EUCLIDEAN GEOMETRY:

A FIRST COURSE

EUCLIDEAN GEOMETRY: a first course

Mark Solomonovich

Grant MacEwan University, Edmonton

Euclidean geometry: a first course

Copyright © 2003, 2004, 2005, 2006, 2009 by Mark Solomonovich

All rights reserved. No part of this book may be used or reproduced by any means, graphic, electronic, or mechanical, including photocopying, recording, taping or by any information storage retrieval system without the written permission of the publisher except in the case of brief quotations embodied in critical articles and reviews.

iUniverse books may be ordered through booksellers or by contacting:

iUniverse
1663 Liberty Drive
Bloomington, IN 47403
www.iuniverse.com
1-800-Authors (1-800-288-4677)

Because of the dynamic nature of the Internet, any Web addresses or links contained in this book may have changed since publication and may no longer be valid. The views expressed in this work are solely those of the author and do not necessarily reflect the views of the publisher, and the publisher hereby disclaims any responsibility for them.

ISBN: 978-1-4401-5348-8 (soft)
ISBN: 978-1-4401-5349-5 (ebook)

Printed in the United States of America

iUniverse rev. date: 8/19/2010

iUniverse, Inc.
New York   Bloomington

*In loving memory of my mother, who always insisted
that I should teach mathematics.*

Preface

He had to write this out two or three times before he could get the Rissolution to look like what he thought it was going to when he began to spell it: but when at last it was finished, he took it round to everybody and read it out to them.
- A.A. Milne, *The House at Pooh Corner*

This book was begun a long time ago and re-written more than three times before I decided it looked right. The first version was written when my daughter was about to start secondary school, for her and for the little private school that she was going to attend. These first notes were based on my recollection of my own secondary school geometry. Then the text survived quite a few changes, especially in the sections concerning the choice of axioms. It has become more rigorous (sometimes I think maybe too much so), which increased both my satisfaction with it and some students' suffering. The former was because I handcrafted a rather rigorous set of axioms that emerge naturally, which I believe is important for those who encounter strict mathematical discussion for the first time. As for the students' suffering, it is probably an inevitable part of studying geometry; however, I did observe that in many instances the excitement of discovering "new worlds" alleviates the pain.

This latest version, which seems satisfactory to me, has been taught both as a high school course and as an introductory college course. In each of these cases, the course demanded a significant effort from the students and the instructors, many of whom were home-schooling parents.

A detailed account of how this textbook was written, how it compares to other (old and contemporary) textbooks, and a review of some of these textbooks and their common features can be found in the essay "Why this Textbook is Written the Way It Is," which is included in the appendix of this book. I placed it in the appendix for two main reasons. First, it's too long to be a preface, and second, it is not essential for students to read it. As for the instructors, whether they be college professors, school teachers, or home-schooling parents, I recommend that they read this essay, even if some parts of it may seem too obscure for those who are not professional mathematicians.

This book can be taught as a one-year course for high school students. For instance, it has been taught as such at the Regina Coeli Academy for the last two years. Still, I believe it should be taught at a slower pace, over two or even three years, starting at grade eight or nine. This is the way it was taught to me when I was a secondary school student about forty years ago in the former Soviet Union. (A standard algebra course was taught independently during the same school years.) This course can also be taught in a shorter period, for example, in one term, for post-secondary students who *may have seen* some geometry during their school years. (If you would like to know what the above *may have seen* stands for, please see the aforementioned essay "Why this Textbook is Written the Way It Is").

The textbook consists of eleven chapters, each accompanied by a set of problems and questions. There are over 1200 problems and questions in the book. Many of these problems and questions are supplied with answers, hints to solutions, or detailed solutions.

The solutions section of the book consists of two parts:

(i) Answers, Hints
This part contains the answers to all problems that allow just short answers, particularly the numerical ones. Also there are hints that will help the students to compose their own solutions.

(ii) Solutions, Hints
This part includes detailed solutions to some problems. I would suggest that students should read these solutions, especially those of construction problems, even if they have already found their own solutions, in order to get accustomed to the culture of discussing and proving statements. Also, there are some hints, usually more detailed than in the first part of the solution manual and some additional theoretical material which is not included in the main part of the text since it would be too much of a digression. Some of these theoretical excerpts are rather advanced topics, which are optional for secondary school students.

Currently, I am preparing a companion to this text, which will contain solutions to additional problems (which may be used for tests and home assignments) and some supplementary materials for instructors. It is planned that this companion be issued almost simultaneously with the book.

I am grateful to many people whose support and encouragement has helped me to write this book. My daughter Sonya was the first student of this course, who has shown me that it can be studied in secondary school and that the process of studying is not necessarily painful. I always recall with gratitude the late Mrs. Pibrovec, the principal of the Bilingual Montessori Learning Centre, who encouraged me to write a Euclidean Geometry course for secondary schools. I am very thankful to Stuart Wachowicz, Director of Curriculum of Edmonton Public Schools, for his enthusiastic support.

I really appreciate the courage and effort of all the Grant MacEwan College students who dared to take the Euclidean Geometry course, with special thanks to Adrian Biglands, Rachel Kok, and the late Guo Yi. Another student who deserves a special acknowledgement is Steven Taschuk. The many discussions I have had with him have significantly improved this book.

I would like to thank all the home-schooling parents and all the teachers who have been using this text, with special thanks to Sandra Harrison, who teaches geometry at the

Regina Coeli Academy and Tina MacLennan, a passionate supporter of classical tradition in education.

Professor Robin Hartshorne from the University of California at Berkley has written a very thorough and very encouraging review of this book, which I really appreciate.

Whatever I know about teaching mathematics the way that makes it inspiring and enjoyable, I have learned from the late Efim Fainstein, my mathematics teacher in high school, a brilliant mathematician and a very kind man.

I am grateful to my wife Elena and my son Philip for giving me the motivation and the moral support in this endeavor.

CONTENTS

1. Introduction 1
   1.1 Introductory remarks on the subject of geometry
   1.2 Euclidean geometry as a deductive system. Common notions
   1.3 Undefined terms. Basic definitions and first axioms

2. Mathematical Propositions 31
   2.1 Statements; Sets of Axioms; Propositions; Theorems
   2.2 Theorems: the structure and proofs
   2.3 Direct, converse, inverse, and contrapositive theorems

3. Angles 46
   3.1 Basic Notions
   3.2 Kinds of angles. Perpendiculars
   3.3 Vertical angles
   3.4 Central angles and their corresponding arcs. Measurement of angles

4. Triangles 64
   4.1 Broken line; polygons
   4.2 Triangles; the basic notions
   4.3 Axial Symmetry
   4.4 Congruence of triangles. Some properties of an isosceles triangle
   4.5 Exterior angle of a triangle
   4.6 Relations between sides and angles in a triangle
   4.7 Comparative length of the straight segment and a broken line connecting the same pair of points
   4.8 Comparison of a perpendicular and the obliques
   4.9 Congruence of right triangles
   4.10 The perpendicular bisector of a segment and the bisector of an angle as loci

5. Construction Problems 101
   5.1 The axioms and tools allowing geometric constructions
   5.2 Basic construction problems
      a) To construct an angle equal to a given angle, with the vertex at a given point and one side lying in a given line
      b) To construct a triangle having given the three sides (SSS)
      c) To construct a triangle having given two sides and the included angle (SAS)
      d) To construct a triangle having given two angles and the included side (ASA)
      e) To construct a triangle having given two sides and an angle opposite to one of them (SSA)
      f) To bisect an angle (draw the symmetry axis of an angle)
      g) To erect a perpendicular to a given line from a given point in the line
      h) To drop a perpendicular onto a line from a given external point
   5.3 Strategy for construction: analysis, construction, synthesis, investigation

6. Parallel Lines 112
   6.1 Definition. Theorem of existence. .
   6.2 Angles formed by a pair of lines and their transversal. Three tests for the parallelism. Parallel postulate (Playfair's formulation)

- 6.3 Tests for the nonparallelism of two lines. Euclid's formulation of Parallel Postulate
- 6.4 Angles with respectively parallel sides. Angles with respectively perpendicular sides.
- 6.5 The sum of the interior angles of a triangle. The sum of the interior angles of a polygon. The sum of the exterior angles of a convex polygon
- 6.6 Central Symmetry

7. PARALLELOGRAM AND TRAPEZOID 134
- 7.1 Parallelogram
- 7.2 Particular cases of parallelograms: rectangle, rhombus (diamond), square. Symmetry properties of parallelograms
- 7.3 Some theorems based on properties of parallelograms
- 7.4 Trapezoid
- 7.5. a) Translations
  b) Translations and symmetry in construction problems

8. CIRCLES 162
- 8.1 Shape and location
- 8.2 Relations between arcs and chords and their distances from the centre
- 8.3 Respective position of a straight line and a circle
- 8.4 Respective position of two circles
- 8.5 Inscribed angles, some other angles related to circles. Construction of tangents
- 8.6 Constructions using loci
- 8.7 Inscribed and circumscribed polygons
- 8.8 Four remarkable points of a triangle

9. SIMILARITY 196
- 9.1 The notion of measurement
  a) The setting of the problem
  b) Common measure
  c) Finding the greatest common measure
  d) Commensurable and incommensurable segments
  e) Measurement of segments. Rational and irrational numbers
  f) The numerical length of a segment. The ratio of two segments

- 9.2 Similarity of Triangles
  a) Preliminary remarks
  b) Corresponding sides; definition and existence of similar triangles

- 9.3 Three tests for the similarity of triangles

- 9.4 Similarity of right triangles

- 9.5 Similarity of polygons

- 9.6 Similarity of general plane figures
  a) Construction of a figure similar to a given one
  b) Similarity of circles

- 9.7 Similarity method in construction problems

- 9.8 Some theorems on proportional segments
  a) Segments related to parallel lines
  b) Properties of bisectors

- 9.9 Metric relations in triangles and some other figures
  a) Mean proportionals in right triangles
  b) The Pythagorean Theorem

      9.10   Proportional segments in circles

      9.11   Trigonometric functions of acute angle
- a) Definitions
- b) Basic trigonometric identities
- c) Construction of angles given their trigonometric functions
- d) Behaviour of trigonometric functions of acute angles
- e) Solving triangles using trigonometry

      9.12   Applications of algebra in geometry
- a) Construction of the golden ratio
- b) General principles of algebraic methods for solving geometric problems
- c) The construction of some elementary formulae

10. REGULAR POLYGONS AND CIRCUMFERENCE 271

      10.1   Regular polygons

      10.2   The circumference of a circle

      10.3   The circumference as a limit

11. AREAS 308

      11.1   Areas of polygons
- a) Basic suggestions (principles)
- b) Measurement of area
- c) Area of a rectangle
- d) Areas of polygons
- e) Pythagorean Theorem in terms of areas

      11.2   Areas of similar figures

      11.3   The area of a circle

12. ANSWERS, HINTS (1) 330
         SOLUTIONS, HINTS (2) 337

APPENDIX: Why this Textbook is Written the Way It Is.............................. 378

1. INTRODUCTION

"We are going to discover the North Pole."

"Oh!" said Pooh again. "What *is* the North Pole?" he asked.

"It's just a thing you discover", said Christopher Robin carelessly, not being quite sure himself.

A.A. Milne, *Winnie-the-Pooh*.

1.1 Introductory remarks on the subject of geometry

The word *geometry* descends from two Greek words, γεα - [gea] meaning "earth" and μετρειν - [metrein] meaning "measurement". It is evident from these words as well as from numerous historical documents that geometry originated as a practical art of measuring land.

The origins of geometry can be traced to early Babylonians and Egyptians, for whom it was an *empirical* (i.e., experimental) science dealing with practical mensuration used in construction and agriculture. Particularly, ancient Egyptians used measurements and some empiric results for the repeated surveying of land in the Nile valley: every year the boundaries between land plots were washed out by a flood.

The aforementioned results, which were usually presented as a sequence of arithmetic instructions, were derived empirically and were often incorrect. For example, the Egyptians used the formula $A = \frac{1}{4}(a+c)(b+d)$ to calculate the area of an arbitrary quadrilateral with successive sides of length a, b, c, and d. It is easy to show (try to do it!) that this formula is correct for a rectangle but not for an arbitrary quadrilateral. Yet, since the formula was applied to evaluate areas of approximately rectangular pieces of land, no one noticed that it was flawed.

Of course, any empirical result can be established only to a certain degree of accuracy, and thus cannot be correct *in general*. Therefore, mere observation and measurement of real (*physical*) objects cannot provide us with *precise* results applicable in other situations.

How would it be possible to obtain such results, i.e. ones that would be universal or *absolutely correct*? Gradually, people came to the conclusion that such results cannot be attributed to real objects, but rather to their *idealizations*. That is how the notion of a *geometric figure* emerged: such a figure is not a physical object but rather an image of one, and interests us only from the point of view of its *geometric properties*.

From this point of view a stretched string and the edge of a soccer field are straight lines, a sheet of paper and the top of a table are both rectangles, an orange as well as a tennis ball can be called spheres. The objects in each pair have the same *shape*, and the latter two may be also of the same *size*, in which case they are *identical* from a geometric viewpoint.

It is not easy to define the notions of shape and size (students are encouraged to attempt to do this). That is why these (and many other) examples do not provide us with a precise definition of *geometric properties* and of the *subject of geometry* but they do clarify something. Firstly, in this subject, we are not going to deal with physical objects but rather with the *idealized* representatives of these objects, called *geometric figures*. Secondly, these figures may differ in their *geometric properties*, which are the subject of our interest.

Now, if we properly define what it means for figures to be *geometrically identical*, we shall have a satisfactory definition of the *geometric properties*, and thus we will be able to define the *subject of geometry*. We shall postpone the formulation of this definition until section 1.3, because it will require from us a certain level of maturity. It should be noted here, for our justification, that the history of geometry began more than 4000 years ago, and the first *rigorous* definition of the subject was given only at the end of the nineteenth century.

1.1 Questions

1. Does geometry study *real* or *ideal* objects?

2. What, in your opinion, are geometric properties of figures?

3. In what case would you say that two figures are identical from the geometric viewpoint?

4. Can absolutely precise results be attributed to real objects? (Adduce examples).

1.2 Euclidean geometry as a deductive system. Common notions.

a) Greeks' contribution to the development of geometry.

Ancient Greeks contributed by transforming geometry into a theoretical science, a *deductive system* based upon a few basic statements, *axioms*. All the other results, called *propositions*, were derived from these axioms by means of logic.

The term "*axiom*" was introduced by Aristotle (c.a. 384 – 322 B.C.), an eminent scholar from Plato's school of philosophy. In Greek, the word $\alpha\xi\iota o$ [axio] means "authority", "respect", "importance". One can say that *axioms are important facts believed to be true without questioning.*

Aristotle constructed a theory of statements that begins with *common notions* or *axioms*, *special notions* or *postulates* and underlying *definitions*. This theory, together with *deductive logic* (reasoning from general laws to a particular case) formed a basis for Greek mathematics.

Nowadays, any field of knowledge that claims the status of a *theoretical science* is required to be represented as an axiomatizable system. This system may be incomplete in the following sense: at this point it does not explain all phenomena in the given field. Then scholars search for new axioms that would complete the system by being added to the old ones or by replacing some of them.

The first, and, for thousands of years, the only example of an axiomatized field of knowledge was Euclidean geometry. In his monumental work *The Elements*, Euclid (c.a. 330 – c.a. 270 B.C.) built geometry as a formal axiomatic system based on a few fundamental definitions, axioms concerning *things in general* (he called them *common notions*) and axioms dealing particularly with geometric features (these were named *postulates*).

Though in modern science all fundamental principles are called axioms or postulates, and these two words are considered as complete synonyms, we find Euclid's reference to *common notions* convenient and we shall use it in our discussion.

In his *ELEMENTS* Euclid did not list all axioms he had really used to obtain his results. Some of them were used *implicitly*, i.e. without making an explicit reference. We shall do the same. For example, we will accept as self-evident that *of any three points in a line, exactly one of them lies between the other two* and will not make a reference to this fact, though this is an axiom in some modern *rigorous* systems. A mature scholar can easily complete our system of axioms with the missing axioms of *incidence* and *betweenness*, e.g. from Hilbert's system of axioms to fill the "gaps" in our discussion.

Also, in modern rigorous systems common notions are not used: all axioms or postulates, which are one and the same, must be explicitly described. We shall use common notions in order to make the discussion as elementary and clear as possible. For example, we will not introduce the *axioms of segments and angles addition*, which are used in Hilbert's system: Common Notion 1 (see below) will be used instead. Yet, we shall apply the common notions only to segments and angles; thus we can easily replace them with additional axioms concerning the congruence of segments and angles.

b) <u>Common notions</u>

Now we are going to list the *common notions* that will be used in the course. We shall use the symbol "$\Rightarrow$" as a short replacement for the words "it follows". For example, we can use this symbol in the following *syllogism*.

$$\left.\begin{array}{l}\text{I love animals}\\ \text{Crocodile is an animal}\end{array}\right\} \Rightarrow \text{I love crocodiles}.$$

In future we shall also use the symbol "$\Leftrightarrow$" that means the *equivalence* of two statements. Two statements are said to be *equivalent* if each one follows from the other.

For example: n is an even number $\Leftrightarrow$ n^2 is an even number.

This symbol ($\Leftrightarrow$) is read *if and only if* or, for brevity, *iff*.

Common Notions (general axioms).

(1) *If equals are added to equals, the sums are equal*:

$$\left.\begin{array}{l}a=b\\ c=d\end{array}\right\} \Rightarrow a+c=b+d$$

(2) *If equals are subtracted from equals, the remainders are equal:*

$$\left.\begin{array}{c}a=b\\c=d\end{array}\right\} \Rightarrow a-c=b-d$$

(3) *If equals are added to unequals, the results are unequal in the same order:*

$$\left.\begin{array}{c}a>b\\c=d\end{array}\right\} \Rightarrow a+c>b+d$$

(4) *If equals are subtracted from unequals, the results are unequal in the same order:*

$$\left.\begin{array}{c}a>b\\c=d\end{array}\right\} \Rightarrow a-c>b-d$$

(5) *If unequals are subtracted from equals, the results are unequal in the opposite order:*

$$\left.\begin{array}{c}a=b\\c>d\end{array}\right\} \Rightarrow a-c<b-d$$

(6) *Things equal to the same quantity, are equal to each other, i.e. equals may be **substituted** for equals. (This property is called Transitivity of equality)*

$$\left.\begin{array}{c}a=b\\b=c\end{array}\right\} \Rightarrow a=c$$

(7) *If one thing be greater than a second and the second greater than a third, the first is greater than the third. (Transitivity of order)*

$$\left.\begin{array}{c}a>b\\b>c\end{array}\right\} \Rightarrow a>c$$

(8) *The whole is greater than any one of its parts.*

(9) *The whole is equal to the sum of its parts.*

Also, it follows from the above common notions that

(i) *multiples of equals are equal*:

$a=b, \Rightarrow na=nb$ for any natural number n

(ii) *equal parts of equals are equal*:

$a=b, \Rightarrow \dfrac{a}{n}=\dfrac{b}{n}$ for any natural number n.

This set of common notions (hereafter we shall refer to them as CN) or general (not specifically geometric) axioms is close to the one used by Euclid. Still there are some deviations. Let us discuss them.

Instead of applying our common notions to *things* in general, we restrict ourselves to the more narrow and specific category of those *things* that can be *measured – magnitudes*. As you will see later, the only figures to which we shall apply our common notions will be *segments*, *angles*, and *arcs of circles*. We shall always compare magnitudes of the same kind. This way we can avoid ambiguous questions like: who is greater – an elephant or a giraffe? – In terms of volume – the first one, and in terms of height – the latter.

In the original Euclidean discussion, one of the common notion (*the things are equal if they coincide*) was applied to various geometric figures, such as polygons, and therefore was in fact a geometric *postulate*. We do not include this common notion in our set: we shall replace it with geometric axioms concerning the equality, or *congruence*, of geometric figures.

1.2 Questions and Problems

GREEK ALPHABET:

A α alpha	B β beta	Γ γ gamma	Δ δ delta	E ε epsilon	Z ζ zeta
H η eta	Θ θ theta	I ι iota	K κ kappa	Λ λ lambda	M μ mu
N ν nu	Ξ ξ ksi	O o omikron	Π π pi	P ρ rho	Σ σ sigma
T τ tau	Y υ upsilon	Φ φ phi	X χ chi	Ψ ψ psi	Ω ω omega

1. Modern English alphabet originates in Greek alphabet (which, in its turn, took origin in Ancient Hebrew and Phoenician). List English letters that are explicitly reminiscent of the corresponding Greek letters.

2. Some modern words and names are of Greek origin. For example, *geometry, axiom, logic, sophisticated, Alexander, Sophia, Philip, Jason..* Find more examples. What do these words mean? Try to write the roots of these words in Greek.

3. Ancient Egyptians used the formula $A = ¼(a+c)(b+d)$ to calculate the area of a quadrilateral with successive sides of lengths *a,b,c,d.* You have an intuitive idea of the area of a plane figure as the amount of space covered by the figure and know that the area of a rectangle is evaluated as the product of the lengths of its sides.
   a) Show that this formula is correct for rectangles.
   b) Is it correct for an arbitrary (non-rectangular) quadrilateral?
   c) Why do you think ancient Egyptians successfully used this formula for measuring land?

4. What, in your opinion, is the main difference between an empirical and theoretical field of knowledge?

5. What is the main difference between the pre-Greek and Greek (Euclidean) geometries?

6. Four definitions of the word ***axiom*** found in different sources follow:
(i) a self-evident truth;
(ii) a universally accepted principle or rule;
(iii) a proposition evident at first sight, that cannot be made plainer by demonstration;

(iv)   a proposition which is assumed without proof for the sake of studying the consequences that follow from it.

Consider the applicability of each of these definitions for the following cases:
a) *Axioms of the norms of behaviour*, e.g. traffic regulations (move along the right side on a two-way highway; do not proceed when the light is red, etc…).
b) *Arithmetic axioms*, e.g. Peano axioms for natural numbers.
c) *Scientific axioms*. Examples:
- An axiom of ancient astronomy: The Earth is immovable, and the Sun and stars rotate about it.
- Newton's first law of motion: If an object is not subjected to any forces, it moves with a constant speed along a straight line.

7. Create an example of a syllogism and write it using the $\Rightarrow$ symbol.

8. Create an example of two equivalent statements and write it down using the $\Leftrightarrow$ symbol.

9. Consider quantities such as *length, area, volume*. What kind of numbers expresses these quantities?

10. Check whether all of our common notions are valid for quantities expressed in negative numbers.

11. Can you define the notion of *set*?

12. What are the undefined notions of arithmetic?

13. Adduce an example of an undefined term from any field of knowledge.

14. What kind of declaration can serve as a definition of a new term?

15. What is the role of axioms with respect to the *undefined* terms?

1.3 Undefined terms. Basic definitions and first geometric axioms

Another distinction between our discussion and the original Euclidean one concerns the introduction of primary geometric terms such as *point, line, surface,* and *plane*. Euclid used definitions like *"a line is a breadthless length"* or *"a straight line is a line that lies evenly with the points on itself"*. As for the first of these "definitions", it uses two new fundamental terms, *breadth* and *length* to introduce the most fundamental term *line*. Therefore it cannot be a definition. As for the second "definition", one can easily see that it can also be related to a circle; hence it can hardly be a definition of a straight line. These definitions are rather explanations.

Thus, we shall accept that there are some *undefined* basic notions, and we shall only attempt to describe and explain them. This will be done by means of axioms. All secondary notions (such as *angle, polygon*, etc.) will be defined in terms of these fundamental *undefined notions*.

In this section we shall introduce the *undefined notions* of *point, line, plane,* and *congruence of segments*. These notions (sometimes called *the primitive terms*) are the underlying objects of Euclidean geometry and therefore cannot be defined. However, their properties, which are essential for the theory, will be *postulated* or described in *axioms*.

a)   Geometric figures

Part of the space bounded from the ambient space is called a *geometric solid*. Solids are bounded by *surfaces*. Part of a surface is separated from the adjacent part by a *line* (either a *curve* or a *straight line*.)

Part of a line is separated from the adjacent part by a *point*.

In real (physical) world, solids, surfaces, lines, points do not exist independently of one another. Still we can abstract from our experience with physical objects and consider these notions as idealizations. Then we can discuss surfaces independently of solids they enclose, lines – independently of surfaces, and points – independently of lines. We can imagine that a surface has no thickness but has length and width; a line has length but has no width and thickness; and a point has no magnitude.

By a *geometric figure* we mean a configuration of points, lines, surfaces and solids.

b)   Planes

A *plane* is a *flat* surface. The test for flatness is that *if any two points are taken in the surface, the straight line that connects them lies wholly in that surface*. This is, actually, one of the plane axioms used in *solid geometry*. In the nearest future we shall consider only *plane geometry*, where all figures lie in one plane.

We suggest that any part of a plane or the plane itself can be moved, or subjected to a *rigid motion*. As a result of such a *motion*, a plane or its part is evenly (without distortion) superimposed on another plane or its part or onto itself. With that the part of the plane that is being superimposed may be turned upside-down (see Figure 1.3.1).

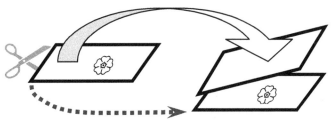

Figure 1.3.1

Rigid motions, also called *isometries*, play a very important (one could say: the most important) role in Euclidean geometry. They will be defined more specifically, or as mathematicians say *rigorously*, at the end of this section.

c)   Straight line; ray; segment

A line may be *straight* or *curved*. One can imagine a *straight line* as a trajectory of a point moving without changing its direction. A *curved line* (or a *curve*) is the trajectory of a particle that moves changing its direction continually from point to point.

Hereafter we shall use the term "line" to denote a straight line, unless it is specified otherwise. For a curved line we shall reserve the word "curve". If a geometric figure does not include curves it is called a *rectilinear* figure. We shall start our discussion of rectilinear figures by *postulating* properties of straight lines and their parts. These properties are described in the corresponding *axioms*.

AXIOM 1 *Given any two points in space, one and only one line can be drawn through these points.*

Let us notice that this axiom contains two statements: (1) *for any two points there exists a line that passes through them*, and (2) *there is only one (not more than one) such line*.

In mathematics, one can often encounter such statements that claim both the *existence* of an object and its *uniqueness*. When formulating them, it is convenient to use the phrase *exactly one*, which means the same as *one and only one*. For example, by saying "Earth has *exactly one* natural satellite", we claim two things: (1) Earth does have a natural satellite, and (2) this satellite is unique, that is: Earth does not have another one. In comparison, by saying "Earth has one natural satellite" (which sounds very similar, but the word *exactly* is missing) one would assert only the existence but not the uniqueness of the satellite.

It follows from Axiom 1 that *if any two points of a line coincide with two respective points of some other line, then these lines coincide in all their points.* (Otherwise it would be possible to draw two distinct lines through a given pair of points).

Therefore, two distinct lines may intersect only at one point.

A few lines that pass through a point are said to be *concurrent*. A set of such lines is called a *pencil*, and the point is called the *vertex* of the pencil.

Let us emphasize that AXIOM 1 asserts not only the *existence* of a line through a pair of points but also the fact that such a line *can be drawn*. This reflects the *constructive* nature of Euclidean geometry as laid down by ancient Greek scholars: the existence of an object implies our ability to draw it.

Thus AXIOM 1 implies the existence of an instrument that allows us to draw a straight line through any pair of points. Such an instrument is called a *straightedge*. That is why AXIOM 1 is also called the STRAIGHTEDGE POSTULATE.

We denote lines by small Latin letters, usually from the middle of the alphabet (e.g., line *m*, line *l*), or by pairs of capital letters denoting two respective points determining lines (e.g., line *AB* in Figure 1.3.2).

Three points do not, in general, lie in one line. If three (or more) points accidentally lie in one line, they are called *collinear*.

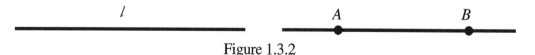

Figure 1.3.2

Part of a line bounded on one side by a point lying in the line is called a *ray emanating* from this point. The point is called the *origin* or *vertex* of the ray (Fig.1.3.3.)

A ●────────────────────────────

Figure 1.3.3

Part of a line bounded from two sides by two points lying in the line is called a *segment of the straight line* (Fig. 1.3.4), or a *straight segment*. (For the sake of brevity we shall use the word "segment" as a substitute for "straight segment". For any other kind of segments we shall provide explaining supplements, e.g., *a segment of a circle, a segment of a curve, etc*).

────●─────────────────────────●────
    A   *B*

Figure 1.3.4

The bounding points are called the *endpoints* or *extremities* of the segment. A segment is denoted by a pair of capital letters naming its endpoints, e.g., "segment *AB*". The same segment can be denoted "*BA*". Sometimes one small letter is used, e.g. "segment *a*", "segment *c*", etc.

Since every straight segment is part of the straight line that passes through its endpoints, one can use a straightedge to extend a given segment into a ray or into a straight line. According to AXIOM 1, such a line is unique for each given segment.

Some segments are said to be *congruent* or equal to each other. In practical terms it means that the segments do not differ from one another, they may just be located in different places. The relation of congruence is usually denoted by the "=" sign, for instance, *AB=MN*, or $a = b$. (Other symbols that are commonly used to denote congruence are ≃, ≅, and ≡). The relation of *congruence of segments* is a fundamental undefined notion that obeys the following two axioms – AXIOMS 2 and 3.

<u>AXIOM 2</u>

(i) *Each segment is congruent to itself: a = a (reflexivity);*

(ii) *If one segment is congruent to the other, then the second segment is congruent to the first one: a = b ⇒ b = a (symmetry);*

(iii) *If each of two segments is congruent to the third segment, they are also congruent to each other:* $\left.\begin{array}{l}a = c\\b = c\end{array}\right\} \Rightarrow a = b$ *(transitivity).*

The latter axiom says that the *congruence* of segments is a so-called *equivalence relation*. (A few examples of equivalence relations are discussed in the problems for this section). This relation splits the whole set of segments in a plane into families (also called *equivalence classes*) of congruent segments. It can be proved (students should attempt!) that two segments are congruent if and only if they are members of the same class. It means that two congruent segments always belong to the same class and two segments from the same class are necessarily congruent. In particular, each segment is a member of exactly one class (since by property *(i)* it is congruent to itself).

The second axiom of congruence of segments helps us to visualize the notions of congruent and non-congruent segments.

<u>AXIOM 3</u> *One and only one segment congruent to a given segment can be laid off on a ray starting from the origin.*

If two segments laid off on a ray from its vertex are *non-congruent*, their second endpoints do not coincide, and the segment whose second endpoint lies within the *interior* (that means: between the endpoints) of the other segment is said to be *the least* of the two. The other segment is *the greatest* of the two. For instance, in Figure 1.3.5, *AC is less than AB*: *AC<AB*, or *AB is greater than AC*: *AB>AC*.

Now let us **define** the sum of two segments: *if C is an interior point of AB* (Fig. 1.3.5), *then AC + CB = AB*.

For the same triple of segments we shall say that *AC* is the *difference* between *AB* and *CB*: *AC = AB – CB*.

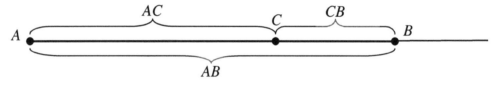

Figure 1.3.5

If there exists such a point that divides a segment into two congruent segments, this point is called a *midpoint* of the segment. A line passing through such a point is said to *bisect the segment*. In Chapter 4 we shall prove that every segment has exactly one midpoint.

One can easily extend this definition for the sum of three or more segments and check that the sum of segments possesses all the properties of the sum of numbers (a good exercise for students).

The above definition of the sum can be used to define the *multiplication* and *division* of a segment by a *natural number* (positive integer).

The product of some segment, *AB*, by a natural number, *k*, (*k-multiple of AB*) is the sum of *k* segments, each of which is congruent to *AB* (Fig. 1.3.6 a). *One n-th* of segment *MN* is any of *n* congruent segments whose total sum is *MN* (Fig. 1.3.6 b.)

A certain segment can be chosen as a *unit (of measurement)*. Then, as it will be proved later on (Chapter 9), all other segments can be *measured* in terms of this segment: for each given segment one determines the *number of units* contained in the segment. This number is called *the length* (or *the measure*) of the segment in given units. If, for instance, in Figure 1.3.6, *AB* is a unit of measurement, then the length of *CD* equals 4.

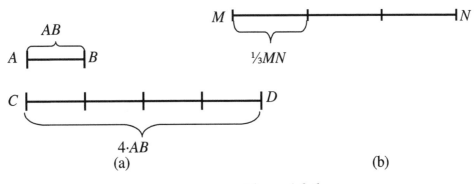

(a)   (b)

Figure 1.3.6

In some cases a unit does not fit an exact number of times on a segment. Then the length is represented by a non-integer number. If, for example, in Figure 1.3.7, *MN* is the unit, the length of *AB* equals $2\frac{1}{3} = \frac{7}{3}$. If one chooses $PQ = \frac{1}{3}MN$ as a unit, the length of *AB* will be equal to 7, an integer.

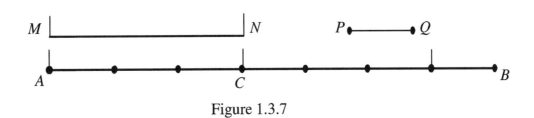

Figure 1.3.7

As we shall see later, in Chapter 9, it is not always possible to find a unit for which the length of a given segment is an integer or a fraction. It will be shown, however, that, when a unit is chosen, there always exists a *real number* (not necessarily an integer or a fraction) that represents the length of a segment. To prove this we shall also need two *axioms of continuity*: the Archimedes' axiom and Cantor's principle of nested segments. Meanwhile we shall consider only integer or fractional lengths.

For the segments whose lengths are integers or fractions, one can use common notions to show that *congruent segments have equal lengths and vice versa* (a good exercise for students!). This statement can be proved in general (and we will be able to do this after the notion of length has been introduced). That is why in some texts *congruent segments* are called *equal segments*.

Also, it can be proved that the length of the sum of two segments equals to the sum of the lengths of the two segments.

For example, if *PQ* is the unit in Figure 1.3.7,
length (*AB*) = length(*AC+CB*) = 7;   on the other hand,
length (*AC*) + length (*CB*) = 3+4=7.

It should be noticed, that all segments with distinct endpoints will have positive lengths. *Length zero* will be assigned only to the segments with coinciding endpoints. We shall say in cases like this that a *segment degenerates into a point*.

Regular (non-degenerate) segments can be measured by means of a *ruler*, a *straightedge* with an appropriate scale. For instance, in Figure 1.3.8, where the segments are measured by a ruler with the unit equal to 1cm, *PN* is congruent to the unit, and *MN* = 5 cm (or its *length* is 5).

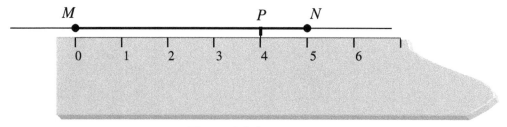

Figure 1.3.8

If two segments are not congruent, *the greater one has the greater length*. Really, if two segments are laid off starting from the vertex on the same ray, the lesser segment will be part of the greater one, and, thus, will contain fewer units (compare segments *AB* and *AC* in Figure 1.3.9) and, respectively, will have lesser length. One can also prove the *converse*: if the lengths of two segments are not equal, *the segment with the greater length is greater than the one with the lesser length*. (Suggest for simplicity that both lengths are expressed by integers and prove the converse statement).

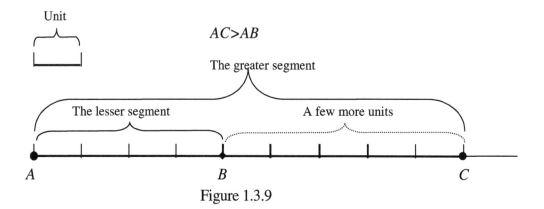

Figure 1.3.9

Let us emphasize that length is a numerical characteristic of a segment. It is convenient for stating and solving practical problems and illustrating examples, – that is why we talk about it here at all. Later on in the course, in Chapter 9, we shall need lengths in order to compare the so-called similar figures. Yet the lengths as numerical representations of segments *do not play any role in our discussion of geometric properties of figures* in Euclidean plane.

a) Distance between two points.

According to AXIOM 1, given a pair of points, there is one and only one straight segment joining them. This segment is called *the distance between the two points*.

If a unit of measurement is chosen, the length of this segment can serve as a uniquely defined numerical characteristic of the respective position of the two points. This numerical representation of the distance is also often called the *distance between two points*. We shall use the term *distance* in this meaning only in practical problems and applications: it is easier to say "the distance between two cities is 20 km" than "the distance between two cities is a segment whose length is 20 if one kilometer is the unit of length." Otherwise (in all theoretical discussions), the distance between two points is the *segment* connecting them.

b) Rigid motion (isometry). Congruence of geometric figures. Subject of Geometry.

In the very beginning of this course we had attempted to describe the subject of geometry, and then we postponed our discussion, being unable to define what is meant by "geometric properties of figures".

We could only adduce a few examples allowing us to develop our intuition and to realize that in geometry we deal with *ideal* images of real physical objects. As for the question, what are "geometric properties", so far we can answer only that they are "those properties studied in geometry". Thus we have gone around the vicious circle: we defined geometry as the science that studies geometric properties of figures, and geometric properties as being those properties studied in geometry. In order to break the circle we must define *geometric properties*.

The definition must be such that all identical, or indistinguishable, from the point of view of our geometry figures will have the same geometric properties. Such figures are called *congruent*. Thus, we are going to define what we mean by *congruence* of geometric figures in our geometry.

Now, being armed with the notion of the distance between two points, we are able to do this. At first we shall introduce the idea of *rigid motion* or *isometry*.

The words *rigid motion* are intuitively clear: a figure subjected to such a motion behaves as a rigid object. This means that it is not stretched or compressed or distorted in any other way: the figure as a whole moves (in a plane), while the positions of different points of the figure with respect to each other do not change. The latter condition will be satisfied if the distance between every pair of points of the figure does not change, i.e. each segment joining two points of the figure is transformed into a congruent segment.

Figure 1.3.10 illustrates such a motion. One can think of an iron (a rigid object) moving along an ironing board (a plane). The shaded figure, which looks like the base of the iron, moves in a plane in such a way that the distance between any two points, for example, A and B, of the figure, remains the same for any new position of the figure: $AB = A'B'$.

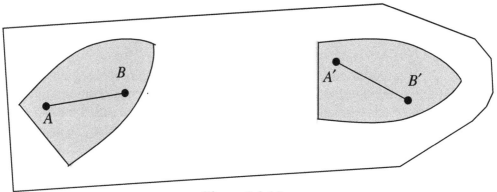

Figure 1.3.10

Keeping in mind this example, we can give the general definition of an *isometry* (*rigid motion*).

An isometry is a distance–preserving transformation of figures in the plane. That means: some figure F is said to be transformed into a figure F' by means of an isometry, if each pair of points, A and B, of F will be transformed into such a pair of points, A' and B', of F' that the distance between the points of each pair is the same, that is the segments AB and $A'B'$ are congruent: $AB = A'B'$.

We also say that the figure F is *superposed* (*superimposed*) on(to) F', and all their *corresponding* segments, such as AB and $A'B'$, are congruent. In other words, F and F' coincide through superposition. In Euclidean geometry we identify such figures as indistinguishable or having the same *geometric properties*. In *The Elements* Euclid called them *equal*. We shall call such figures *congruent*, though the word *equal* will also be used for segments and angles (measurable things). The notion of congruent figures is very important since it determines the subject of Euclidean geometry.

<u>Definition</u>. ***Two geometric figures are said to be congruent if one of them can be made to coincide with the other by means of an isometry.***

In other words, one of the congruent figures can be *superimposed onto the other* in such a way that they will totally coincide. Now we are able to give the required definition of geometric properties of figures in Euclidean Geometry:

Geometric properties of figures are those properties that are the same for all figures that can be obtained from one another by means of isometries.

Finally, we can define our subject: *Euclidean Geometry is the science that studies those properties of geometric figures that cannot be changed by rigid motions (isometries).*

Mathematicians prefer to speak of *transformations of the plane* over the transformations of figures lying in the plane. When a plane is subjected to a transformation, each point A of the plane is transformed into a point A' of the plane. A' is called the *image* of A under the transformation, and A is called a *pre-image* of

A'. We shall consider only those transformations for which every point is transformed into exactly one image, and every image has exactly one pre-image. Then *isometry* is defined as a *transformation of the plane (or space) for which the distance between points is **invariant*** (which means: it does not change). In other words, a transformation of the plane is called an isometry if for every pair of points A and B of the plane and their respective images A' and B', the distance between the images is congruent to the distance between the points: $A'B' = AB$.

It should be noted that the above definition of congruence and the subject of geometry itself are meaningful only if we know that isometries exist, but their existence does not follow from AXIOMS 1 – 3 (and so far, we have no other axioms!). We can imagine isometries as motions of rigid physical objects; however this intuitive model does not prove the existence of such motions as transformations of geometric figures. Hence, we have to *postulate* the existence of isometries, i.e. to introduce another axiom that will state their existence.

When formulating such an axiom, we shall use our experience with physical motions as a starting point. One can imagine a few pictures that are exact copies of each other (our model of congruent figures), each cut from a flat piece of cardboard lying on the table, i.e. in the same plane. See, for example, Figure 1.3.11a below.

Now let us try to imagine all the possible motions that we may need in order to superimpose each of these figures (flags) onto any of the others. We will call these flags AB, CD, EF, and GH, according to the letters standing near the bases and tops of the flagpoles.

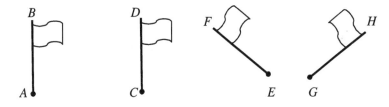

Figure 1.3.11(a)

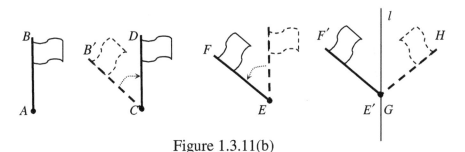

Figure 1.3.11(b)

In order to superimpose flag AB onto flag CD, we move the first flag in such a way that A falls on C. Later on we shall call such a motion *translation along a line*: we *translate* point A along the line AC. If we are lucky, AB will stay vertical during this motion, and will take the position CD immediately after point A has arrived at C. However, we cannot count on it: it may happen that B will not fall on D, and the flag will take some position *CB'* (the dotted flag with its base at *C* in Figure 1.3.11(b)), in which *B'* does not coincide with *D*.

Then, in order to make them coincide, we shall *rotate* the flag about point *C* until the flag completely coincides with *CD*.

In a similar manner we can superimpose flag *AB* onto *EF*. We shall start by imposing the base of the first flag, *A*, onto *E*. As a result of such a translation, *AB* will take some position, for instance, the one shown by the dotted line in Figure 1.3.11b. Then we shall *rotate* the flagpole of the dotted flag about *E* until it coincides with the flagpole of *EF*. The required motion is a combination of *translation along a line* and *rotation*.

We can use motions similar to the above in order to put any of the first three flags into position *E'F'*. However, there does not exist a rotation that will impose the latter flag onto *GH*. Now we shall need a transformation that will change the *orientation* of the part of the plane in which *E'F'* lies. One of the ways of performing such a transformation is the following. We shall fold the sheet of cardboard (i.e. our plane) along the line *l*, which passes exactly halfway between the two flagpoles (Figure 1.3.11(b)) and impose the part of the plane containing *E'F'* onto the other part, containing *GH*. Thus the left half of the plane has been turned upside-down and imposed onto the right half-plane by means of a *reflection in line l*.

Now that we know in practical terms what kinds of motions, or isometries, we may need, we shall postulate their existence.

<u>AXIOM 4 (Existence of Isometries).</u>

- *(i)   For any pair of points, there exists an isometry that carries one of these points into the other.*

- *(ii)   For any pair of rays with a common vertex, there exists an isometry that carries one of these rays into the other.*

- *(iii) For any line in a plane, there exists an isometry that leaves every point of the line unchanged and maps a half-plane lying on one side of the line onto the half-plane on the opposite side of the line.*

In Figure 1.3.12 the actions of the three isometries postulated above are illustrated:

- (i)   Translation of a point into a point carries a point (*A*) into another point (*B*).

- (ii) Rotation about a point (*O*) imposes a ray (*r*) onto another ray (*k*).

- (iii) Reflection in a line (*l*) imposes a figure in the left half-plane onto the right half-plane without moving the points of the line. Obviously, this reflection also imposes the right half-plane onto the left one; that is why we can say that a transformation that reflects in a line interchanges the two half-planes that lie on the opposite sides of this line while keeping the points of the line stationary.

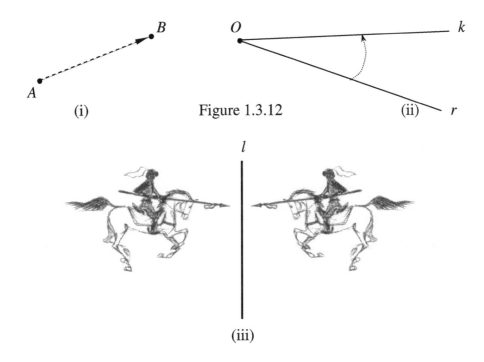

Figure 1.3.12

Figure 1.3.12

Thus we have postulated the existence of three elementary isometries; all other isometries that we may need, will be obtained as combinations, or, as mathematicians say, *compositions* of these. For example, in Figure 1.3.11 we used a composition of a translation and a rotation in order to impose flag *AB* onto *CD*: a translation has been followed by a rotation.

In the following chapters (Chapters 3 and 4) we shall use the existence of isometries to establish the criteria for congruence of various geometric figures.

> Let us notice here that the definition of congruence is in some sense "more fundamental" than others: it determines the subject of studies in Euclidean geometry: *those properties of figures that do not change (are invariant) under isometries*. At the same time, the replacement of this definition by another definition of congruence (another equivalence relation on the set of geometric figures), would engender a new geometry.
>
> This new geometry would study *invariant properties* of figures under some other *transformations (not isometries)*. Such transformations have certain algebraic properties: they form a group. The latter means that *the product*, or composition, of two transformations (one of them is followed by the other) is a transformation of the same kind; also, each transformation has *the inverse*, which, if being applied, abolishes the action of the original transformation.
>
> Geometry, in general, is defined as the study of invariant properties of elements called geometrical figures under specified groups of transformations. Such definition of the subject of geometry has been proposed in the late 19th century by Felix Klein, an eminent mathematician, in his Erlangen Programme.

f) <u>The Compass. Circles. Circles as loci. Arcs.</u>

According to AXIOM 3, one can always lay off on a given ray a segment congruent to any given segment. Thus the axiom postulates the existence of a special instrument that makes this operation possible in practice. Such an instrument is called a *compass*. One can say that compasses are used for *transferring segments*. That is why AXIOM 3 is also called the COMPASS POSTULATE.

The tool itself consists of two rigid legs joined by a hinge. One of the legs ends with a sharp needle, the other with a lead. If the needle is stuck into a given point in the plane, and the respective position of the legs is fixed, the lead will reach all the points in the plane that stand a given distance from the endpoint of the needle. A compass is shown in Fig. 1.3.13.

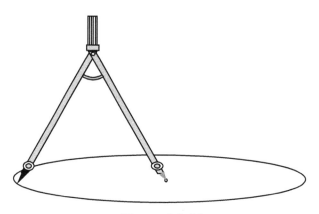

Figure 1.3.13

If the leg equipped with a lead performs one complete revolution about the other leg of a compass, it will draw a *closed plane curve* called a *circle*. The words *closed curve*, though intuitively clear, do require some explanation. We mean that when the lead is describing a circle, it *continuously* moves and returns to the point from which it departed, so there are no "gaps" in the curve. The latter can be fulfilled if we require that every ray emanating from the centre intersects the circle. (Such a requirement follows from the so-called continuity axioms that will be studied in Chapter 9). Thus every point whose distance from the centre is congruent to the distance between the lead and the needle of the compass will lie on the circle. Now we can define the circle.

A circle is the set of all points that are equally distant from a point called the centre. Any of congruent segments connecting the centre with a point lying on the circle is called a radius (plural – *radii*) *of the circle*.

Using this terminology, one can reformulate the COMPASS POSTULATE as follows: *a circle may be described with any point as a centre and any segment as a radius*. This formulation was actually used by Euclid; if it is assumed that every ray emanating from the centre of a circle intersects the circle, this formulation implies AXIOM 4. Indeed, given a ray (*l* in Figure 1.3.14), one can describe a circle of a given radius with its centre at the vertex (*O*) of the ray. The segment (*OB*) connecting the vertex with the point (*B*) of the circle-ray intersection, will be congruent to the radius. Any other segment (*OA*), which joins the vertex with some other point on the ray, will not be congruent to the radius, since it does not lie on the circle.

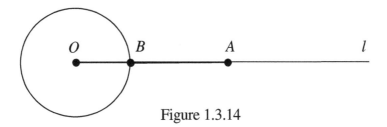

Figure 1.3.14

The definition of the circle suggests that a circle of radius *r* with the centre at *O* consists exclusively of those points whose distance from *O* is *r*, and only points lying on the circle are located at the distance *r* from *O*. Mathematicians say that *a circle is the **locus** of points equidistant from a given point called the centre of the circle.*

It is clear from the previous discussion, what the word *locus* means.

A locus is a set of points such that

(i) *all points in the set satisfy a given condition, and*

(ii) *all points that satisfy the given condition belong to this set.*

We shall meet many other examples of *loci* in this course.

It is easy to prove that the following two statements are true (exercises for students):

1) *A circle cannot have more than one centre.*

2) *Two circles are congruent if and only if their radii are congruent.*

Before turning to the discussion of arcs of circles, we should make a brief note on the terminology. The word *circle* is loosely used both for the curve as well as for its interior. Yet in more rigorous texts the word *disk* is used to denote the part of the plane enclosed by a *circle*. Sometimes the word *circumference* is used as a synonym to the *circle*. We shall think of *circumference* as the boundary of a circle and define it rigorously in Chapter 10.

Two circles that have the same centre are said to be *concentric*; an example of such circles is shown in Figure 1.3.15a.

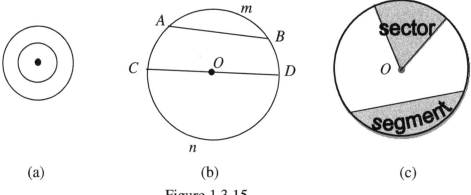

(a)　　　　　　　　(b)　　　　　　　　(c)

Figure 1.3.15

Any portion of a circle is called an *arc*. A *chord* is a straight segment connecting the endpoints of an arc. A chord (as any other segment) is denoted by two capital letters representing its endpoint. An arc *subtended* by the chord is denoted by the same two letters, labelling its *endpoints* and the symbol of arc $\cup$ put before or above the letters. For example, $\cup AB$ is subtended by AB.

Sometimes, in order to specify, which of the two possible arcs, *minor* (the lesser arc) or *major* (the greater one) is considered, other letters, labelling *exterior* points of the arc or one small letter between them are added to the notation. For example, in Figure 1.3.15b, chord AB subtends arcs AmB (minor) and AnB (major); we can also denote them $\cup AB$ and $\cup ACDB$, respectively (C and D are exterior points of the major arc).

A chord passing through the centre of a circle (e.g., CD in Figure 1.3.15b) is called a *diameter*.

A figure bounded by two radii and a subtended arc (Fig. 1.3.15c) is called a *sector*. A *segment* of a circle is bounded by a chord and a subtended arc (Figure 1.3.15c).

A straight line (or a part of a straight line) passing through any two points on a circle is called a *secant* of the circle. A line which touches a circle in but one point is called a *tangent*; this common point is called a *point of tangency* (sometimes – *a point of contact*).

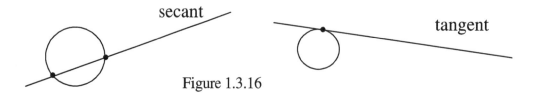

Figure 1.3.16

The definitions of the sum and the difference of two arcs, a multiple and a fraction of an arc are totally analogous to those for segments, e.g., in Figure 1.3.17,

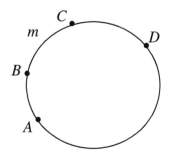
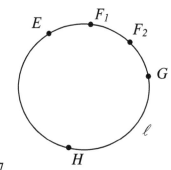

Figure 1.3.17

$$\cup AmD = \cup AB + \cup BC + \cup CD; \qquad \cup BC = \cup BD - \cup CD;$$

$$\cup EF_1 = \cup F_1F_2 = \cup F_2G = (1/3) \cup EG; \quad \cup E\ell H = 2\cup GH = 2\cup EG.$$

It can be shown that these operations of addition and multiplication by numbers are also subject to commutative, distributive and associative laws.

1.3 Questions and problems.

1. One may think of curves, surfaces, solids as formed, or *generated,* by motion:
   a) If a point moves, its path is a curve.
   b) If a curve moves, it generates a surface.
   c) If a surface moves, it generates a solid.
   Adduce examples for a), b), and c).

2. Yet in some cases the motion of a curve does not generate a surface, the motion of a surface does not generate a solid. Adduce examples of such situations.

3. Describe, how one should move a straight line in order to generate
   a) a plane;   b) a cone;   c) a cylinder.

4. In this course we study *plane geometry.* What does that mean?

5. Given a point in a plane, draw a straight line passing through this point. How many straight lines that pass through that point can you draw?

6. Draw a straight line that passes through two given points. How many straight lines that pass through a given pair of points can be drawn? How many curves can be drawn through a given pair of points?

7. Compare the statements "I have one dog" and "I have exactly one dog". Is there a difference in their meaning? Which of these two means the same as "I have one and only one dog"?

8. Propose an example illustrating the difference between the meanings of the phrases *one* and *exactly one*.

9. How many common points may two distinct straight lines have?

10. How many rays can emanate from one point?

11. How many rays emanate from one point and pass through some other point?

12. Name all the rays obtained as a result of the intersection of two lines, *AB* and *CD*, at some point, *O*. (Draw the figure).

13. There are four points, *A,B,C,D*, lying in one plane. How many lines do we need to connect each of these points with all the others in each of the following cases:
    a) all four points are collinear;
    b) *C* is collinear with *A* and *B*, but *D* is not collinear with them;
    c) none of possible triples of the given points is collinear.

14. There are four points, *A,B,C,D*, lying in one plane. *A,B,C* are collinear, and *B,C,D* are collinear. Prove that all four points are collinear.

15. Lines *a, b, c,* and *d* lie in one plane. Lines *a, b,* and *c* are concurrent. Also, lines *b, c,* and *d* are concurrent. Prove that all four lines are concurrent.

16. The straight line created by the extension of segment *AB* passes through some point *O*. The extension of segment *AC* into a straight line also passes through *O*. Are *A,B,C* necessarily collinear? – Prove or disprove (by a counterexample).

17. Visually compare segments *a* and *b* presented in Figure 1 (in each of the four cases). Then use a ruler or a compass to check your suggestion. What would you conclude based on this observation?

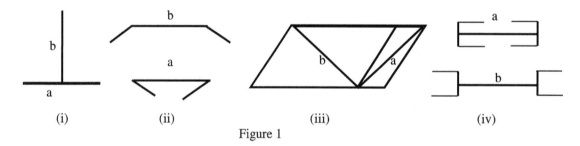

Figure 1

18. Segments *a,b,c* (Fig.2) are given. Construct the following segments:
    $a+b$; $a-c$; $a+b-c$; $2a$; $3c$; $a+2c-b$; $2a+2b$; $2a+b-c$.

Figure 2

19. Point *M* is located on the straight line *PQ*, passing through the points *P* and *Q*.
    a) Find *MP* if *MQ=16cm, PQ=31cm*.
    b) Find *MQ* if *PQ=0.42m, PM=3.8 dm*.

20. *A, B, C, D* are four collinear points taken in consecutive order. Determine segment *AD* if *AC=8 cm, BD=12 cm,* and *BC=5 cm*.

21. Two segments, *a* and *b*, are situated in the same line. Their common part equals *c*. Determine the segment covered by *a* and *b* together.

22. Points *M, N, S, R* are collinear. It is known that $MR > SM$, $NS < NR$, $MN < MS$.
   a) Draw figures showing all possible respective locations of the points in the line (we assume that two configurations of points provide distinct solutions if their respective sequences of the letters read as different "words"; for instance, *RNMS* in which *N* is very close to *M* and *RNMS* in which *NM* is much greater, represent the same solution).
   b) How many solutions would the problem have if the last inequality was not given?

23. Imagine you are a consulting mathematician kidnapped by pirates. They want you to determine whether a piece of a page from the diary of some captain (shown in Figure 3) provides enough information to enable them to find the buried treasure (otherwise they will not embark on a journey to that remote and notoriously unhappy archipelago). You have to prove that there is enough information to find the treasure (or you will not likely survive the adventure).

Remember that pirates are mean and perfidious; they do not trust anyone. However, they believe in postulates of Euclidean geometry. Thus, your argument must be based on these postulates.

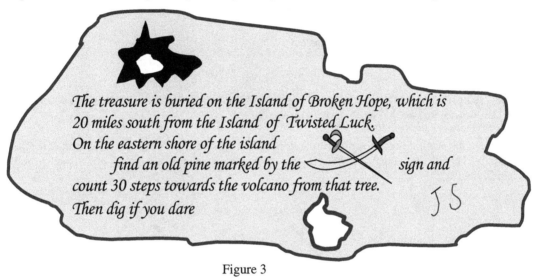

Figure 3

24. Children play soccer on a grass field. There is no lines or any other marks on the field, - only a pair of goals faced to each other. Children did not actually need any marks until one of the goalkeepers beat off the ball directed in his goal over the cross-bar (Figure 4). Now the other team must serve a corner.

Develop an *algorithm* (i.e., propose a sequence of actions) that will allow children to find the exact location of the corner of their field, if the width of the goal is 6 steps and they decided the width of the field to be 50 steps. Use the axioms of Euclidean geometry to substantiate each point of your algorithm.

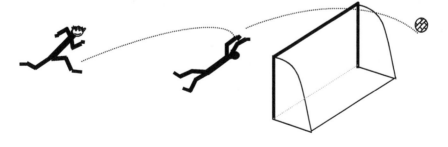

Figure 4

25. Use a compass only to *bisect* a segment approximately, i.e. to divide a segment approximately into two equal segments. In other words, propose an *algorithm* (sequence of steps) that will allow you to find the *midpoint* of a segment approximately within any given accuracy with the help of a compass only.

For the rest of the section use the result of Problem 25 whenever you need to find the midpoint of a segment.

26. Find segments a and b if their sum and difference are given (Fig.5).

Figure 5

27. Use a ruler to estimate the thickness of a page of your textbook. Is your estimate more accurate for a thicker or a thinner book? Substantiate your answer.

28. Compare the *distance* between points A and B with the distance between points C and D (Fig.6) visually. Then use a compass to check if your suggestion is right. What is your conclusion from this experiment?

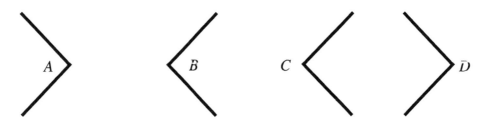

Figure 6

29. Two segments, $AB=44$ cm and $AC=72$ cm, are laid off on the same ray emanating from some point A. Find the distance between the midpoints of the two segments, AB and AC.

30. Two segments, $OA = a$ and $OB = b$, are laid off on the same ray emanating from some point O.
    a) Find the distance between the midpoints of the two segments.
    b) Find the distance between O and the midpoint of AB.

31. Use the map in Figure 7 and a ruler with the centimeter scale to find
    a) the width of the bridge, b) the length of the bridge, c) the distance between the house and the entrance to the bridge (located on the same shore as the house).
    Which of your estimates would be more accurate?

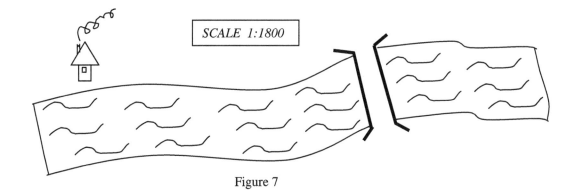

Figure 7

32. In a map of Europe using a scale of *1: 16 000 000 (how many kilometers to one centimeter?)* the distance between Paris and Barcelona equals *5cm* , and the distance between London and Dresden equals *54mm*. Find the actual distances in kilometers. Which of your estimates represents the distance along the earth surface (neglect the distortions caused by the relief) more accurately?

33. Suppose the scale of the map presented in Figure 7 has been changed to 1:900. What will be the width and the length of the figure depicting the bridge on the new map? What kind of transformation will every figure on the map experience as a result of such rescaling?

34. Explain the term "transformation". Propose a synonym of the word *transformation*. (Hint: in the preceding problems we used geographical maps in order to obtain results concerning parts of the Earth's surface. Can we see drawing a map, or mapping, as a transformation?).

35. Figure 8 shows a few figures in the plane. For each of the following situations propose an example or, if possible, a few examples of transformations of the plane.

    (a) A transformation that changes every figure.
    (b) A transformation that changes some figure or figures and leaves the rest unchanged.
    (c) A transformation that does not change any of the figures shown although it changes the positions of some points of the plane.

36. What kind of transformation is called an *isometry*? Why is it also called a *rigid motion*?

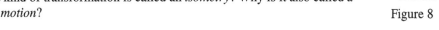

Figure 8

37. Draw a few figures in the plane. Now subject the plane to an isometry; describe your isometry. (i) Can you perform such an isometry that the figures will change their shapes? (ii) Can you propose an isometry that will change the respective positions of the figures in the plane? (iii) Can you propose an isometry that will not change the respective positions of the figures in the plane?

38. Use any available tools to check if the segments *a, b,* and *c* in Figure 9 are congruent. Explain your actions. Is *b* an isometric image of *a?* Is *c* an isometric image of *a?* If it is possible, describe isometries that will transform *a* into *b*; *a* into *c*.

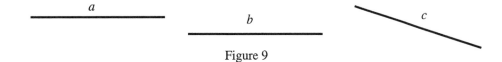

Figure 9

39. Specify, which of the transformations among the ones you have proposed in your solutions of problem #35 are isometries. For each of them determine what type of isometry it represents: (i) translation along a line; (ii) rotation about a point; (iii) reflection in a line.

40. Some Australians facetiously insist that the world map must look as it is shown in Figure 10. Propose a transformation or a sequence of transformations that will return the map to its regular view. What type of isometry do you absolutely have to use in order to solve the problem?

Figure 10

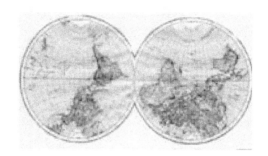

41. Describe the types of isometries that will carry the caravel in Figure 11 from its original position labeled (a) into each of the positions (b), (c), and (d). Did you have to use all three types of isometries whose existence has been postulated?

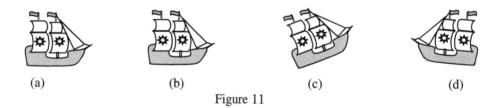

(a)   (b)   (c)   (d)

Figure 11

42. Describe the types of isometries that will carry the space ship in Figure 12 from its original position labeled (a) into each of the positions (b), (c), and (d). Did you have to use all three types of isometries whose existence has been postulated?
Why do you think you do not need all these types of isometries? What is the feature of the shape of the space ship that the caravel does not possess?

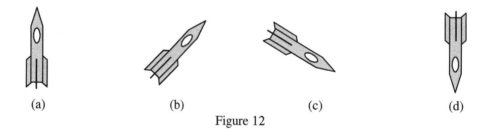

(a)   (b)   (c)   (d)

Figure 12

43. A barefoot hiker had not noticed the sign and crossed the border to the property of an evil magician. You can guess (look at Figure 13) what has happened to him.

a) Had his left footprints on the sand been isometric images of one another before he crossed the border?
b) Had his left footprints been isometric images of his right footprints before he crossed the border?
c) Is the transformation that magician applied to the hiker an isometry?

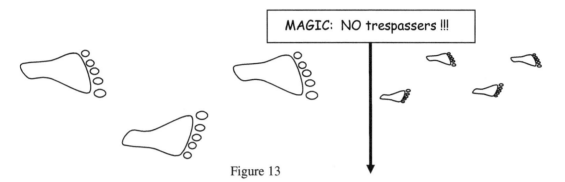

Figure 13

The following discussion and problems 44-47 are optional.

Transformation obtained as a result of two (or more) consecutive transformations is called a *composition of the transformations*.

For example, the first small footprint of the right foot in Figure 11 is obtained from the big right footprint by means of the following composition: *translation* of the big footprint in the forbidden area is followed by its *contraction*. If we denote the resulting transformation by T, we can present the composition by the following formula:
$$T = contraction \circ translation.,$$
which means that
$$T(\text{footprint}) = contraction(translation(\text{footprint})).$$

If some transformation, τ, produces figure B out of figure A, then the transformation that makes A out of B (if such a transformation exists!) is said to be the *inverse* of τ and is denoted τ^{-1}.

For example, if τ transforms a fly into an elephant, τ^{-1} transforms the elephant back into the original fly. We can use formulae to write that:

$$\tau(\text{fly}) = \text{elephant}; \qquad \tau^{-1}(\text{elephant}) = \text{fly}.$$

Thus a composition of a transformation and its inverse (in any order) is a transformation that does not change the original object, It is called an *identity transformation* and denoted I. For the above example:

$$I(\text{fly}) = (\tau^{-1} \circ \tau)(\text{fly}) = \tau^{-1}\big(\tau(\text{fly})\big) = \tau^{-1}(\text{elephant}) = \text{fly}.$$

44. Describe the following transformations illustrated in Figure 13, first with words and then with formulae (introducing appropriate notations):
   a) A transformation that produces a big left footprint out of the preceding big left footprint.
   b) A transformation that produces a big right footprint out of a big left footprint.
   c) A transformation that turns the last big left footprint before the magic line into the first small right footprint after the line.
   d) The transformations that are inverse to each of the above three (a, b, c).

45. Is a composition of isometries an isometry? Is the transformation inverse to an isometry an isometry? Substantiate your answers.

46. The clown's face in Figure 14 is subjected to four different transformations, α, β, γ, and δ, which are shown by arrows.
    a) Explain, why do you think that these transformations are isometries?
    α, β, γ represent three basic elementary types of isometries. These transformations are called: α – *reflection (in a line)*, β – *rotation*, γ – *translation*. How would you call their inverses?
    b) Describe transformation δ as a *composite* isometry (i.e. as a sequence of some isometries). Write a formula that represents δ as a composition of some other isometries shown in the figure or/and their inverses.
    c) For the three transformations, β, γ, δ, represent each of them in terms of the other two (and, maybe, their inverses).

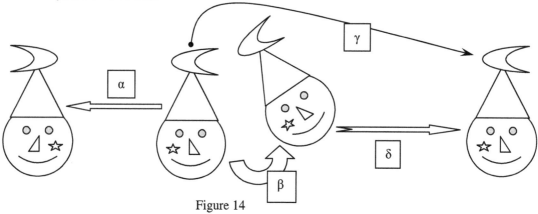

Figure 14

47. Draw a figure (you may also use the clown's face in Figure 14) and subject it to a translation and, independently, - to a rotation. (Use a ruler and a compass).
    a) Represent the translation as a composition of a few reflections.
    b) Represent the rotation as a composition of a few reflections.

In the following three problems, use isometries to prove the properties of *congruence of figures*.
(It is convenient to use capital English letters from the end of the alphabet, e.g., W, X, Y, Z to denote figures and small Greek letters for isometries, where appropriate).

48. Any figure is, obviously, *congruent* to itself. (This property of congruence is called *reflexivity*). What is the isometry that transforms a figure into itself? (It is called the *identity transformation*).

49. If a figure is congruent to some other figure, then the second figure is congruent to the first one. (This property of congruence is called *symmetry*). How are the isometries that transform the first figure into the second and the second figure into the first related? (Such transformations are called *the inverse* of each other).

50. If a figure is congruent to some other figure, and this other figure is congruent to a third figure, then the first figure is congruent to the third one (this property is called *transitivity*).

The following discussion of equivalence relations and problems 51 -56 are optional.

It follows from problems 48 – 50 that congruence of figures is an *equivalence relation*: it is a *binary relation*, which means it is a relation connecting two objects, that possesses the three properties listed in the above problems.

In general: a relation ~ , defined on some set, is called an *equivalence relation* if it is a binary relation and for all elements (X, Y, Z, T,...) of the set the following conditions hold:
(1) Reflexivity: X ~ X.
(2) Symmetry: If X ~ Y, then Y ~ X.
(3) Transitivity: If X ~ Y and Y ~ Z, then X ~ Z.

It can be proved that an equivalence relation splits the whole set on which it is defined into non-intersecting *subsets* called the *classes of equivalence* (see problem 53).

51. Explain why *being a relative* is an equivalence relation (on the set of all humans). Is *being an enemy* an equivalence relation? Substantiate your answer.

52. Show that the equivalence relation *being a relative* splits the set of all humans into subsets that do not have common elements (*non-intersecting* subsets). In other words, if we call the set of all relatives of a person a *family*, prove that every human is a member of exactly one family. That means: (1) every person is a member of some family; (2) one person cannot be a member of two or more families.

53. Let an equivalence relation be defined on the elements of some set S. The subset of S that consists of all elements equivalent to a given element x is called *the equivalence class* of x and denoted $[x]$. Prove that two equivalence classes either coincide completely or do not have common elements, and therefore S is a union of non-intersecting equivalence classes of its elements.

54. If two integers, m and n, produce the same remainder when divided by some integer k, they are called *congruent modulo k*, and this relation is denoted: $m \equiv n \pmod{k}$. For example, $6 \equiv 1 \pmod 5$ since $6 = 5 \cdot 1 + 1$ (the remainder is equal to 1) and $1 = 5 \cdot 0 + 1$ (the remainder is 1 as well). Also, $-4 \equiv 1 \pmod 5$ since $-4 = 5 \cdot (-1) + 1$ (again, the remainder is 1), etc.
   a) Prove that the described relation is an equivalence relation.
   b) Consider the relation of *congruence modulo 5* on the set of integers.
      In the above example we found that the numbers 6, 1, and -4 are congruent to 1 modulo 5, that is they belong to [1] (of course, this class could be denoted [6] or [-4] as well). In order to list all the elements of this class we can write: $[1] = \{..., -14, -9, -4, 1, 6, 11, 16, ..., 5n+1, ...\}$.
      Describe all the other equivalence classes of this relation. How many of them have you found?

55. Adduce examples of equivalence relations.

56. Adduce examples of *binary* relations (i.e. relations involving pairs) that are not equivalence relations.

57. Euclid formulated the COMPASS POSTULATE as follows: *A circle may be described with any point as its centre and any segment as its radius.* Assume that every ray emanating from the centre of a circle will have a point of intersection with the circle (i.e. lines and circles are continuous) and explain why this postulate is equivalent to AXIOM 3: *One and only one segment congruent to a given one can be laid off on a ray starting from the origin.*

58. What is the *locus* of the points collinear with two given points, A and B, and removed from A by no more than 5 cm? Substantiate your opinion.

59. Construct the locus of points collinear with given points A and B and located exactly 6cm from A. Prove that this is a locus.

60. Draw two concentric circles with their radii equal 1 cm and 3 cm respectively. Draw if possible the locus of points equidistant from these circles along the rays emanating from their centre.

61. How many chords can be drawn through a point taken inside (in the interior of) a circle?

62. How many diameters can be drawn through a point lying in the interior of a circle? Substantiate your answer.

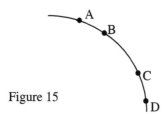

Figure 15

63. In Figure 15, an arc, *ABCD,* of some circle is shown.
    a) Given ∪AB=∪CD, prove that ∪AC=∪BD.
    b) Given ∪AC=∪BD, prove that ∪AB=∪CD.

64. Prove that two circles of equal radii are congruent.

65. Prove that a circle cannot have two centers.

2. MATHEMATICAL PROPOSITIONS

> The first person he met was Rabbit.
> "Hallo, Rabbit," he said, "is that you?"
> "Let's pretend it isn't," said Rabbit, "and see what happens."
> A. A. Milne, *Winnie-The-Pooh*.

2.1 Statements; Sets of Axioms; Induction and Deduction; Propositions; Theorems

a) <u>Formal statements</u>

We have so far introduced basic geometric notions and, where necessary, described their properties by means of axioms (we *postulated* the properties). Now we are ready to turn to the derivation of geometric results. Before we commence we must agree about "the rules of our game", i.e. we have to describe what is suggested to be a rightful way of obtaining results. Results are usually formulated as *statements –declarative sentences that are either true or false* (the latter two notions are undefinable). Not all declarative sentences are statements. For example, sentences such as "I am tired" or "Calgary is far away" could be true or false depending upon the context or circumstances. Thus they are not statements as we have formally defined this term, even though they are statements in the everyday meaning of that word. We shall denote statements by capital Latin letters.

b) <u>Connectives; elementary notions of logic</u> (**This subsection is optional**)

Statements may be used to form new statements by means of operations that are called *sentential connectives*. For the sake of brevity, each *connective* is denoted by a certain symbol. These notations are not universal, i.e. they may differ from text to text. We shall use the following very common set:

Connective	Name	Symbol	Other symbols
not (S)	*negation (of S)*	$\text{bar}(\bar{S})$	¬ ~
or	*disjunction*	∨	∪ +
and	*conjunction*	∧	∩ ·
it follows	*implication*	⇒	→
if and only if	*biconditional*	⇔	↔

Let us describe these connectives and their properties and give some examples.

For a given statement **S**, its *negation* is a statement meaning (consisting in) the falsity of **S**. Usually it is denoted $\overline{S}$ or $\neg S$; we shall use the more compact $\overline{S}$ notation.

For example,   **S**: it is raining $\overline{S}$: it is not raining

                       F: *m* is even $\overline{F}$: *m* is not even

Disjunction is the statement formed from two given statements by the connective *or*, thereby asserting the truth of at least one of the given statements (compare to the notion of *union* or *sum* of sets in the set theory).

Conjunction is the statement formed from two given statements by the connective *and*; it is true if and only if both given statements are true (compare to *intersection* or *product* of sets).

Let us consider some examples of disjunctions and conjunctions.

<u>Example (a)</u> **A**: this flying object is an airplane;   **B**: this flying object is a bird.

$A \wedge B$: This flying object is a bird and an airplane (at the same time). It is clear that this statement is *false* or *contradictory* independently of the falsity or truthfulness of **A** and **B**. Such a statement is called a *contradiction*.

$A \vee B$: this flying object is an airplane or a bird or both. We realize that the latter is impossible since an airplane cannot be a bird; however, the operation of disjunction suggests this option to be included since for some other statements it can be true (this kind of disjunction is called inclusive).

It is convenient to illustrate statements by figures as it is done in set theory. Such figures are called *Venn diagrams*.

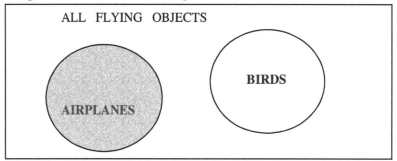

Figure 2.1.1

For instance, the last example is illustrated by the diagram in Figure 2.1.1. Each point inside the rectangle represents the statement: *this is a flying object*; the points in the grey and white circles represent respectively the statements: *this is an airplane* and *this is a bird*. These circles do not intersect since none of the known birds is an airplane and an airplane is never a bird: the conjunction of the two statements cannot be true.

Example (b)   *T*: *k* is divisible by 2;   *R*: *k* is divisible by 3.
   $T \wedge R$: *k* is divisible by 2 and by 3.
   $T \vee R$: *k* is divisible by 2 or by 3 or by both 2 and 3.

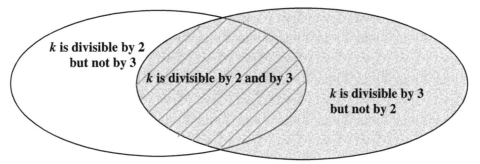

Figure 2.1.2

In this example, the conjunction may be true since there are numbers that are divisible by both 2 and 3. The statements representing the conjunction belong to the common part of the oval in the above diagram. It can be shown that the conjunction is equivalent to the statement *k is divisible by* 6:
$T \wedge R \Leftrightarrow$ *k is divisible by 6*.

The following two *principles* (the word *principle* is another synonym of *axiom*) of logic determine the relations between a statement and its negation.

(i) Law of the excluded middle
For every statement S, either S is true or $\overline{S}$ is false.

In other words, at least one of the statements S or $\overline{S}$ is true, – there is no "middle" option (hence the name of the principle). One can also formulate the principle as an assertion that the *disjunction* (*logical sum*) of a statement and its negation $S \vee \overline{S}$ *is universally true*. Such a statement, which is true independently of the truthfulness of the elementary statements constituting it, is called a *tautology*. For example, the following statements are tautologies: it is either raining or not raining; the number *m* is even or it is not even.

Thus, at *least one* of S or $\overline{S}$ is true. Then the natural question arises: can they both be true? The next principle answers this question.

(ii) Law of contradiction
A statement cannot be both true and false at the same time.

In other words, *at most one* of S or $\overline{S}$ is true.

One can express the same by saying that it is false that both S and $\overline{S}$ are true:
$S \wedge \overline{S} = \varnothing$ - *the conjunction* (*logical product*) of a statement and its negation is *universally false*, i.e. it is false independently of the truthfulness of S. For example, it cannot be true that some number *m* is even and not even at the same time. A universally false statement is called a *contradiction*.

According to the law of contradiction, for any statement S, at most one of the statements S or $\overline{S}$ is true. On the other hand, the law of the excluded middle asserts that *at least one* of these two statements is true. Therefore, these two laws imply that for any statement S exactly one of the statements S or $\overline{S}$ is true. This property is called the *complementarity* of a statement and its negation.

Two statements are *complements* of each other if their union is a *tautology* (i.e. it is universally true), and their intersection is a *contradiction* (i.e. it is universally false). For a statement and its negation:

$S \vee \overline{S} = I$ where I stands for *tautology*;

$S \wedge \overline{S} = \emptyset$ where $\emptyset$ denotes *contradiction* (in set theory this symbol is used to denote an *empty set*).

Complementarity is easy to illustrate:

In Figure 2.1.3, the whole bounded domain (tautology) is divided into two non-intersecting regions, – a shaded one (representing S) and a white region (representing $\overline{S}$). Every point of the whole bounded region belongs to one and only one of the regions representing S or $\overline{S}$.

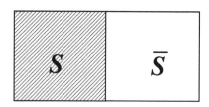

Figure 2.1.3

In particular, complementarity means that a statement S is true if and only if its negation $\overline{S}$ is false, that is S is a negation of $\overline{S}$. This important consequence of complementarity is called the *law of double negation*:

The negation of the negation of a statement is equivalent to the original statement: $\overline{\overline{S}} \Leftrightarrow S$.

For the examples considered in the previous section this property can be illustrated as follows: $\overline{\overline{S}}$ $\Leftrightarrow$ S

     it is not true that it is not raining $\Leftrightarrow$ it is raining
     it is not the case that m is not even $\Leftrightarrow$ m is even

The law of double negation is often used for so-called indirect proofs: instead of proving directly that some statement S is true we can prove the falsity of $\overline{S}$. If $\overline{S}$ is false, then it follows from the law of double negation that S is true. We shall discuss this kind of proofs in detail in the next section (2.2).

c)   Inductive and deductive reasoning; axioms and propositions

The next question is: how can we establish whether a statement is true or false?

First of all we can use *induction – a method of reasoning, which obtains general laws from the observation of particular facts or examples*. This is how we have established our *axioms*, statements that are *believed* to be true. It would be appropriate to remind here that axioms are introduced to the theory to describe properties of our *undefined* notions, basic notions of the given field of knowledge.

There is a set of requirements imposed on systems of axioms in fundamental sciences. Of course, these requirements are axioms themselves; they may be called *logical axioms*. Most commonly these requirements are *consistency, independence, completeness*.

A set of axioms is *consistent* if the axioms do not contradict each other and it is impossible to derive from these axioms results that will contradict each other.

An axiom of a set is said to be *independent* if it cannot be concluded from some other axioms of the set. This property ensures that a theory is built on a minimal number of axioms. It is important for the clearness of the theory and for the real understanding of what the theory is built on.

The *completeness* guarantees that a set of axioms is sufficient to obtain all possible results concerning a given field of knowledge. The latter means the aggregate of all statements concerning the *original set of undefined terms*. In other words, a set of axioms is *complete* if it is impossible to add an additional consistent and independent axiom without introducing additional *undefined terms*.

Once a set of axioms is established, all statements of the theory can be *deduced* from it. For this purpose they use *deduction – reasoning from general laws to particular cases*.

Deduced, or *proved* statements are usually called *propositions*. The most important propositions are called *theorems*. They form a "skeleton" of a theory. Less important and less general statements, derived from the axioms and basic theorems, are called *problems*.

In modern science the boundary between theorems and problems is extremely vague. Still, in well-established areas, such as elementary geometry, it is very clear, which propositions should be called theorems.

If a theorem is auxiliary in its nature, i.e. it is mostly used to simplify the proof of some basic theorem, it is called a *lemma*. A theorem that follows immediately (i.e. its proof is very short and straightforward) from some preceding axiom or proposition is usually called a *corollary*. For example, the fact that two lines can intersect only at one point is a corollary of AXIOM 1.

The division of theorems into theorems themselves, lemmas and corollaries is, of course, conventional.

For instance, such fundamental results as Schur's Lemma in Algebra or Borel's Lemma in Analysis are more important than many theorems following from them. They are called *lemmas* rather by historical reasons.

2.1 Questions and Problems.

1. Determine, which of the following declarative sentences are *formal statements*, and, if possible, which of the latter are true:
   a) Mt. McKinley is high.
   b) Mt. McKinley is over 20000 feet high.
   c) I am old.
   d) I am 325 years old.
   e) The way from my house to my friend's house is long.
   f) It is nice to live in Toronto.
   g) Ontario is the deepest lake in the world.
   h) Iron is heavy.
   i) Every positive number has two square roots.

2. Adduce examples of a declarative sentence that
   a) is not a formal statement;
   b) is a formal statement that is true;
   c) is a false formal statement.

3. What kind of reasoning is called *inductive*? Adduce examples.

4. A few samples of inductive reasoning and inductive *inference* follow. In each case determine whether the conclusion is true or false. In every case explain why inductive reasoning did or did not result in the right conclusion.
   a) I have seen a lot of birds; all of them can fly. Hence, all birds can fly.
   b) There is nothing but sand in a desert.
   c) This fellow always smiles to me and praises me, so he is my friend.
   d) If I eat this cereal, I will become a famous basketball player.
   e) This teacher is mean: he gave me a low mark.
   f) All bad movies are produced in Hollywood.
   g) The Sun moves around the Earth, which is immovable.
   h) If an object floats, it is made of material that is lighter than water.
   i) Bodies fall quicker in exact proportion to their weight. (Aristotle).
   j) All bodies fall at the same rate. (Galileo).
   k) If a moving object is not subjected to a force, it will come to rest. (Aristotle).
   l) If a moving object is not subjected to a force, it will continue to move with the same speed. (Newton).
   m) Total energy of a closed system is a constant.

5. Bring examples of poor inductive reasoning from, newspapers, magazines, TV or radio broadcasts.

Sometimes, inductive reasoning results in axioms that serve for developing a theory (where else axioms may come from?). One of such axioms, called *the principle of mathematical induction*, plays an important role in the theory of natural numbers (do you know any other axioms for natural numbers?).

THE PRINCIPLE OF MATHEMATICAL INDUCTION:

If a set of natural numbers includes 1 and also includes $n+1$ whenever it includes n, then the set includes every natural number.

This principle is one of axioms for natural numbers, and is, of course, an inductive statement following from our experience with an *undefined notion* of natural number.
It is often used for proving various statements concerning sequences of statements. For such occasions it is usually reformulated as follows:
   Suppose there is an infinite sequence of (mathematical) statements
   $$S_1, S_2, S_3, S_n, ...,$$

which altogether form some general statement S.
If
    (i) S_1 is true;
    (ii)   It is shown that S_{n+1} is true if S_n is true,
Then each of the statements of the sequence is true, and, thus, *S* is true.

6. Use the principle of mathematical induction to prove the following statements:
   a) *n* distinct concurrent lines in a plane divide the plane into $2n$ regions.
   b) *n* lines lying in one plane divide the plane into regions whose number does not exceed 2^n.
   c) $1+2+3+...+n = \dfrac{n(n+1)}{2}$
   d) $1^2 + 2^2 + ... + n^2 = \dfrac{n(n+1)(2n+1)}{6}$
   e) $(1+p)^n \geq 1+np$ for $p \geq -1$.

7. What kind of reasoning is called deductive?

8. Are axioms obtained by means of induction or deduction? Adduce examples.

9. Which of the following examples represent deductive inference? Explain and comment on the truthfulness or falsity.

   a) This solid is made of wood, so it will float.
   b) This segment is the sum of the segments *a* and *b*, so it is greater than *a*.
   c) Look at odd numbers: 3, 5, 7, 9, 11,... Odd numbers are either prime or divisible by 3.
   d) Perpetual motion is impossible.
   e) The trajectory of this particle is not a straight line, thus some forces have been applied to the particle.
   f) Albert has been receiving only low marks, so he is not a smart fellow.
   g) UFO (unidentified flying objects) do not exist.
   h) These two segments are laid off from the same vertex on the same ray, and their second endpoints do not coincide, hence these segments are not equal.
   i) These two circles have equal radii, so they are congruent.
   j) Number 121 is not divisible by 2, by 3, by 5, by 7... It must be prime.
   k) A body moving along a circle is subjected to a force.

10. Adduce examples of deductive reasoning from:
    a) everyday life or literature;
    b) natural science;
    c) mathematics.

11. Theories are built on axioms.
    a) What do axioms describe?
    b) Which sets of axioms are said to be (i)consistent, (ii)independent, (iii)complete?

12. Propose a set of axioms that would enable us to operate with natural numbers the way we do. (If you find this assignment difficult find in mathematical literature Peano Axioms).

13. What kind of statement is called a proposition?
    Which of the following are propositions?
    a) $(x+y)^2 = x^2 + 2xy + y^2$
    b) $\dfrac{a}{c} + \dfrac{b}{c} = \dfrac{a+b}{c}$
    c) A solid floats if it is made of material that is lighter than water.
    d) A perpetual engine does not exist.

e) $2^3 = 2 \cdot 2 \cdot 2$.
f) A circle is the locus of points equidistant from a given point called the centre.

14. Explain the differences (if there are any) between the following terms: theorem, lemma, problem, corollary. Adduce supportive examples.

2.2 Theorems: the structure and proofs.

a) <u>Conditional form; direct proofs</u>

Each theorem consists of two parts: the *hypothesis (or condition)* and the *conclusion*. The *hypothesis (condition)* determines what is suggested as given. The *conclusion* is something that will be shown to follow from the hypothesis, or to be an *implication* of the condition. In other words, a theorem asserts that, given the hypothesis, the conclusion will follow.

Thus, the structure of a theorem may be represented by the diagram:
If [hypothesis] *then* [conclusion], or
[hypothesis] $\Rightarrow$ [conclusion].

In some cases a theorem may state only the conclusion; the axioms of the system are then implicit (assumed) as the hypothesis. If a theorem is not written in the aforesaid *conditional form*, it can nevertheless be translated into that form. For example,

Theorem 2.2.1 *Congruent arcs (of the same circle) are subtended by congruent chords* can be formulated as

if two arcs of a circle are congruent, then the chords that subtend them are congruent as well.

(Apparently, the word *then* can be omitted in this formulation, as in many other cases). Let us prove Theorem 2.2.1 (hereafter for the sake of brevity we shall write "Th" instead of "Theorem" when referring to some theorem).

Proof. Suppose there is a circle (Fig. 2.2.1), and two arcs in this circle are congruent:
$\cup AmB = \cup CnD$.

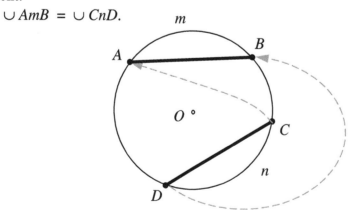

Figure 2.2.1

According to the definition of congruence, we can use a rigid motion (isometry) to superimpose $\smile CnD$ on $\smile AmB$ so that these two arcs will totally coincide. Suppose (for certainty) that point C coincides with point A, and D coincides with B.

Then, by AXIOM 1, there is one and only one straight segment that connects points A and B. Thus, as a result of this isometry, segment CD will coincide with AB, in other words, the two segments are congruent, Q.E.D. (Q.E.D. is the abbreviation of *Quod Erat Demonstrandum* meaning "what was to be shown"; it signifies the end of a proof).

This is a typical *direct proof*. It consists of a chain of statements *inferred* one from another, starting with the hypothesis and ending with the conclusion of the theorem. Each step in the proof (each "link of the chain") is justified by means of axioms or definitions. In other cases (e.g., more advanced theorems) justifications may also rely on previously proved propositions, theorems, lemmas or preceding steps in the proof.

b) Proof by contradiction (RAA)

Among *indirect proofs* the most frequently used one is the *proof by contradiction*.

It is also called *Reductio Ad Absurdum* (abbreviated RAA).

To describe this method we have to consider the notion of *the negation* of a statement. For a given statement S its negation is a statement meaning (consisting in) the falsity of S. Usually it is denoted $\overline{S}$.

A few examples of a statement (left column) and its negation (right column) follow below:

S	$\overline{S}$
This animal is a dog	This animal is not a dog
X is a negative number	X is positive or zero
These figures are not congruent	These figures are congruent

One of the *principles* (axioms) of logic, called the *law of the excluded middle*, asserts that for any statement (consider, e.g., the above examples), either the statement itself or its negation is true. Another principle, called the *law of contradiction*, claims that a statement and its negation cannot be true simultaneously. Therefore, according to these two principles, for every statement S, exactly one of the statements S or $\overline{S}$ is true..

In particular, if $\overline{S}$ is not true, then S is true.

The latter law of logic serves as the basis for the *proof by contradiction* (RAA): instead of proving directly the truth of a statement, we prove the falsity of its negation.

To prove the conclusion of a theorem using RAA, one assumes the *negation* (denial) of the conclusion to be true and deduces an absurd statement using the hypothesis of the theorem in the deduction. Thus, the proof shows that the hypothesis and the negation of the conclusion are contradictory. Then, the negation of the conclusion is false, therefore the conclusion is true.

To demonstrate this method we shall prove the following theorem:

Theorem 2.2.2 *A circle cannot have more than one centre.*
(In conditional form it would be: ***If** a figure is a circle, **then** it cannot have more than one centre.*)

Proof. Let us suggest that the conclusion of the theorem is false; in particular the circle shown in Figure 2.2.2 has two or more centres. Let us denote two of the centres O and P.

According to AXIOM 1, there is exactly one line that passes through the points O and P. Let M and N be the points of intersection of this line with the circle.

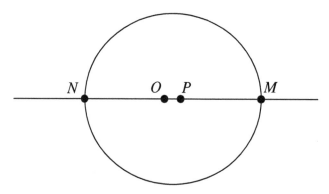

Figure 2.2.2

Since we have suggested that O is a centre of the circle, the segments NO and OM are congruent as two radii of the circle centred at O: $NO = OM$. Similarly, $NP = PM$ as two radii of the circle centred at P.

Point O lies in the interior of NP, hence $NO < NP$; similarly, $PM < OM$. Then we can write: $NO < NP = PM < OM = NO$, which means that $NO < NO$. This result contradicts to the axiom of congruence of segments which asserts that each segment is congruent to itself. Thus the suggestion of the existence of two (P and O) or more distinct centres is false; therefore a circle cannot have more than one centre, Q.E.D.

2.3   Direct, converse, inverse, and contrapositive theorems.

As we have already mentioned, each theorem can be formulated in the *If... then...* form that is also called an *implication* or a *conditional statement*.

Denoting the hypothesis (condition) of a theorem by ***H*** and the conclusion by ***C*** (hereafter we shall use capital Latin letters to denote statements), we can symbolically represent the *If **H**, then **C*** theorem as ***H*** $\Rightarrow$ ***C***. It is read as ***C** follows from **H*** or ***H** implies **C***. Mathematicians also say: ***H** is sufficient for **C*** or ***C** is necessary upon **H***.

The *if* clause, ***H***, is called the *antecedent* and the *then* clause, ***C***, the *consequent*.

The theorem formed by interchanging the hypothesis (antecedent) and the conclusion (consequent) of a given *direct* theorem, is called the *converse* of the given theorem. For example, one can formulate the theorem "if chords are congruent $\Rightarrow$ the respective arcs are congruent", which would be the converse of Theorem 2.2.1,

"if two arcs are congruent $\Rightarrow$ the respective chords are congruent".

It is not always the case that the converse of a true statement (theorem) is true. For example, let us consider the following conditional statement:

I live in Edmonton $\Rightarrow$ *I live in Alberta.*

The converse statement would be:

I live in Alberta $\Rightarrow$ *I live in Edmonton,*

that is evidently wrong (also, see the picture at the end of the section).

If the converse of a true theorem is true, as for instance, in the case of Th. 2.2.2 and 2.2.1, one says that the hypothesis and the conclusion of the theorem are logically *equivalent*:

$H \Leftrightarrow C$, or H iff C

(the latter is a short notation for: *H is true if and only if C is true*).

The *inverse* of a given theorem is formed by negating both the hypothesis (antecedent) and the conclusion (consequent): $\overline{H} \Rightarrow \overline{C}$, that means: if the hypothesis is false, then the conclusion is false.

For the conditional statement: *I live in Edmonton* $\Rightarrow$ *I live in Alberta,* the inverse would be: *I do not live in Edmonton* $\Rightarrow$ *I do not live in Alberta,* that is, of course, false. This example shows that in general **the inverse of a true statement is not true.**

It should be noticed here that when mathematicians say "is not true", they mean "is not true in general" or "is not necessarily true". However, "is not true" in mathematical sense does not exclude the option "sometimes is true". Thus, some inverses of valid statements are true. For instance, the inverse of Th. 2.2.1 is *unequal arcs in the same circle are subtended by unequal chords*, that is, of course, true (one can prove). Another example: a mathematician would say that *an integer is not even* (in general), however one would agree that some integers are even.

One can also observe another *important* feature of deductive (logical) systems: a statement can be disproved by a single counterexample (that is how we have shown that the inverse [also the converse] of a valid statement is not valid). Still one cannot prove a statement by adducing examples in favour of this statement. For instance, one can show people who do not live in Edmonton and do not live in Alberta, and yet there are quite a few of those who live in Alberta but not in Edmonton.

It is easy to see that for a given conditional statement (no matter whether it is true or false) its **converse and inverse are logically equivalent**, i.e. if the converse is true, then the inverse is true and vice versa. Really, let us suppose (using our standard notation) that for some proposition $H \Rightarrow C$ (that may be true or false), its converse is true: $C \Rightarrow H$. Then, if H is not true, C cannot be true, because C

necessarily implies H (H is necessary if C is true). Translating the latter into formal notation, we obtain: H is not true $\Rightarrow C$ is not true,

or $\overline{H} \Rightarrow \overline{C}$, the inverse statement., Q.E.D. Similarly, one can prove that the inverse implies the converse, and thus, the inverse and the converse are equivalent.

Finally, the *contrapositive* of a given conditional statement $H \Rightarrow C$ is $\overline{C} \Rightarrow \overline{H}$ formed by interchanging the hypothesis (antecedent) and the conclusion (consequent) and negating both of them.

For our example *I live in Edmonton* $\Rightarrow$ *I live in Alberta,*

the contrapositive will be : *I do not live in Alberta* $\Rightarrow$ *I do not live in Edmonton.*

One can easily show that **a direct statement is logically equivalent to its contrapositive**.

This property of conditional statements is called *contrapositive reasoning*. This is the kind of reasoning we use when proving conditional statements *by contradiction* (RAA *method*). Let us see how it works. Suppose, we have to show that if H is true, then C is true: $H \Rightarrow C$. We suggest that $\overline{C}$ is true, which is equivalent to C *is false* and arrive to the conclusion that H cannot be true, i.e. $\overline{H}$ is true; thus $\overline{C}$ implies $\overline{H}$. Since the contrapositive of $H \Rightarrow C$ is true, the direct statement is also true.

Contrapositive reasoning is often unconsciously used in our everyday life. For example, when looking on a shelf for a certain book with a green cover, one would skip (deny) all books of other colours (non-green).

Let us denote the statements:

H: this is the required book; $\qquad C$: this book has a green cover.

The respective negations are:

$\overline{H}$: this is not the required book; $\qquad \overline{C}$: the cover of this book is not green.

It is known that $H \Rightarrow C$ (if it is the needed book, then it is green). Then we start searching. The cover of the first book is blue (not green), hence it is not what we need: $\overline{C} \Rightarrow \overline{H}$; the second book is red (not green), so it is not the one we need: $\overline{C} \Rightarrow \overline{H}$; ... the twenty seventh book is green, – let us look at the title, it may be the one we are looking for.

Let us notice that the converse of $H \Rightarrow C$ is not necessarily true, therefore the green book is not necessarily the one we need.

The properties of conditional statements are conveniently visualized by means of *Venn diagrams*, where statements are represented by points in the corresponding plane regions.

For our "territorial" example: *If I live in Edmonton, then I live in Alberta*, the diagram emerges naturally (see Figure 2.3.1 below).

A plane bounded domain represents Alberta, and everyone within this domain lives in Alberta. Edmonton is shown as a rectangle located approximately in the middle of the province. A person inside this rectangle illustrates the direct statement, the inhabitants of the Northern and Southern Alberta disprove, respectively, the converse and inverse, and a resident of the North West Territories supports the contrapositive statement.

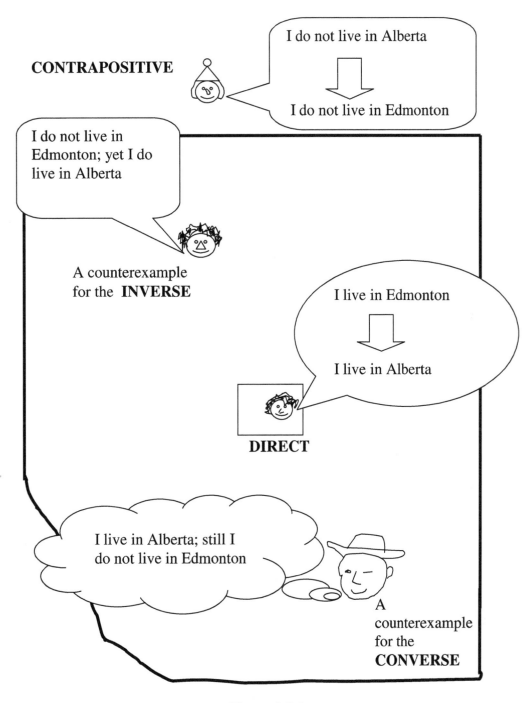

Figure 2.3.1

We shall conclude this section with the summary of logical relations between a direct and the related conditional statements.

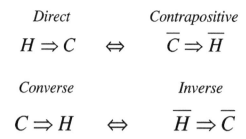

2.2-3 Questions and Problems.

1. In each of the following cases put a statement in the conditional form. Underline the hypothesis (condition) by a straight line and the conclusion – by a wavy line.
   a) Stop on red light.
   b) Wooden objects float.
   c) An isometry transforms a figure into a conruent figure.
   d) Two circles of equal radii are congruent .
   e) A statement based on empiric induction is not necessarily true.
   f) Axiom that follows from some other axioms of the same set, violates the independence property.
   g) A city located in Hungary is a European city.
   h) Decent people do not cheat.
   i) Two distinct lines have at most one common point.

2. Explain, what kind of proof is called direct. What kind of proof did you use in problems 12, 13, 42, 44, 52 of Section 1.3?

3. For each of the following statements in the left column choose its negation from the statements in the right column. Substantiate your choice based on properties of complements.
   a) This animal is a dog.
      (i) This animal is a cat
      (ii) This is not an animal
      (iii) This animal is not a dog

   b) The expression in brackets is positive.
      (i) The expression in brackets is negative
      (ii) The expression in brackets is either negative or zero.
      (iii) The expression that stands outside the brackets is positive.

   c) This number is divisible by 5.
      (i) This number is prime
      (ii) This number is divisible by any number except 5.
      (iii) This number is not divisible by 5.

   d) Segment a is congruent to segment b.
      (i) Segment a is greater than segment b.
      (ii) Segments a and b do not lie in the same line.
      (iii) Either $a < b$ or $a > b$.

4. Propose an example of a statement and its negation. Propose also a few false (confusing) negations for your example.

5. Use RAA (proof by contradiction) to prove that two distinct lines may have at most one common point.

6. Use RAA to prove that a circle has exactly one centre.

7. For each of the following statements do the following: (i) put it in the conditional form; (ii) write its converse, inverse, and contrapositive statements (iii) determine if the converse is true; (iv) illustrate the statements by a Venn diagramm.
   a) All humans have one nose.
   b) Congruent circles have equal radii.
   c) An even number that is divisible by 3, is divisible by 6.
   d) A moving object that changes the direction of motion, is subjected to a force.

8. Make a true statement whose converse is false. Is the inverse false? Is the contrapositive false?

9. Make a statement whose converse is true. Reformulate the statement using the expressions *if and only if, and conversely, and vice versa* to show that the converse is also true. Is the inverse true? Is the contrapositive true?

10. For each of the following pairs of statements answer the following questions:
    Is (a) necessary for (b)?
    Is (a) sufficient for (b)?
    Are (a) and (b) equivalent?
    (i)   (a) Number $k > 2$ is prime.   (b) Number $k > 2$ is odd.
    (ii) (a) This is a mammal.   (b) This is a whale.
    (iii) (a) Each of these transformations is an isometry.   (b) The composition of these transformations is an isometry.
    (iv) (a) $x^2 = 25$ (b) $x = 5$
    (v)   (a) This natural number is odd and divisible by 5.   (b) The last digit of this natural number is 5.

3. ANGLES

> "That's funny," said Pooh. "I dropped it on the other side ,"said Pooh, "and it came out on this side! I wonder if it would do it again?"
>
> A.A.Milne, *The House at Pooh Corner*.

3.1 Basic Notions

An *angle* is a figure formed by two rays that emanate from a common point and do not lie on the same line. The point is called the *vertex* of the angle, and the rays are called the *sides* or *arms* of the angle. The sides of an angle divide the whole plane into two parts, called the *interior* and *exterior* of the angle.

Usually (if it is not specified otherwise) the *interior* is the region containing any segment that connects two points on the sides of an angle (Fig. 3.1.1). Such a region is called *convex*.

Figure 3.1.1

Some examples of convex and non-convex regions are shown in Figure 3.1.2.

In general, a *convex region is a subset (part) of the plane that contains the entire segment connecting any pair of points lying in the subset.*

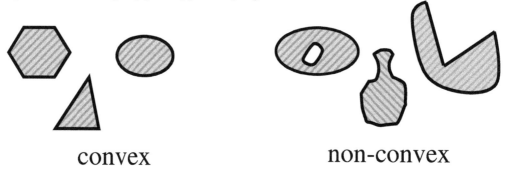

Figure 3.1.2

There are a few different ways of naming angles. An angle may be denoted by the angle sign "$\angle$" and a capital letter labelling the vertex. In Fig. 3.1.3, the angle formed by rays l and m emanating from A may be denoted as $\angle A$, or $\angle(l, m)$, or $\angle(m, l)$.

Figure 3.1.3

However, in some cases, when a few angles have a common vertex, such a notation may lead to ambiguities. Then it is convenient to label a point on each side of an angle and to denote the angle by three letters. It is important that the letter naming the vertex of the angle be placed in the middle. Numbers with the angle sign and small Greek letters without any signs are also used to denote angles. For example, in Fig. 3.1.4, $\angle BAC$ is also denoted $\angle 1$ or α; $\angle CAD$ is also $\angle 2$ or β; $\angle EAD$ may also be called $\angle 3$ or γ; δ is another name for $\angle DAB$.

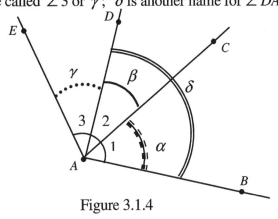

Figure 3.1.4

We do not distinguish between $\angle BAC$ and $\angle CAB$ or, e.g., $\angle CAD$ and $\angle DAC$.

According to the general definition of congruence, two angles are *congruent* if they coincide as a result of superposition, or, in other words, one of them can be transformed into another by means of an isometry. In order to compare two angles, e.g. $\angle AOB$ and $\angle A'O'B'$ in Figure 3.1.5, one should superpose them so that the vertices, O and O' and two corresponding sides, e.g. OA and $O'A'$, coincide, and the interiors of the angles are located on the same side of the coinciding arms (OA & $O'A'$). If the other two corresponding sides (OB & $O'B'$) also coincide, then the angles are congruent (Fig. 3.1.5).

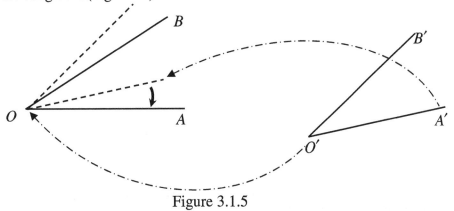

Figure 3.1.5

For any pair of congruent angles, such as $\angle AOB$ and $\angle A'O'B'$, one can describe the sequence of isometries that will superimpose one of these angle onto the other. At first we apply a (type (*i*)) isometry that will carry O' onto O. Then, if the arm obtained from $O'A'$ of $\angle A'O'B'$ does not fall on OA, we shall apply a rotation (type (ii) isometry) that will remedy this. If the image of $O'B'$ lies on the same side of *OA* as *OB*, the angles will coincide (Fig.3.15a). Otherwise, we shall reflect the plane in *OA* (type(iii) isometry) to make the angles coincide (Fig.3.15b).

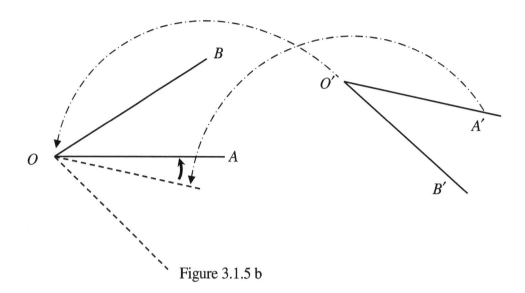

Figure 3.1.5 b

If two angles are not congruent, the interior of one angle, e.g. $\angle O'A'B'$, will constitute, maybe as a result of a motion, part of the interior of the other angle ($\angle OAB$). Then we shall say that the first angle is *less* than the second: $\angle O'A'B' < \angle OAB$ (Fig.3.1.6). One can also say that the second angle is *greater* than the first one.

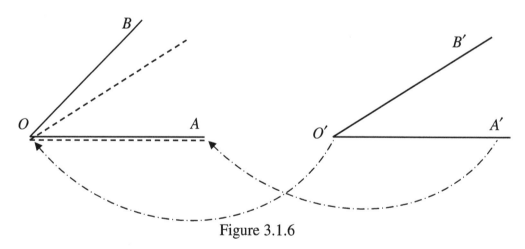

Figure 3.1.6

Two angles are called *adjacent* (in some texts *adjoining*) if they share a common vertex and one arm, and their interiors are located on opposite sides of the common arm. For instance, in Fig. 3.1.7 (b), $\angle AOB$ and $\angle BOC$ are adjacent.

The *sum of two angles* is the angle formed by the sides that are not common if the two angles are made adjacent to each other. Thus, in Figure 3.1.7, in order to add angles *LMN* and *PQR*, we built an angle *AOB* congruent to *LMN* and, adjacent to it, ∠BOC congruent to ∠PQR. (It will be shown later how to perform this construction). Then, ∠AOC = α + β = ∠LMN + ∠PQR.

It can be shown that addition of angles is *commutative* and *associative*.

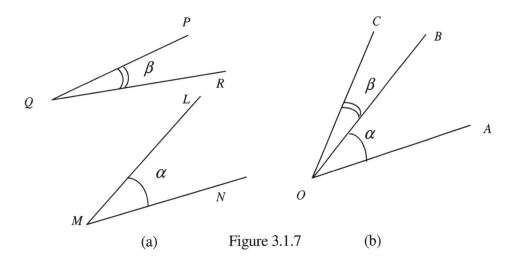

(a) Figure 3.1.7 (b)

The other arithmetic operations are naturally defined as follows. The difference ($\gamma - \beta$) of angles γ and β, is such an angle α that $\alpha + \beta = \gamma$.

In Fig. 3.1.7, ∠AOB = α = ∠AOC − ∠BOC.

If some angle α is added to a congruent angle, the sum is called *double* the original angle, 2α. Then another angle congruent to α can be added, and the resulting angle can be called triple the original. The sum of k angles congruent to α is called a *k-multiple* of α and denoted $k\alpha$. The original angle α will be called *one k^{th}* ($1/k$) of the obtained angle. The definitions of multiples and fractions of angles are illustrated in Figure 3.1.8, where ∠DOH = 4α = 4∠ABC, and ∠ABC = $\frac{1}{4}$∠DOH.

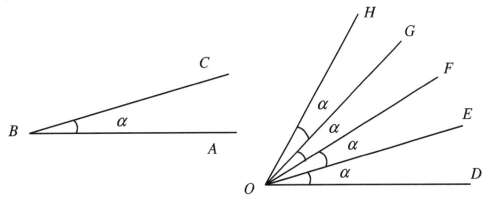

Figure 3.1.8

Later (in Chapter 4) we shall prove that each angle can be *bisected – divided into two congruent* angles. A ray emanating from the vertex of an angle and dividing the angle into two congruent angles is called a *bisector* of the angle. For instance, in Figure 3.1.8 $\angle DOE = \alpha$, and $\angle EOF = \alpha$, hence these angles are congruent, and we can say that *OE* is a *bisector* of $\angle DOF$. We shall also prove that *an angle can have only one bisector.*

3.2. Generalization of the notion of an angle. Types of angles. Perpendiculars.

a) <u>Generalization of the notion of an angle</u>

Our definition of angles suggests that the interior of an angle is convex. Yet, when adding two sufficiently large angles, one can obtain a figure whose interior is not convex. (see, for example, Figure 3.2.1).

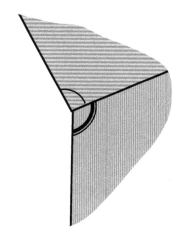

Figure 3.2.1

Although such figures are not angles in the sense of our original definition and we will not use them in geometric constructions, it is convenient to deem them as *generalized angles*. (These angles will be needed in the future when introducing *trigonometric functions* of angles). Both ordinary and generalized angles may be obtained as results of revolution. Let us consider this point of view.

An angle may always be thought of as generated by a ray which rotates about its origin (which is fixed). If ray *AB*, emanating from *A* with some point *B* on it (Fig. 3.2.2), rotates about *A*, it successively takes positions AB^I, AB^{II}, AB^{III}, AB^{IV}, AB^V, AB^{VI}, where B^I, ..., B^{VI} are *images* of *B* as results of respective rotations and AB^V coincides with *AB*.

Thus a sequence of angles: $\angle BAB^I$, $\angle BAB^{II}$, $\angle BAB^{III}$, ..., and $\angle BAB^{VI}$ is generated. With that, each successive angle of the sequence contains the preceding angle as a part of it, and, according to the common notion *the whole is greater than any of its parts*, one can say that the angle *increases* as the rotation goes on.

$\angle BAB^I$, and $\angle BAB^{II}$ are angles in the sense of the original definition: their interiors are convex. All the other angles: $\angle BAB^I$, ..., $\angle BAB^{VI}$, are generalized angles. There are some special terms reserved for the latter, and we shall mention them even though generalized angles will not be used until the very end of the course, when trigonometric functions will be introduced.

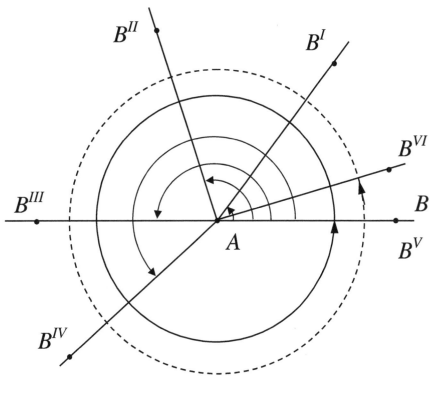

Figure 3.2.2

If a ray is rotated about its vertex until it coincides with its original position, then the generalized angle obtained (e.g., $\angle BAB^V$ in Fig. 3.2.2) is called a *perigon* or a *complete revolution* (sometimes just a *revolution* for brevity). A figure formed by two rays that extend each other into a straight line (e.g, $\angle BAB^{III}$ in Fig. 3.2.2) is called a *straight angle*. It is easy to prove that all straight angles are congruent. A generalized angle that is less than a perigon but greater than a straight angle, e.g. $\angle BAB^{IV}$, is called a *reflex angle*.

b) <u>Types of angles. Perpendiculars</u>

If the non-common sides of two adjacent angles extend each other into a straight line, the angles are called *supplementary*, or *supplements*. (Using the terminology of the previous subsection, one can say that two angles are *supplements* of each other if their sum is a straight angle.)

Examples of supplements are shown in Figure 3.2.3. *AD* and *FH* are straight lines; then ∠*ACB* and ∠*BCD* are supplementary to each other, as are ∠*FGE* and ∠*EGH*.

It looks like the latter two angles (∠*FGE* and ∠*EGH*) are congruent to each other.

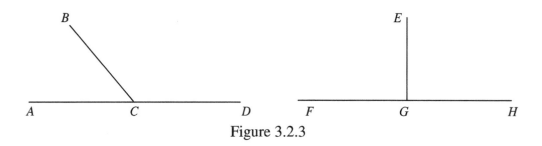

Figure 3.2.3

We say "it looks like" since we have not been informed of the way these angles were constructed. Can these two angles be made to coincide by superposition? We can imagine performing such a superposition physically, by folding the plane (the sheet of paper with our figure) along *EG*.

If, as a result of such folding, the ray *GH* falls onto *GF*, then we can say that the angles ∠*FGE* and ∠*EGH* can be made to coincide by superposition, and therefore they are congruent.

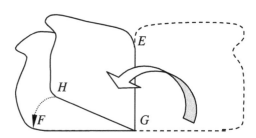

Figure 3.2.4

Such a physical superposition, which has been performed as a folding along a line, can be identified as a type (*iii*) isometry, whose existence is postulated. Now, keeping in mind such a physical operation with a sheet of paper, let us use the corresponding isometry to show that there exist angles that are supplementary and congruent. Such angles are very important and thus they deserve a special name: *right angles*. Thus we define a *right angle* as an *angle that is congruent to its supplement*. Now let us show that such angles exist.

<u>Theorem 3.2.1</u>   *There exists an angle that is congruent to its supplement, i.e. right angles exist.*

<u>Proof.</u> Let *AB* be a straight line and *M* a point not lying on *AB* (see Figure 3.2.5).

Let us subject the plane to a type (*iii*) isometry: reflection in *AB*. That means the isometry will carry every point of the plane lying on one side of *AB* into a point lying on the other side of this line, and every point of *AB* will remain stationary, that is it will not move. (Compare this to the physical action: folding the figure along *AB*).

As a result of this transformation, point *M* will be carried into some point *N* lying on the other side of *AB*. (*N* is called the *image* of *M* whereas *M* is a preimage of *N*).

Let us join *M* with *N* by means of a straight line. This line will intersect *AB* at some point, which we shall denote *C*.

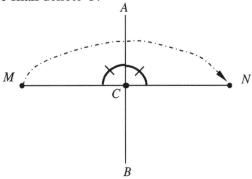

Figure 3.2.5

Since *CA* does not move as a result of the isometry performed, ∠*MCA* is transformed into ∠*NCA*, which means that these two angles are congruent: ∠*MCA* = ∠*NCA*. At the same time, since the two sides (*MC* and *CN*) of these angles extend each other into a straight line (*MN*), these angles are supplementary. Therefore ∠*MCA* and ∠*NCA* are congruent supplementary angles, which means such angles exist, Q.E.D.

Now let us consider some important properties of right angles and lines that form such angles.

Theorem 3.2.2   *All right angles are congruent.*

Proof. Let us prove the statement by contradiction (RAA).

Suppose some right angles are not congruent. In particular, let us assume that in Figure 3.2.6, ∠*ACB* = ∠*BCD* = α are right as are ∠*FGE* = ∠*EGH* = β, and $\alpha \neq \beta$. For total certainty, let us suggest that $\alpha > \beta$, or ∠*BCD* > ∠*EGH*.

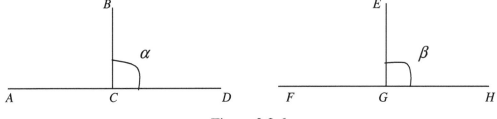

Figure 3.2.6

For convenience, we can suggest that *GH* = *CD* (by AXIOM 3, we can lay off a segment congruent to *CD* on any ray emanating from *G*). By the definition of isometry, one can apply an isometry to superimpose *GH* onto *CD*. Then, by AXIOM 1, the whole line passing through *G* and *H* will coincide with the line through *C* and *D*; hence *HF* will lie on *DA* (Figure 3.2.7).

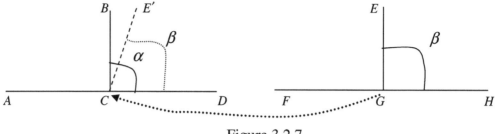

Figure 3.2.7

According to our suggestion, $\angle BCD > \angle EGH$, therefore *GE* will take some position *CE'* lying in the interior of $\angle BCD$. Then, $\angle BCD$ will be located in the interior of $\angle ACE'$, which is obtained by superposing $\angle EGF$, and, therefore, $\angle BCD < \angle ACE' = \angle EGF$. Also, we know that $\angle EGF$ is congruent to $\angle EGH$ as its supplement, since both angles are right.

Thus, we have obtained that $\angle EGH = \beta < \angle BCD = \alpha < \angle ACE' = \angle EGF = \beta$. The latter is a contradiction, and therefore our suggestion that $\beta < \alpha$ was false. Similarly, α cannot be less than β. Hence, $\alpha = \beta$, Q.E.D.

Since all right angles are congruent, we can use a common notation for all of them. We shall denote right angles *d* (the first letter of the French word *droit*, which means *right*). For example, we can use this notation to say that supplements are two angles whose sum is $2d$.

If the sum of two angles is a right angle, the angles are called *complements* of each other (or *complementary* angles).

An angle less than a right angle is called an *acute angle*. An angle greater than a right angle is called an *obtuse angle*. In Figure 3.2.8, $\angle BOC = \alpha$ is acute, and its supplement $\angle AOB = \beta$ is obtuse.

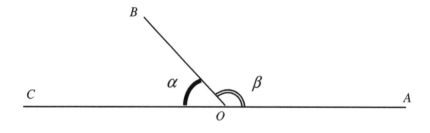

Figure 3.2.8

The arms of a right angle are said to be *perpendicular* to each other. In general, two straight lines (rays, segments) are *perpendicular* to each other if they form a right angle. Thus, one can say that a common side of two congruent supplementary angles is *perpendicular* to the line in which the other two sides lie.

Otherwise, (if the supplements are not congruent) the common side is called an *oblique* to the line that is formed by the two non-common sides. In Figure 3.2.9, *AB* is a perpendicular, and *CD* is an oblique to the line *l*.

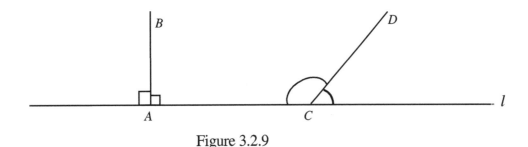

Figure 3.2.9

There is a special symbol, ⊥, which denotes perpendicularity. For example, one can write $AB \perp AC$ (segment *AB* is perpendicular to *AC*), or $AB \perp l$ (*AB* is perpendicular to the line *l*). If a perpendicular to a given line is drawn starting from a point lying in the line, one says that a perpendicular to the line is *erected*. If a perpendicular is drawn to a line from an *exterior* point (point that does not belong to the line), one says that it is *dropped* onto the line; in this case the word *perpendicular* denotes the segment joining the exterior point with the line. This segment is also called the *distance from the point to the line*.

The point in the line from which a perpendicular is erected or upon which a perpendicular that is dropped upon the line falls, e.g., point *A* in Fig. 3.2.9, is called the *foot of the perpendicular*. Similarly, the point where an oblique intersects a line, such as point *C* in the above figure, is called the *foot of the oblique*.

Theorem 3.2.3 *One and only one perpendicular can be erected from a given point in a given line.*

The proof of this theorem follows immediately from Theorem 3.2.2 (prove by contradiction).

Theorem 3.2.4 *A perpendicular can be dropped onto a given line from a given exterior point.*

Proof. First of all let us notice that the words *exterior point* mean a point that is not located on the line.

Suppose some line, *AB*, and a point, *M*, located outside of this line, are given (see Figure 3.2.10). It is required to prove that a perpendicular can be dropped from *M* onto *AB*.

In order to draw a perpendicular to *AB* from *M*, we can perform the same isometry that we have used to prove the existence of right angles. This will be a reflection of the plane in *AB*, which is a type *(iii)* isometry. Such a transformation interchanges the points of the two half-planes lying on the opposite sides from *AB* while keeping the points of *AB* stationary.

As a result of this isometry, *M* is transformed into some point *N* lying on the side of *AB* that is opposite from *M*. (In physical terms one can say that we fold the plane along *AB* so that this line does not move, and the left half-plane falls onto the right one and vice versa. Then *M* falls on some point *N*. Let us mark point *N* in the right half-plane and (completely) unfold our figure.) Then we connect *M* and *N* with a straight segment. It will intersect *AB* at some point; let us name this point *C*.

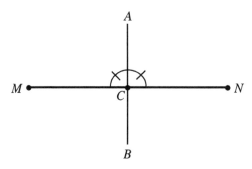

Figure 3.2.10

Since ∠*NCA* is obtained from ∠*MCA* by means of an isometry, the two angles are congruent. Also, these angles are supplementary, as two of their sides lie on the same line, *MN*. Hence, each of these angles is right: ∠*NCA* = ∠*MCA* = *d*, and thus *MN* ⊥ *AB*, or *MC* ⊥ *AB*. Therefore a perpendicular can be drawn to any line *AB* through any of its exterior points, Q.E.D.

It can also be shown that such a perpendicular is unique for a given line and a given point. However, the proof will require a few auxiliary statements, which will be established in due sequence in Chapter 4; therefore we shall postpone the proof till Section 4.5 of that chapter.

To complete the discussion of perpendiculars we introduce another useful notion: the projection of a segment. The *projection of a segment upon a straight line* is the portion of the line that lies between the feet of the perpendiculars dropped from the endpoints of the segment onto the line. Thus, in Figure 3.2.11, *MN* is, in each instance, the projection of *AB* upon *CD*.

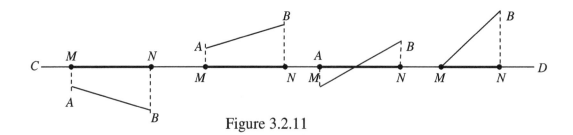

Figure 3.2.11

3.3 Vertical angles

If two lines intersect forming four angles, any two angles that are not adjacent, for instance, ∠AOB & ∠COD, or ∠BOC & ∠DOA in Figure 3.3.1, are called *vertical* angles.

The arms of one of two vertical angles are the arms of the other angle extended through the vertex.

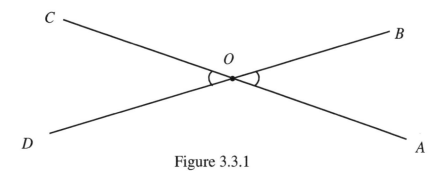

Figure 3.3.1

Theorem 3.3.1   *Any two vertical angles are congruent.*

Proof. Any two vertical angles are supplements of the same angle, e.g., both ∠COD and ∠AOB are supplements of ∠COB. Thus, ∠COD + ∠COB = ∠AOB + ∠COB, hence ∠COD and ∠AOB are congruent, according to Common Notions, Q.E.D.

3.1-3. Questions and Problems.

1. Two angles, α and β, are shown in Figure 1. (i) Use only a sheet of hard paper, a tack, a pencil, a straightedge, and scissors to make moulds of these angles. Substantiate each step of your procedure.   (ii) Use the moulds to draw the following angles: $\alpha + \beta$, $\alpha - \beta$, 3α, 3β, $3(\alpha + \beta)$, $3\alpha + \beta$.   (iii) Which of the six constructed angles have non-convex interiors?

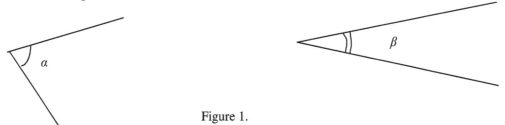

Figure 1.

2. Cut of an angle from a piece of paper. Use the minimal number of instruments to draw its bisector. (Of course, you will need a pencil and a straightedge. Do you need anything else?) Substantiate your actions.

3. Which of the six angles you have constructed for Problem1 are acute? obtuse? reflex?

4. Take a regular page (like one from your notebook) and use only scissors to make a mould of a right angle. Substantiate your actions.

5. (i) In Figure 2, ∠AOB = ∠COD. Prove that ∠AOC = ∠BOD.
   (ii) Prove the converse to (i).

6. In Figure 2, *OM* is the bisector of ∠*BOC*. Given that ∠*DOB*=∠*AOC*, prove that *OM* bisects ∠*AOD*.

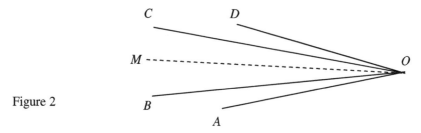

Figure 2

7. List all pairs of *adjacent* angles in Figure 2. Is any of these pairs a pair of *supplements*? A pair of *complements*?

8. In Figure 3, points *A, B, O* are collinear. Which of the angles presented in the figure are *supplementary* to each other?

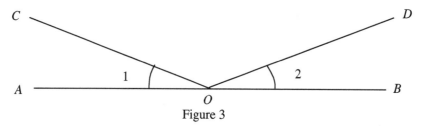

Figure 3

9. Prove that all straight angles are congruent.

10. *Right angle* is defined as *an angle that is congruent to its supplement*. Prove that all right angles are congruent.

   Hereafter, based on the result of Problem 10, we denote *right angles* by *d*.
   For example, ∠*MPQ* = *d* means that angle *MPQ* is *right*; ∠*ABC*=2*d* means that angle *ABC* is *straight*, etc.

11. Angles *MON* =α and *NOP*=β are adjacent. Find the angle between their bisectors. Apply the result for the case when α and β are supplementary angles.

12. Angles *MON*=α and *NOP*=β are adjacent, and so are ∠*NOP* and ∠ *POQ*=γ. Find the angle between the bisectors of the angles *MON* and *POQ*.

13. Suppose in Figure 3, ∠*1*=∠*2*, and ∠*COD*=⅝*d*. Find ∠*1* (or ∠*2*) in terms of *d*.

14. Angles α, β, γ are expressed as fractions of a right angle: α = ⅞*d*, β= 0.3*d*, γ=0.75*d*. For each of the following angles determine whether it is acute, obtuse or reflex: δ=α + β + γ;   θ= 2α −β +3γ.

15. Suppose in Figure 3 angles *1* and *2* are congruent and each of them equals one quarter of angle *COD*. Determine ∠ *1*.

16. Two segments are equal. Are their *projections* upon the same line equal? Formulate *sufficient* conditions under which the projections of two equal segments upon the same line are equal. Are these conditions *necessary*?

17. Prove that the bisectors of two *vertical angles* extend each other into a straight line.

3.4   Central angles and their corresponding arcs. Measurement of angles.

An angle formed by two radii of a circle (for instance, ∠ AOB in Figure 3.4.1.) is called a *central angle*; such an angle *intercepts* (*or cuts off*) an arc (∪ AmB) of the circle. (From the Latin "intercepio" which means "cutting off") The central angle and the intercepted arc are said to *correspond* to each other. The following property relates central angles with their corresponding arcs.

Theorem 3.4.1. *In the same circle or in congruent circles:*
(i)   *if two central angles are congruent, then the corresponding arcs are congruent;*
(ii)   *if two arcs are congruent, then the corresponding central angles are congruent.*

Proof. Let ∠ AOB = ∠ COD (Fig. 3.4.1); we will show that ∪ AmB = ∪ CnD.

Suppose sector AOB is rotated about O in the direction shown by the arrow, so that OA coincides with OC (this is a type (*ii*) isometry). Then, because of the congruence of the angles, OB will fall upon OD, and since each of these segments is a radius, they will coincide.

Thus, as a result of the above isometry, sector AOB is transformed into a sector bounded by the radii OC and OD. Let us prove (by contradiction) that the arcs of the sectors will also coincide as a result of the isometry performed.

Suppose ∪AmB has been transformed by this isometry into an arc that has at least one point that does not lie on ∪CnD. Let us denote this point M. Let us draw a straight segment joining M and O. This segment (as it is shown in Figure 3.4.1b) or its extension into a straight line will intersect ∪CnD at some point N.

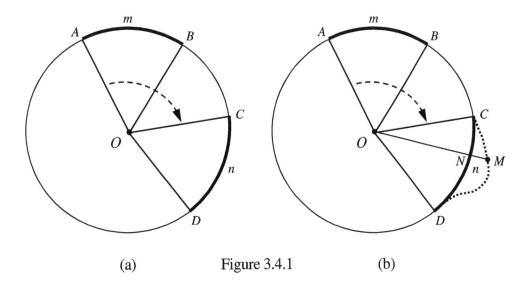

(a)   Figure 3.4.1   (b)

Then each of the segments OM and ON is congruent to the radius of the circle, but they cannot be congruent to each other unless M coincides with N. Thus we have arrived at a contradiction, which shows that every point of ∪AmB will fall onto a point of $\cup CnD$ as a result of the isometry described. The latter means: $\cup AmB = \cup CnD$, Q.E.D.

Students should attempt to prove the converse statement, (ii), after they have covered Chapter 4 (see Problem 18 to the Review Problems for Chapter 4).

It should be noticed that the size of the circle (the radius) did not play any role in the above discussion. Thus, one can compare different angles with a common vertex by measuring their corresponding arcs in a circle of any radius with the centre at the vertex. This property allows us to introduce a characteristic of the size of angles called the *angle measure*.

Suppose some circle is divided into 360 equal parts, and there is a radius drawn to each of the points of the partition. This will create 360 congruent (because they are subtended by congruent arcs) central angles. Each of the above arcs is called an *angular degree*. It is denoted by °. In future we shall say *degree* for brevity. Thus *one degree* is one three hundred sixtieth of one complete revolution. The degree is subdivided into 60 equal parts called *minutes* ('). The minute is divided into 60 equal parts called seconds, (''). Thus, an angle of 20 degrees, 15 minutes, and 6 seconds would be denoted 20° 15´ 6´´.

The number of degrees contained in an angle is called the *angle measure*. It is usually denoted m ($\angle$), e.g., m ($\angle AOB$) = 41, if $\angle AOB$ contains 41 angles of 1°, or the corresponding arc, $\cup AB$, of a circle consists of 41 out of 360 arc degrees.

It can be shown (and it will be done in Chapter 10) that every angle can be assigned a certain numerical measure. The number that represents a measure is a so-called *real number* that not necessarily an integer or a fraction, and the degree measure is not the only one possible. We shall postpone this discussion till the end of Chapter 10, in which it will be shown how to measure arcs of circles.

Using the introduced term of the *measure* one can reformulate Theorem 3.4.1: *Central angles are congruent iff (if and only if) their measures are equal.* Obviously, we can remove the word "central" from the statement. Really, any angle becomes central if a circle is described with the centre at its vertex, and the latter is always possible by the COMPASS POSTULATE.

Then it follows from Theorem 3.4.1 that there exists the following on-to-one correspondence between angles and their measures:

<u>**Theorem 3.4.2**</u>   *Angles are congruent iff (if and only if) their measures are equal.*

Similarly, one can surround angles by circles and relate angle measures to corresponding arcs, to prove the following theorem.

<u>**Theorem 3.4.3**</u>   *For any angles, α, β, and a positive integer, n,*
(i) $m(\alpha + \beta) = m(\alpha) + m(\beta)$;
(ii) $m(\alpha - \beta) = m(\alpha) - m(\beta)$; where $\alpha > \beta$;
(iii) $m(n\alpha) = n\, m(\alpha)$;
(iv) $m(\frac{1}{n}\alpha) = \frac{1}{n} m(\alpha)$.

It follows from the last theorem that since a complete revolution (perigon) contains 360°, the measure of a straight angle is 180°, and a right angle contains 90°.

Let us notice that even though the degree measure of angles is a convenient tool for solving some practical problems and provides convenience in numerous applications, it does not play any role in establishing geometric properties of the Euclidean plane. Do not forget that the degree measure is just a numerical characteristic of angles, whereas the geometric properties of figures that are the subject of our studies, do not depend on any numbers. When establishing the geometric results, we can do just as well without numerical measures of angles (or segments, by the way). Instead, we can just compare the angles with the right angle. We will do so in all our derivations, and for convenience we shall use the special notation "d" to denote a right angle.

An instrument used for measuring angles is called a *protractor*. It is a half-circle made of wood, metal, or plastic, with the marked centre (at O, Fig. 3.4.2) and with the half-circumference divided by fine lines into 180 equal arcs. Each of these arcs corresponds to 1°, if the vertex of an angle is placed at the centre (at O).

To measure an angle, one should place the protractor upon it so that one side of the angle lies along the radius OA, with the vertex of the angle at O. Then the other side of the angle will fall in some position, such as OP, and the number of degrees and fractions of a degree can be read off directly from the scale. For example, in Figure 3.4.2, $\angle AOP = 57°$; $\angle BOP = 123°$.

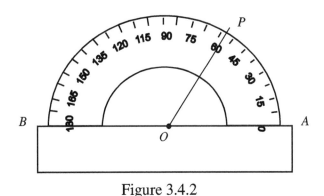

Figure 3.4.2

Remark. "Modern" Geometric Notation for Lines, Rays, Segments, Arcs, and Congruence.

In this book we use the so-called *traditional* geometric notation. This notation has been used, for example, in the translation of Euclid's ELEMENTS by sir Thomas Heath.

In many modern textbooks a slightly different notation (we shall call it *modern*) is used. Since *modern* notation is often used in tests (such as SAT), students should be aware of it. Those symbols of the *modern* notation that are different from the *traditional* ones are listed below.

$\overleftrightarrow{AB}$ represents the line that passes through points A and B, and $\overrightarrow{AB}$ denotes the ray beginning at point A and passing through point B. The line segment with points A and B as endpoints will be represented by $\overline{AB}$ and AB will represent the length of $\overline{AB}$.

The congruence is usually denoted by the symbol $\cong$.

$\overset{\frown}{ACB}$ denotes the arc that joins A and B and passes through C. Also, $\overset{\frown}{AB}$ can be used if it is clear which of the arcs joining A and B is meant.

$m\angle ABC$ stands for the measure of $\angle ABC$ (usually the degree measure is suggested).

EXAMPLES.

1). If one centimeter is chosen as the unit of length, and some straight segment, joining points A and B, is 5 cm long, in the *modern* notation we should express this by writing either $\overline{AB} \cong 5cm$ or $AB = 5$.

2). If $\angle ABC$ contains 72 degrees, in the *modern* notation it can be written as $\angle ABC \cong 72^0$ or $m\angle ABC = 72$.

3.4 Questions and Problems.

1. Prove that if two arcs of a circle are congruent, the corresponding *central angles* are congruent.

2. In Figure 1 a circle with the centre at O is shown. AA', BB', and CC' are three diameters of the circle, and $\angle 1 = \angle 2$. Show that $\cup AB = \cup BC$.

Figure 1

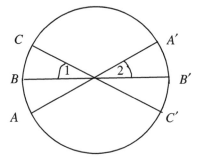

3. Through how many degrees does the minute hand of a clock turn in (i) fifteen minutes? (ii) six minutes? (iii) one hour? (iv) one hour and twenty minutes?

4. Through how many degrees does the hour hand of a clock turn in (i) one hour? (ii) 360 minutes? (iii) 540 minutes? (iv) 36 hours?

5. Find the lesser angle formed by the hands of a clock at (i) 4:00; (ii) 11:30; (iii) 2:45.

6. If a wheel makes ten revolutions per minute, through how many degrees does it turn in a second?

7. What fraction of a circumference does the arc measuring (i) 90°; (ii) 30°; (iii) 27°; (iv) 135° make up?

8. What part of a straight angle makes up an angle of (i) 45°? (ii) 30°? (iii) 72°? (iv) 150°?

9. The Earth turns on its axis once in 24 hours. How many degrees of longitude correspond to one hour?

10. Is one fifth of a revolution an acute or an obtuse angle?

11. Three concurrent lines form six consecutively adjacent angles. Three of these angles are θ, 2θ, 3θ. Find the measure of θ (in degrees).

12. For each of the following angles draw an angle as nearly equal to the given one as you can judge by your eye. Then measure your angle with a protractor. How much error did you make? What fraction and what percentage of the given value of the angle is your error? a) 45°; b) 30°; c) 120°.

13. Can it happen that the sum of three consecutive angles formed by two intersecting lines equals exactly 150°?

14. Two adjacent angles measure 100°20´ and 79°40´ minutes, respectively. Are these angles supplementary?

15. Two adjacent angles measure 61°22´ and 29°38´ minutes, respectively. Are these angles complementary?

16. Find the supplement of each of the following angles: a) 72°; b) 65°28´; c) 171°28´20´´.

17. The supplement of α equals 4α. How many degrees are there in α?

18. The complement of α equals 74α. Find the value of α (in degrees, minutes, seconds).

4. TRIANGLES

> Piglet came a little closer to see what it was. Eeyore had three sticks on the ground, and was looking at them. Two of the sticks were touching at one end, but not at the other, and the third stick was laid across them. Piglet thought that perhaps it was a Trap of some kind.
>
> A.A. Milne, *The House at Pooh Corner*.

4.1. Broken line; Polygons

A *broken line* is a figure formed by straight segments located so that they are not lying in one straight line and each of them, starting from the second, originates at the endpoint of the preceding one. For example, *ABCDE*, in Fig. 4.1.1 is a broken line. The segments forming a broken line (*AB*, *BC*, *CD*, and *DE*) are called its *sides*, and the vertices of the angles formed by neighbouring segments (*A*, *B*, *C*, *D*) are called its *vertices*.

A broken line is called *convex* if it is located entirely in one half-plane obtained by extending any *side* of the broken line into a straight line. For example, broken line *MNPQR* in Figure 4.1.2, is convex, whereas *ABCDE* in Figure 4.1.1 is not (in order to see this, extend *BC*).

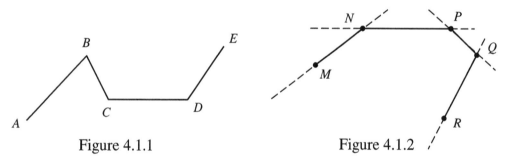

Figure 4.1.1　　　　　Figure 4.1.2

A broken line is said to be *closed* if the origin of the first side coincides with the terminal point of the last side. A closed line, when traced from any point lying in it through its entire extent, is found to return to the starting point. For instance, both *ABCDE* and *MNPQRS* in Fig. 4.1.3 are closed lines.

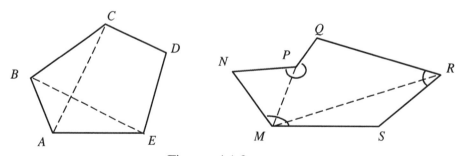

Figure 4.1.3

- 64 -

A figure formed by a closed broken line is called a *polygon*, and the part of the plane bounded by this line, is called the *interior of the polygon*. (In some texts, by a polygon they mean the part of the plane bounded by a closed broken line, which is called the *boundary* or *contour* of the polygon). The sides, vertices, angles of the line are called respectively the *sides, vertices,* and *angles of the polygon*. The segment equal to the sum of all sides of a polygon is called its *perimeter*.

A polygon is *convex* if it is formed by a convex broken line (it is easy to check that this definition conforms to the general definition of convexity given in section 3.1). Polygon *ABCDE* in Figure 4.1.3 is convex, and polygon *MNPQRA* is *non-convex*.

A segment connecting two vertices that are not the endpoints of one side in a polygon, is called a *diagonal* of the polygon. In Fig. 4.1.3, *AC, BE, MP, MR* are diagonals.

It is easy to see that the number of sides of a polygon is the same as the number of its angles (trace the boundary associating each vertex with the following side). A polygon is named according to its number of sides (or angles), e.g., *triangle, quadrilateral* (less common name – *quadrangle*), pentagon, hexagon, etc.; some examples are shown in Figure 4.1.4.

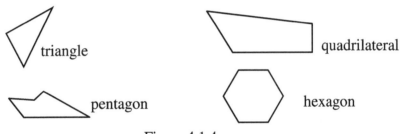

Figure 4.1.4

A polygon is said to be *equilateral* when all its sides are equal, *equiangular* when all its angles are equal, *regular* when it is both equilateral and equiangular. In the nearest future we shall restrict ourselves to the discussion of triangles.

4.2. Triangles; the basic notions.

We shall denote triangles by the "Δ" symbol, e.g. ΔABC. It is also convenient to denote the sides opposed to given vertices by the corresponding small Latin letters, as in Figure 4.2.1: $BC = a$; $AC = b$; $AB = c$.

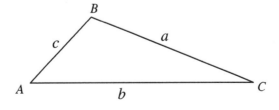

Figure 4.2.1

Triangles are classified with regard to their sides. A triangle is called *scalene*, when none of its sides are congruent; *isosceles*, when two of its sides are congruent, and (as any other polygon) *equilateral*, when all its sides are congruent (Figure 4.2.2).

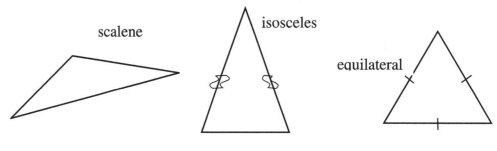

Figure 4.2.2

It can be shown that equilateral triangles exist. More specifically, we will show that there exists an equilateral triangle of any given size, i.e. with any given segment as a side.

Theorem 4.2.1 *An equilateral triangle with a side congruent to a given segment can be constructed.*

Proof (construction) Let $AB = a$ be a given segment (Figure 4.2.3). Describe a circle of radius a from A as a centre, and a circle of radius a from B as a centre (it is possible by AXIOM 3, or, to be more specific, by the COMPASS POSTULATE). The circles will intersect at some point C. Join C with the points A and B (AXIOM 1). $CA = a$ as a radius of the circle described from A; similarly, $CB = a$. Hence, each side of $\triangle ABC$ is congruent to a; the required triangle is constructed and thus its existence has been proved, Q.E.D.

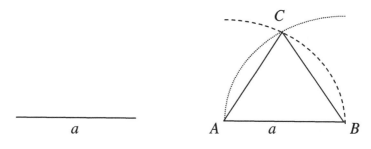

Figure 4.2.3

In an isosceles triangle the term *vertex* is usually applied to the point of intersection of the congruent sides, and the side that is not congruent to any of the others is, respectively, the *base*. The *lateral sides* of an isosceles triangle are congruent. It can be shown that under a certain condition (what is this condition?) an isosceles triangle with a given base and a lateral side can be constructed. The proof is similar to the latter one.

Theorem 4.2.2 *Given segments l and b such that b < 2l, an isosceles triangle with b as a base and l as a lateral side can be constructed.*

Triangles can also be classified with regard to their angles (Fig. 4.2.4). A triangle is said to be *right*, when one of its angles is right; *obtuse-angled*, when one of its angles is obtuse; *acute-angled*, when all three of its angles are acute. It will be proved later on that in any triangle at least two of its angles are acute.

In a right triangle, the side opposite to the right angle is called the *hypotenuse*, and the other two sides, the ones that include the right angle, are usually called *legs*.

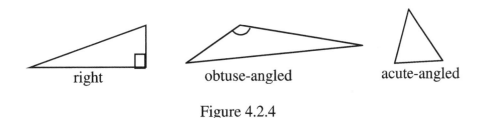

right   obtuse-angled   acute-angled

Figure 4.2.4

In a triangle, a perpendicular dropped from any of its vertices onto the opposite side (such as *BD* in $\triangle ABC$, Fig. 4.2.5a) or its extension (*NR* in $\triangle MNP$, Fig. 4.2.5b) is called an *altitude*. We know that such a segment exists since we have already proved in Th.3.2.3 that a perpendicular can be dropped onto a line from any exterior point.

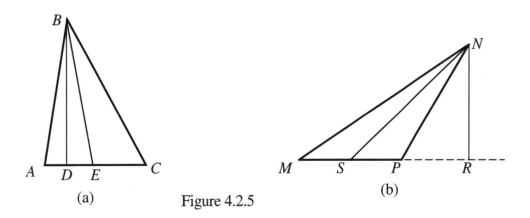

(a)   Figure 4.2.5   (b)

Later in this chapter we shall prove the existence of the other two important segments: a *median* and a *bisector*. A straight segment joining a vertex of a triangle with the midpoint of the opposite side is called a *median*. In Figure 4.2.5, *AE = EC*, and *MS = SP*, thus *BE* and *NS* are medians). A bisector of an angle of a triangle is called a *bisector* of the triangle. Apparently, each triangle has three altitudes, three medians and three bisectors.

A segment connecting the midpoints of two sides of a triangle will be called a *midline* or a *midjoin* of a triangle. (The latter terms are not common, and, yet, convenient; thus we shall use them.)

4.1-2 Questions and Problems.

1. Which of the broken lines shown in Figure 1 are convex? Which of the lines in Figure 1 are closed?

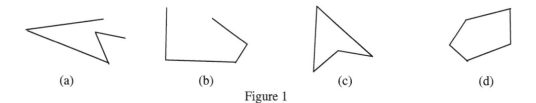

(a) (b) (c) (d)

Figure 1

2. Prove that a polygon bounded by a convex curve, is a convex figure (according to the general definition of convexity).

3. How many diagonals emanate from one vertex of a a) quadrilateral? b) pentagon? c) hexagon?

4. How many diagonals emanate from one vertex of a polygon with n sides (n-gon)? Make a suggestion and prove it using the principle of mathematical induction.

5. Into how many triangles is a *convex hexagon* divided by all diagonals drawn from one vertex? Solve the same problem for a *convex heptagon* (7-gon); *convex octagon* (8-gon).

6. Into how many triangles is a convex polygon of n sides (n-gon) divided by its diagonals drawn from one vertex? Make a suggestion and prove it by the principle of mathematical induction.

7. Draw an equilateral quadrilateral. (Use only a straightedge and a compass). Is it necessarily equiangular?

8. Draw a regular pentagon (use a straightedge, a compass, and a protractor). Is it necessarily convex? *Can you use your result to construct a regular *decagon* (a polygon of 10 angles)?

9. Use a regular pentagon you have constructed for Problem 8 to construct an equilateral non-convex pentagon.

10. Use a straightedge, a compass, and a protractor to construct a regular hexagon. Then use the hexagon you have constructed to construct a non-convex equilateral hexagon.

11. Classify the triangles presented in Figure 2 with regard to their sides and with regard to their angles.

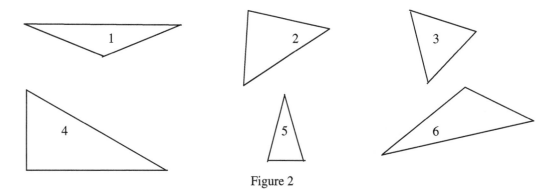

Figure 2

12. Prove that in a triangle there is exactly one bisector, one median, and one altitude starting from each vertex.

13. Draw three scalene triangles: an acute-angled, right, and obtuse-angled. In each of these triangles draw the bisector, the median, and the altitude to the (a) greatest side, (b) least side as a base.

14. Use only a compass and a straightedge to construct an equilateral triangle. Measure its angles. What do you observe?

15. Use a drawing triangle, and a compass to construct two isosceles right triangles. Measure their acute angles. What do you observe?

16. Prove that the median drawn to the base of an isosceles triangle, divides it into two triangles with equal perimeters.

4.3   Axial Symmetry

Two points are said to be *symmetric in a line* if they are located in the opposite half-planes formed by the line, on a perpendicular to the line, and are equidistant from the foot of this perpendicular. For example, points P and Q in Figure 4.3.1 are symmetric in line l if $PO = OQ$ and $PQ \perp l$. Let us see how we can construct a point that is symmetric to a given point in a given line.

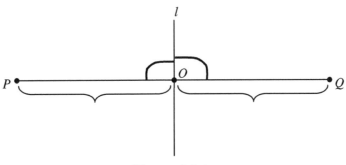

Figure 4.3.1

Let A be a point in the plane, and MN a line that does not contain A. MN, like any other line, divides the plane into two *half-planes*. If the plane is subjected to a type (*iii*) isometry that keeps MN unchanged (we call such an isometry a *reflection* in MN), A will be transformed into some other point A', lying in the opposite half-plane (Figure 4.3.2). A' is called the *reflection image* of A.

Let us join A with A' by a straight line. This line intersects MN at some point, which we label B.

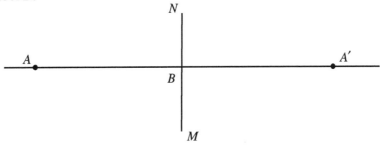

Figure 4.3.2

Since ∠ABN is transformed into ∠A'BN by means of an isometry, these two angles are congruent, and since they are also supplementary to each other, then each of them is right. Thus, $AA' \perp MN$. Also, AB is transformed by this isometry into A'B (we say that A'B is an isometric image of AB), and so these two segments are congruent.

Therefore point A and its reflection image, A', are *symmetric in MN*: they are located in opposite half-planes formed by this line, on a perpendicular to the line and are equidistant from the foot of the perpendicular, B.

Thus, a reflection in a line (type (iii) isometry) transforms points that are symmetric in the line into each other.

Some figures consist entirely of points that are symmetric to each other in some line. Then, under a reflection in this line such a figure is transformed into itself. Then this line is called an *axis of symmetry* of the figure and the figure is said to possess *axial symmetry*.

It is clear why the word *axis* emerges: in order to perform a symmetry (reflection) transformation one actually rotates the plane about the chosen line until the points of one half-plane fall onto the points of the opposite half-plane. As a result of such a transformation, the part of the figure lying on one side of the axis is superimposed onto the part lying on the other side, and they coincide. Thus the symmetry axis divides a figure into two congruent parts.

Axial symmetry is very common in nature and in everyday life. Some examples of this kind of symmetry are shown if Figure 4.3.3, where each figure is divided by a dashed line into two parts that are symmetric to each other in this line.

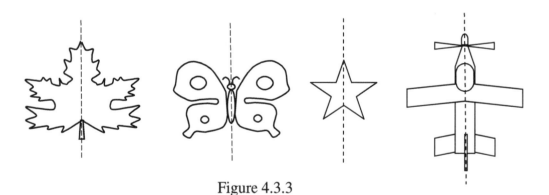

Figure 4.3.3

4.3 Questions and Problems.

For all constructions in this subsection use a straightedge, a compass, and, if you need to draw a perpendicular to a line, – a drawing triangle.

1. Given a line, construct a *symmetric in this line image* of a point exterior to the line.

2. Given a line, construct a *symmetric* (in this line) *image* of a given segment. Consider two cases: a) The segment does not have common points with the line; b) The segment has a common point with the line.

3. Given a line and a triangle, construct a reflection image of the triangle in the line. Is that image congruent to the original triangle? Substantiate your answer.

4. Determine the number of axes of symmetry for each of the following figures: a) a regular triangle; b) a square; c) a rectangle; d) a regular pentagon; e) a regular hexagon; f) a circle.

5. Points B and B' that are symmetric (to each other) in line l, are connected to points A and C lying in line l. Prove that: a) $\angle BAC = \angle B'AC$; b) $\angle BCA = \angle B'CA$; c) $\angle ABC = \angle AB'C$; d) $\triangle ABC = \triangle AB'C$.

6. Line AB divides the plane into two half-planes, and points M and N lie in the same half-plane. Find on AB such a point C that $\angle ACM = \angle NCB$.

7. In a circle, its diameter AB is perpendicular to chord CD. Prove that $AC = AD$.

8. In a circle, chords AC and AD form equal angles with the diameter AB. Prove that (i) C and D are symmetric in AB; (ii) $\triangle ABC = \triangle ABD$.

4.4 Congruence of triangles. Some properties of an isosceles triangle.

According to the definition, two figures are congruent if they can be made to coincide as a result of an isometry. In practice it may be difficult and inconvenient to find the appropriate isometry and subject a figure to the isometry each time when the congruence should be established. At least for simple figures, such as polygons, and especially, triangles, such a procedure can be replaced by the comparison of the *corresponding parts* of the figures whose congruence is being tested.

Corresponding parts are two parts of congruent figures that would coincide if the figures were made to coincide by an isometry. In this section we shall prove a few theorems that will provide simple criteria for the congruence of triangles through the equality of their corresponding sides (**S**) and angles (**A**).

***Theorem 4.4.1** (SAS condition). If two triangles have two sides and the included angle of the one congruent, respectively, to two sides and the included angle of the other, the triangles are congruent.*

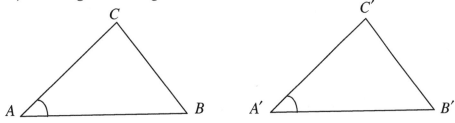

Figure 4.4.1

Proof. Let us suppose that for $\triangle ABC$ and $\triangle A'B'C'$ in Fig. 4.4.1, $AB = A'B'$, $AC = A'C'$, and $\angle A = \angle A'$. Now, since $AB = A'B'$, we can place $\triangle A'B'C'$ on $\triangle ABC$ so that $A'B'$ will coincide with AB, making C' fall somewhere in the same half-plane as C with respect to AB (i.e. the interiors of $\angle A$ and $\angle A'$ will be on the same side of AB).

Then $A'C'$ will lie on AC, because $\angle A' = \angle A$. Moreover, C' will fall exactly on C, because $A'C' = AC$ (by AXIOM 3). Thus, as a result of the performed isometry, B' lands on B, and C' falls on C; then, by AXIOM 1, $B'C'$ will fit exactly upon BC. Therefore, the triangles will completely coincide, hence they are congruent, Q.E.D.

Corollary 1. ***Two right triangles are congruent if two legs of one are congruent respectively to two legs of the other.***

Proof. It follows directly from the theorem and congruence of all right angles.

Theorem 4.4.2 (ASA condition)

If two triangles have two angles and the included side in the one congruent, respectively, to two angles and the included side in the other, the triangles are congruent.

Proof. In Fig. 4.4.2, $\angle A = \angle A'$, $AC = A'C'$, and $\angle C = \angle C'$. Let us prove that $\triangle A'B'C' = \triangle ABC$.

Figure 4.4.2

Let us superpose $A'C'$ on AC so that A' falls on A and C' falls on C (it is possible because the segments are congruent), and the interiors of the triangles fall on the same side of AC.

Then $A'B'$ will fall along AB, since $\angle A = \angle A'$, and $C'B'$ will fall along CB, since $\angle C = \angle C'$. It follows from AXIOM 1 that two distinct straight lines can intersect only at one point, hence B' will coincide with B.

So, the triangles will completely coincide, hence, they are congruent. Q.E.D.

Corollary 1. ***Two right triangles are congruent if a leg and the angle formed by this leg and the hypotenuse in one triangle are congruent, respectively, to a leg and the angle formed by this leg and the hypotenuse in the other triangle.***

Proof. It follows directly from the *ASA* theorem and the congruence of all right angles (Figure 4.4.3).

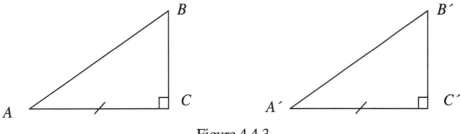

Figure 4.4.3

In order to establish more tests for the congruence of triangles we shall need some properties of isosceles triangles.

Theorem 4.4.3 *In an isosceles triangle, the angles opposite the congruent sides are congruent.*

Proof.   Let $\triangle ABC$ in Figure 4.4.4 be isosceles with $AB = BC$. We are going to prove that $\alpha = \beta$, i.e. the so-called *base angles* are congruent.

The proof of this theorem is comparatively lengthy, and it is convenient to organize it in the so-called *two-column form*: we shall write the statements needed for the proof in the left column, and their substantiations – in the right column.

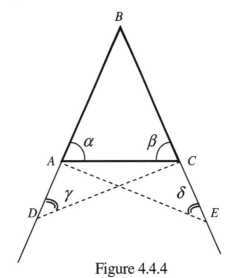

Figure 4.4.4

Extend BA beyond A and BC beyond C	AXIOM 1
On the extension of BA, lay off an arbitrary segment AD	AXIOM 3
On the extension of BC, lay off $CE = AD$	AXIOM 3
$BD = BE$	the sums of pairs of respectively congruent segments : $BD = BA+AD; BE = BC+CE$

- 73 -

$BA = BC$	hypothesis
$AD = CE$	construction
$\Delta DBC = \Delta EBC$	SAS: $\angle B$- common; $BD=BE$ (proved); $BA=BC$ (hypothesis)
$DC = EA$; $\gamma = \delta$	corresponding elements in congruent ΔDBC and ΔEBC
$\Delta DAC = \Delta ECA$	SAS: $DA = EC$; $DC = EA$; $\gamma = \delta$
$\angle DAC = \angle ECA$	corresponding angles in congruent triangles: ΔDAC and ΔECA
$\alpha = \beta$	supplements of congruent angles

Thus, in an isosceles triangle the base angles are congruent, Q.E.D.

Corollary 1 **An equilateral triangle is also equiangular.**

Proof. It follows immediately from Th. 4.4.3.

The converse of this theorem is also true:

Theorem 4.4.4 *If in a triangle two angles are congruent, their opposite sides are also congruent, and thus the triangle is isosceles.*

Proof. Let in ΔABC in Figure 4.4.5, $\angle BAC = \angle BCA$. We shall prove that $AB = BC$. Let us use the RAA method (proof by contradiction).

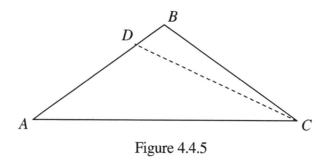

Figure 4.4.5

Suppose, AB and BC are not congruent, then one of them is greater than the other. Let us suggest $AB > BC$. Then, on AB we can lay off $AD = BC$ (AXIOM 3). In ΔDAC and ΔBCA: $AD = BC$ (construction), $\angle DAC = \angle BCA$ (hypothesis), AC is common; hence $\Delta DAC = \Delta BCA$ by SAS, and therefore $\angle BAC = \angle DCA$ as corresponding angles (opposite to congruent sides BC and AD). The latter is absurd since $\angle DCA < \angle BCA$ (the former lies in the interior of the latter) and $\angle BCA = \angle BAC$ (hypothesis).

The obtained contradiction shows that our suggestion $AB > BC$ was false. Similarly, $AB < BC$ would be false, and therefore $AB = BC$, Q.E.D.

One can summarize the results of the last two theorems by stating that *in a triangle congruent sides are opposed to congruent angles and vice versa.*

The following theorem provides another sufficient condition for the congruence of triangles.

Theorem 4.4.5 *(SSS condition). If two triangles have three sides of the one congruent respectively to the three sides of the other, they are congruent.*

Proof. Suppose in Fig. 4.4.6, $AB = A'B'$, $AC = A'C'$, $BC = B'C'$ and AB and $A'B'$ are the largest corresponding sides.

Let us place $\Delta A'B'C'$ in the position ABC'', thus making $A'B'$ coincide with its congruent AB and causing C' to take up some position on the other side of AB than C; this position will be marked C''. Join C with C'' by a segment.

$\Delta C''AC$ and $\Delta C''B'C$ are isosceles, $\Rightarrow$ by Th.4.4.3, $\angle 1 = \angle 2$ and $\angle 3 = \angle 4$; $\Rightarrow \angle ACB = \angle 1 + \angle 3 = \angle 2 + \angle 4 = \angle AC''B$.

That means the triangles ACB and $AC''B$ are congruent by **SAS** condition, and so are the triangles ACB and $A'B'C'$, Q.E.D.

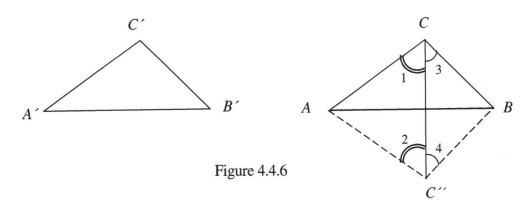

Figure 4.4.6

It follows from the above conditions of congruence that *in congruent triangles, corresponding sides are opposed the corresponding angles and vice versa.*

Let us notice that the notion of corresponding parts can be extended for the case of non-congruent triangles. If an angle of one triangle is congruent to an angle of another triangle, these angles are called *corresponding angles*, and the sides opposed these angles are said to be the *corresponding sides*.

Now we can prove the existence of bisectors of angles.

Theorem 4.4.6 *Every angle has a bisector.*

Proof (construction) Let ABC be an angle formed by some rays BA and BC emanating from some point B (Figure 4.4.7).

On BA, lay off an arbitrary segment BD from point B. On BC, lay off $BE=BD$ from point B. The above constructions are allowed by AXIOM 3. Join D with E by a straight segment DE (AXIOM 1).

On DE as a base, construct an equilateral triangle DEF (Th. 4.2.1). Draw a straight line l through B and F. $\Delta DBF = \Delta EBF$ by **SSS**: BF is common,

BE=BD, and *DF=EF* by construction. Hence, ∠*DBF* = ∠*EBF*, and thus *l* is a bisector of ∠*ABC*. The existence of a bisector of an angle has been proved, Q.E.D.

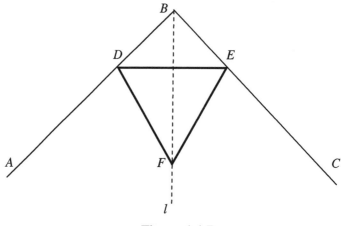

Figure 4.4.7

Theorem 4.4.7   *An angle has exactly one bisector.*

(Since the existence of a bisector has been proved, it is sufficient to show that *an angle cannot have two or more bisectors*)

Proof   We shall prove the statement by RAA (by contradiction).

Suppose ∠*BAC* in Figure 4.4.8, has two bisectors: *AD* and *AE*. Then, ∠*BAD* = ∠*DAC*> ∠*EAC* = ∠*EAB* > ∠*BAD*, which is impossible since an angle (in our case *BAD*) cannot be greater than itself. Hence, an angle cannot have more than one bisector, Q.E.D.

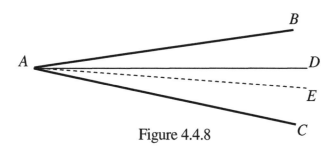

Figure 4.4.8

Theorem 4.4.8   *In an isosceles triangle, the bisector of the angle included by the congruent sides is also a median and an altitude.*

Proof. Let △*ABC* in Figure 4.4.9 be isosceles, and *BD* bisect the vertex angle *B*.

We are going to prove that *BD* is also a median and an altitude.

By the hypothesis, *AB* = *BC*, and ∠1 = ∠2; also, by Th.4.4.3, ∠5=∠6. Therefore, △*ABD* = △*CBD* by **ASA**. As a result of this congruence, *AD=DC* as corresponding sides opposite to congruent angles 1 and 2, and ∠3 = ∠4 as corresponding angles opposed *AB* and *BC*.

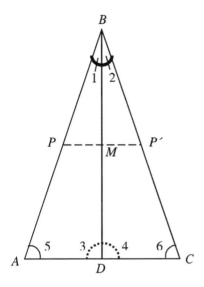

Figure 4.4.9

So: $DA = DC$ and BD is a median; $\angle 3 = \angle 4$ and their sum is two right angles, $\Rightarrow$ each of them is right, $\Rightarrow$ BD is an altitude, Q.E.D.

Corollary 1 **In an isosceles triangle, the perpendicular bisector of the base passes through the vertex and divides the triangle into two congruent right triangles.**

Proof The uniqueness of the bisector of an angle was shown in Th.4.4.7. It has been proved in Theorem 4.4.8 that the bisector, altitude, and median drawn from the vertex of an isosceles triangle, coincide. So, the bisector of the angle at the vertex will be the perpendicular that bisects the base, Q.E.D.

It was shown in the course of proving Th.4.4.8 that $\triangle ABD = \triangle CBD$, Q.E.D.

Corollary 2 **The bisector of the angle at the vertex of an isosceles triangle is a symmetry axis of the triangle.**

Proof. According to AXIOM 3, for any point P lying on AB (see Figure 4.4.9) there is unique point P' lying on CB such that $BP' = BP$. Let us show that P' is the reflection image of P in BD. If we connect P with P' by a straight segment PP', intersecting BD at some point M, we shall create an isosceles $\triangle PBP'$, were BM is the bisector included between the equal sides.

Then, by Th.4.4.8, $PP' \perp BD$ and $PM = P'M$, hence, by the definition, P' is the reflection image of P, Q.E.D.

Corollary 3 **Every segment has one and only one midpoint.**
Proof. First, let us notice that there are two statements in this corollary:
(i) Every segment has a midpoint;
(ii) a segment cannot have more than one midpoint.

Given a segment, let us construct an isosceles triangle with this segment as a base. According to Th. 4.4.8, the bisector of the angle opposed to the base is also a median, thus it will cut the base at the midpoint. The uniqueness of the midpoint follows from the uniqueness of the bisector. Thus, for every segment a midpoint exists and it is unique, Q.E.D.

Corollary 4 **One an only one median can be drawn to each side of a triangle.**
Proof. According to AXIOM 1, there is one and only one straight segment that joins a vertex of a triangle with the midpoint of the opposite side, Q.E.D.

Corollary 5 **A perpendicular can be dropped from any point upon any line that does not contain this point.**
Proof. Construct an isosceles triangle with the vertex at the given point and the base on the given line. In this triangle, according to Th. 4.4.8, the bisector of the angle opposed the base is the altitude drawn to the base, hence it is perpendicular to the given line.

(The result of the last corollary has been proved in Th. 3.2.3 in a different way).

Let us notice, in conclusion of this section, that the ability to recognize congruent triangles is absolutely necessary for solving more complicated geometric problems.

4.4 Questions and Problems.

SAS condition.

1. Use a protractor, a straightedge, and a compass to construct such a triangle that two of its sides, equal, respectively, *6.4 cm* and *4.6 cm*, include an angle of *68°*. Is the triangle defined uniquely by the given data? – Explain.

2. In Figure 1 *AM=AN*, and *D* is an arbitrary point of the bisector of ∠*BAC*. Prove that *DM=DN*.

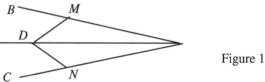

Figure 1

3. △ *ABC* in Figure 2 is isosceles, and *BM=NC*. Prove that (i) vertex *A* is equidistant from *M* and *N*; (ii) ∠*AMN*=∠*ANM*.

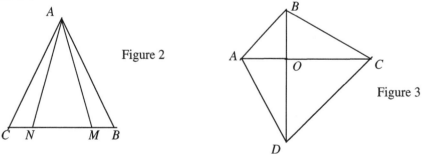

Figure 2

Figure 3

4. In Figure 3, *AC* and *BD* are diagonals in *ABCD*. They intersect at *O,* and *AO=OB; DO=OC*. Prove that *AD=BC*.

5. The median *AD* of △ *ABC* is extended beyond *D* by segment *DE* that is equal to *AD*. Point *E* is connected with *C* by a straight segment. Evaluate ∠*ACE* if ∠*ACD*=56°, and ∠*ABD*=40°.

6. Prove that in an isosceles triangle, medians that are drawn to the equal sides, are equal.

7. In isosceles triangle *ABC*, *BM=BN* (Figure 4). Prove that *CM=AN*.

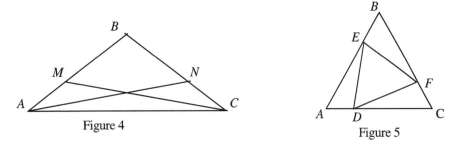

Figure 4   Figure 5

8. △ *ABC* in Figure 5 is equilateral, and *AD=CF=BE*. Prove that △*DEF* is equilateral.

9. △ *ABC* in Figure 6 is equilateral, and *AD=BE=CF*. Prove that △*DEF* would be equilateral.

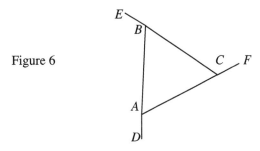

Figure 6

10. In Figure 7, *AB=AC*, *AD=AE*. Show that *BD=CE*.

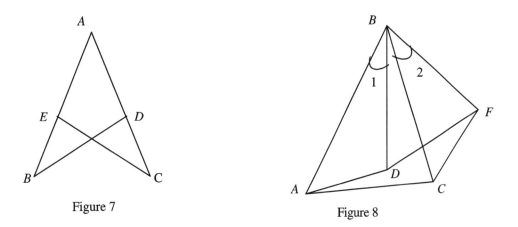

Figure 7   Figure 8

11. In Figure 8 *AB=BC*, *BD=BF*, and ∠1=∠2. Find a pair of congruent triangles in that figure. Show their congruence.

- 79 -

ASA condition.

12. In Figure 9, $AD=CF$, $\angle BAC=\angle EDF$, and $\angle 1=\angle 2$. Prove that $\triangle ABC = \triangle DEF$.

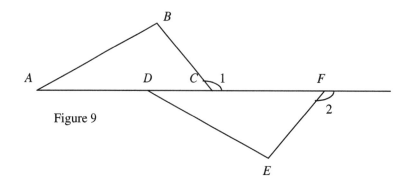

Figure 9

13. Rays AD and BC (Figure 10) intersect at O; $\angle 1=\angle 2$, $OC=OD$. Show that $\angle A=\angle B$.

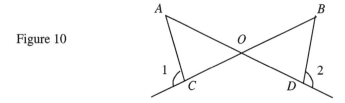

Figure 10

14. In $\triangle ABC$ (Fig.12), $AB=AC$, and $\angle 1=\angle 2$. Show that $\angle 3=\angle 4$.

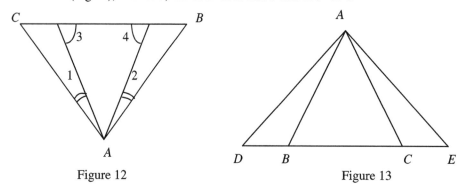

Figure 12   Figure 13

15. In Figure 13, $AD=AE$, $\angle CAD = \angle BAE$. Prove that $BD=CE$.

16. Prove that in an isosceles triangle the bisectors drawn to the lateral sides are congruent.

17. Prove that in congruent triangles the bisectors of corresponding angles are congruent.

18. Prove that two isosceles triangles are congruent if the base and an adjacent angle of the one are congruent, respectively, to the base and an adjacent angle of the other.

19. In quadrilateral *ABCD* (Figure3) $\angle DAB = \angle CBA$, and its diagonals *AC* and *BD* form equal angles with *AB*. Show that the diagonals of the quadrilateral are congruent.

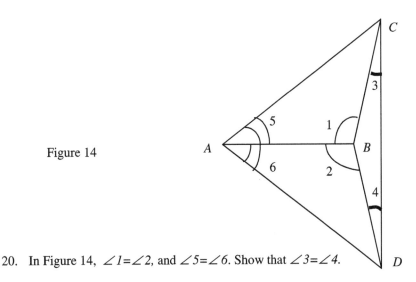

Figure 14

20. In Figure 14, $\angle 1 = \angle 2$, and $\angle 5 = \angle 6$. Show that $\angle 3 = \angle 4$.

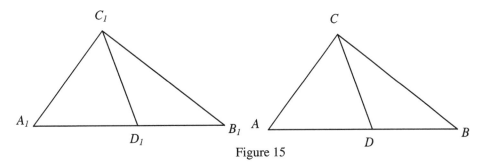

Figure 15

21. In Figure 15, $\triangle ABC = \triangle A_1B_1C_1$, and $\angle BCD = \angle B_1C_1D_1$. Prove that $AD = A_1D_1$.

SSS condition.

22. Three rods are hinged at their endpoints as shown in Figure 16. Can a different triangle be made from them by joining them up in different order? Substantiate your answer.

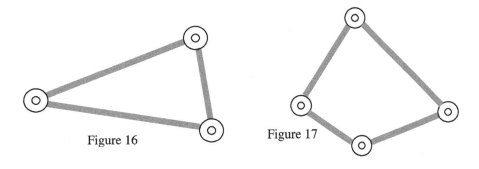

Figure 16   Figure 17

23. Four rods are hinged, as shown in Figure 17. Is the obtained figure rigid? If not, where would you put in the fifth rod to make the figure rigid? Substantiate your solution.

24. In quadrilateral $ABCD$ (Fig. 18) $AB=AD$, and $BC=DC$. Prove that AC bisects $\angle BAD$ as well as $\angle BCD$.

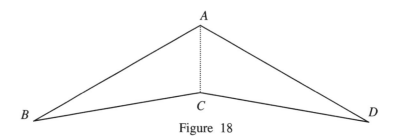

Figure 18

25. Points *A, D, C, F* in Figure 19 are collinear. *AD=CF; AB=EF; BC=DE*. Show that ∠ *ADE* = ∠ *FCB*.

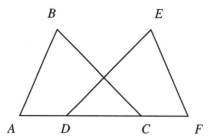

Various problems

26. Two circles, with their centers at *O* and *Q* intersect at points *A* and *B*, as shown in Figure 20. Prove that *AB* ⊥ *OQ*.

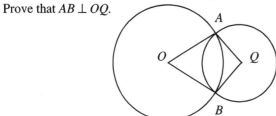

Figure 20

27. Prove that a perpendicular drawn to a bisector of an interior angle of a triangle, cuts off congruent segments on the sides that include the angle.

27. Figure 21 illustrates a method of determining the width of a river. Describe and substantiate the method.

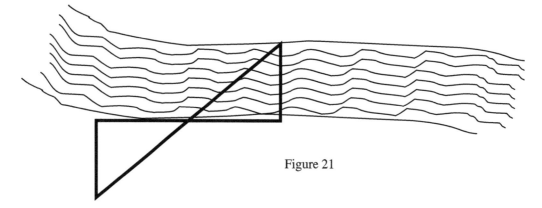

Figure 21

28. Prove that if a median of a triangle coincides with the altitude drawn from the same vertex, - then the triangle is isosceles.

29. Prove that if a bisector of an angle of some triangle coincides with the altitude drawn from the same vertex, - then the triangle is isosceles.

30. Prove that in an isosceles triangle a point in the altitude drawn to the base is equidistant from the vertices including the base.

4.5 The exterior angles of a triangle

When talking about an angle of a polygon, we always assume it to be an *interior angle*. Such an angle contains the interior points of the polygon in its interior. For example, the interior of a triangle is the intersection of the interiors of its angles (Figure 4.5.1).

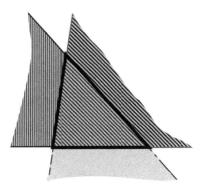

A supplement of an angle of a triangle (or any other polygon), is called an *exterior* angle of the triangle (or a polygon). In Figure 4.5.2, $\angle FAC$, $\angle BCE$, $\angle CBD$, $\angle SRM$, $\angle MNV$, etc. are *exterior*, in contrast with *interior* angles, such as $\angle ACB$, $\angle CAB$, $\angle RMN$, etc. The latter are normally called *angles* of a triangle (polygon).

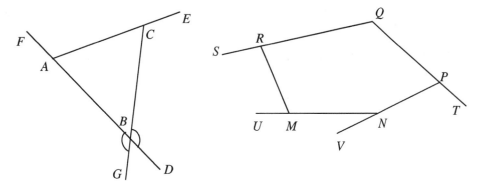

Figure 4.5.2

Each vertex of a triangle is associated with two exterior angles that may be obtained by extending either one of the two sides that form the vertex. These two exterior angles are congruent as verticals. For instance, in Figure 4.5.2, ∠ABG = ∠CBD.

For each exterior angle of a triangle, there are two interior angles that are not adjacent to it; they are called *opposite* to this exterior angle.

Theorem 4.5.1 *An exterior angle of a triangle is greater than either of the opposite interior angles.*

Proof. Let us prove that in Fig. 4.5.3, ∠BCD is greater than either of ∠BAC, ∠ABC. Through point E, the midpoint of BC, draw the median, AE, and prolong it to F, making EF = AE. Then, in △ABE and △FCE, AE = EF, BE = EC, and ∠BEA = ∠CEF (as verticals); therefore △ABE = △FCE, whence ∠ABE = ∠FCE. ∠FCE is only a part of ∠DCE, hence, ∠DCE > ∠FCE = ∠ABC, Q.E.D.

Similar procedure (drawing the median to AC, etc.) shows that

∠ACG = ∠BCD > ∠BAC, Q.E.D.

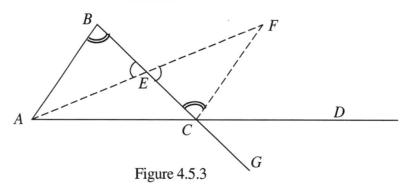

Figure 4.5.3

The exterior angle theorem is one of the most fundamental results concerning the properties of the Euclidean plane. As we shall see in the following propositions, not only is it helpful for establishing the fundamental inequalities involving sides and angles of polygons, but it also implies another test for the congruence of triangles and, even more surprisingly, the uniqueness of a perpendicular through a point to a line and the uniqueness of a reflection in a given line!

Corollary 1 *If in a triangle one of its angles is right or obtuse, then the other two angles are acute.*

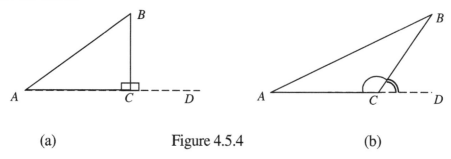

(a) Figure 4.5.4 (b)

Proof. Suppose ∠ACB of △ABC is right (Figure 4.5.4a) or obtuse (Figure 4.5.4b); then the corresponding exterior angle *DCB* will be respectively right or acute. By Th.4.5.1, ∠DCB > ∠A, and ∠DCB > ∠B, whence follows that ∠A and ∠B are acute, Q.E.D.

Corollary 2 **Given a line and a point, there exists only one perpendicular to the given line through the given point.**

Proof. If a point is located on a line, it is called an *interior point of the line*; otherwise it is called an *exterior point of the line*. We have already proved in Theorem 3.2.3 the uniqueness of a perpendicular to a line through a given interior point (it follows from the congruence of all right angles). Now let us prove the uniqueness of a perpendicular to a given line through a given exterior point.

Suppose one can draw two distinct perpendiculars, *AB* and *AC* to a given line *l* through some exterior point *A* (Figure 4.5.5). Then in △ABC two angles, ∠ABC and ∠ACB are right. This contradicts Corollary 1 of the Exterior Angle Theorem; hence our suggestion of the existence of two distinct perpendiculars is false, and a perpendicular through a given line to a given point is unique, Q.E.D.

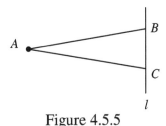

Figure 4.5.5

So far, we have never asked: is it possible to perform two distinct reflections in the same line? First, let us discuss what the words "distinct reflections" mean; then we shall see why it is important to answer this question.

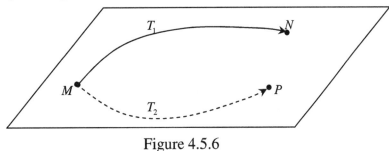

Figure 4.5.6

Suppose T_1 and T_2 are two transformations applied to the same plane. If, as a result of T_1, every point of the plane will take the same position as it takes when T_2 is applied, we shall say that T_1 and T_2 are not distinct (really, how can we distinguish between them if the results of their actions are exactly the same?).

Otherwise, if at least one point *M* of the plane is transformed into two different points, *N*, and *P*, by T_1 and T_2 respectively, the transformations are distinct: they produce different results (see an illustration in Figure 4.5.6). One can express this situation by writing: there exists such a point M that $N = T_1(M)$; $P = T_2(M)$; and $N \neq P$: the images of M under the two transformations do not coincide.

The uniqueness of a perpendicular to a line through a point allows us to prove that reflection in a given line is unique.

Corollary 3   **Given a line, there exists exactly one isometry that reflects the plane in this line. (In other words, for a given line, a type (iii) isometry is unique).**

Proof. Two isometries are different if at least one point of the plane will have two distinct images as a result of applying these two isometries. Let AB be a line in the plane. If M lies on AB, then according to the definition of reflection it remains unchanged, hence its image, which is M itself, is the same for any reflection in AB.

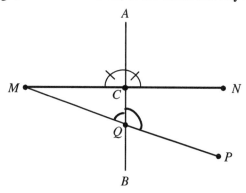

Figure 4.5.7

Now let us consider the case when M does not lie on AB. Suppose N is the image of M as a result of a reflection in AB, and P is the image of M as a result of another reflection in AB.

Let us draw a line through M and N. It will intersect AB at some point C. Then $\angle MCA$ and $\angle NCA$ are congruent as isometric images of each other. Since they are a pair of supplements, that would mean that each of them is right, and MN is perpendicular to AB.

Similarly, MP intersects AB at some point Q, and $\angle MQA = \angle PQA$ as isometric images of each other. Also, they are a pair of supplements, so each of them is right, and MP would also be perpendicular to AB.

The existence of two distinct perpendiculars to AB through M contradicts *Corollary 2* of the Exterior Angle Theorem; therefore point P must coincide with N. Thus we have established that every point of the plane is always transformed into the same image as a result of a reflection in a given line. That means we cannot have two distinct reflections in the same line: a reflection in a line is unique, Q.E.D.

While proving this corollary, we have also established the following important result:

Corollary 4.   **For every point there exists exactly one point symmetric to it in a given line.**

It is quite amazing that even though the Exterior Angle Theorem concerns an inequality, we can use it to prove an important result that establishes equality: another sufficient condition for the congruence of triangles.

Theorem 4.5.2 *(AAS condition) If a triangle has two angles and a side opposite one of these angles congruent respectively to two angles and the corresponding side in another triangle, then the triangles are congruent.*

Proof   Suppose that in triangles *ABC* and *A'B'C'* in Figure 4.5.5, ∠*B* = ∠*B'*, ∠*A* = ∠*A'*, and *AC* = *A'C'*. Let us prove that these triangles are congruent.

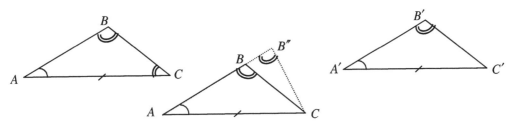

Figure 4.5.7

Let us impose △*A'B'C'* on △*ABC* in such a way that *A'C'* coincides with *AC* (they are congruent), and the line containing *A'B'* falls onto the line of *AB* (this is possible since ∠*A'* = ∠*A*).

If *A'B'* = *AB*, point *B'* will fall on *B*, and as a result the triangles will coincide (why?), which proves their congruence.

Suppose, *A'B'* is not congruent to *AB*, and as a result, *B'* lands at some point *B''* that does not coincide with *B*. If *AB''* > *AB*, then ∠*ABC* is exterior for △*CBB''*, and thus it is greater than ∠*BB''C*. The latter contradicts the hypothesis. Similarly, it can be shown that *B''* cannot lie between *A* and *B*. Therefore, *B'* will fall onto *B*, and by AXIOM 1, *B'C'* will coincide with *BC*. Hence, the triangles are congruent, Q.E.D.

4.5   Questions and Problems.

1. Show that two perpendiculars to the same straight line cannot intersect.

2. Point *M* is located in the interior of △*ABC*. Which of the angles is greater: *ABC* or *AMC*?

3.* Prove the validity of the **SAA** test for the congruence of triangles (*two triangles are congruent if a side, an adjacent to that side angle, and the angle opposed that side of one triangle are congruent, respectively, to a side, its adjacent and opposite angles of the other triangle*).

4. Prove that in an obtuse triangle, an altitude drawn from the vertex of an acute angle, cuts the extension of the opposite side, but not the side itself.

5. Prove that a point has exactly one symmetric image in a given line.

4.6 Relations between the sides and angles in a triangle

Theorem 4.6.1 *In any triangle, (i) if two sides are congruent, the opposite angles will be congruent; (ii) if two sides are not congruent, the angle opposite the greater side will be greater than the angle opposite the smaller side.*

Proof. (i) has already been proved in Theorem 4.4.3.

To prove (ii) we are considering $\triangle ABC$ (Fig. 4.6.1) with $AB > BC$. We are going to prove that $\angle ACB > \angle BAC$.

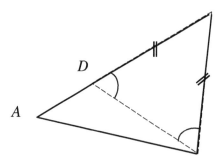

Figure 4.6.1

Let us lay off on AB, starting from B, a segment $BD = BC$. Then connect D with C by a segment. $\triangle DBC$ is isosceles, therefore, by Th. 4.3.1, $\angle BDC = \angle BCD$, and the latter angle is only a part of $\angle ACB$.

On the other hand, $\angle BAC < \angle BDC$, because $\angle BDC$ is exterior for $\triangle ADC$. So, $\angle BAC < \angle BDC = \angle BCD < \angle ACB$, Q.E.D.

The theorems converse to (i) and (ii) of Th. 4.6.1 are also valid (should be formulated and proved [using RAA] by students).

Theorem 4.6.2. *In any triangle, (i) if two angles are congruent, the opposite sides are congruent; (ii) if two angles are not congruent, the side opposite to the greater angle will be greater than the other.*

Proof. (i) Suppose, in $\triangle ABC$ (Fig. 4.6.2a) $\angle A = \angle C$; let us prove that $AB = BC$. Suppose, $AB \neq BC$, e.g. $AB > BC$. Then, by Th.4.6.1, $\angle C > \angle A$. The latter contradicts the hypothesis of our theorem, hence it follows that $AB = BC$, Q.E.D.

(ii) Suppose, in Fig. 4.6.2b $\angle C > \angle A$; let us prove that $AB > BC$.

Let us suggest that $AB \leq BC$. Then, either $AB = BC$, and by Th.4.6.1. $\angle A = \angle C$ (that contradicts the hypothesis), or $AB < BC$, and therefore $\angle C < \angle A$ (that also contradicts to the hypothesis). Then, our suggestion $AB \leq BC$ is false; hence $AB > BC$, Q.E.D.

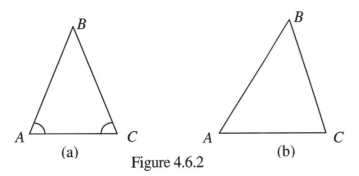

Figure 4.6.2

This theorem immediately implies the following corollaries:

Corollary 1   **An equiangular triangle is also equilateral.**

Corollary 2   **In a right triangle the hypotenuse is the greatest side.**

The proofs of these statements are simple exercises.

4.7   Comparative lengths of the straight segment and a broken line connecting the same pair of points.

Theorem 4.7.1 *(Triangle Inequality). **The sum of any two sides of a triangle is greater than the third side.***

Proof. Suppose, in Fig. 4.7.1, AB is the greatest side of $\triangle ABC$. Then it is enough to prove that $AB < AC + CB$.

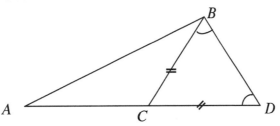

It is natural to start the proof with the construction of a segment equal to $AC + CB$. We shall do this by prolonging AC beyond C till the point D such that $CD = CB$.

Then, $\triangle BCD$ will be isosceles, $\Rightarrow \angle CDB = \angle CBD < \angle ABD$ ($\angle CBD$ is only a part of $\angle ABD$); hence in $\triangle ABD$, $AB < AD$, since AB is opposite to a smaller angle ($\angle CDB$). Thus, $AB < AD = AC + CD = AC + CB$, Q.E.D.

Corollary 1 **Each side of a triangle is greater than the difference between the other two sides.**

Proof. We shall use again Fig. 4.7.1. and the obtained inequality, $AB < AC + CB$.

Subtracting *AC* and then *CB* from the both sides of the inequality, we shall obtain:

$AB - AC < CB$, $AB - CB < AC$.

These inequalities prove the Corollary for *CB* and *AC*, thus it will be valid for the greatest side, *AB*, as well. So, the Corollary is valid for all three sides, Q.E.D.

***Theorem 4.7.2** A straight segment connecting two points is less than any broken line connecting these points.*

Remark. It should be specified that the words *"less than a broken line"* mean *"less than the sum of the sides of a broken line"*.

Proof. For a broken line consisting of two sides, the theorem reduces to Th. 4.7.1. If a broken line consists of a greater (than two) number of segments, the result of Th. 4.7.1 should be applied to prove the statement using the Principle of Mathematical Induction.

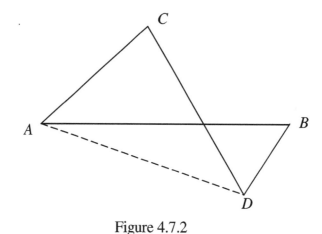

Figure 4.7.2

For instance, for broken line *ACDB* in Fig. 4.7.2, we join points *D* and *A* by a straight segment.

Then, $AB < AD + DB$, and $AD < AC + CD$.

The latter leads to $AB < AC + CD + DB$, Q.E.D.

Hint: In order to complete the proof of the statement by Mathematical Induction, assume that it is true for any broken line that consists of *n* sides and prove that it will be true for a broken line consisting of (*n* + *1*) sides. (The proof is similar to the above, which assumes $n = 2$).

***Theorem 4.7.3** If two triangles have two sides of the one congruent to two sides of the other, then*

(i) if the included angles are not congruent, then the third sides are not congruent, the greater side being opposite the greater angle; and conversely,

(ii) if the third sides are not congruent, then the angles opposite those sides are not congruent, the greater angle being opposite the greater side.

Proof. We shall prove first the direct (i) and then – the converse (ii) statement.

(i) Given (Fig. 4.7.3): $AC = A_1C_1$, $AB = A_1B_1$, and $\angle BAC > \angle B_1A_1C_1$. We are proving that $BC > B_1C_1$.

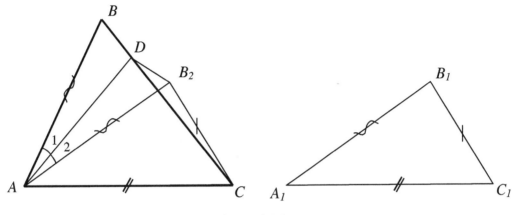

Figure 4.7.3

Let $\triangle A_1B_1C_1$ be placed with A_1C_1 coinciding with its equal AC and A_1B_1 taking the position AB_2 in the interior of $\angle BAC$ (since $\angle B_1A_1C_1 < \angle BAC$). Then $\triangle A_1B_1C_1$ takes the position AB_2C. Point B_2 may get inside $\triangle ABC$, outside $\triangle ABC$, or right on BC; the same kind of proof can be applied to any of these cases. In Fig.4.7.3 point B_2 falls in the exterior of the triangle.

Let us draw AD bisecting $\angle BAB_2$ with D lying on BC. Join D with B_2 by a straight segment. Then $\triangle ABD = \triangle AB_2D$ as having a common side AD, $\angle 1 = \angle 2$, and $AB = AB_2$. Hence, $BD = B_2D$. Now, from $\triangle DCB_2$ we conclude: $B_2C < DC + B_2D = DC + BD = BC$. That means $B_1C_1 < BC$, Q.E.D.

(ii) Consider the same figure (4.7.3) suggesting $AB = A_1B_1$, $AC = A_1C_1$, and $BC > B_1C_1$. We shall prove that $\angle BAC > \angle B_1A_1C_1$.

If we suggest that $\angle BAC = \angle B_1A_1C_1$, i.e. the equality of the angles takes place, then $\triangle ABC = \triangle A_1B_1C_1$, which is impossible, since $BC \neq B_1C_1$.

If $\angle BAC < \angle B_1A_1C_1$, it will lead, by point (i) of this theorem, to $BC < B_1C_1$. The latter contradicts the condition of the theorem. Thus, the only option left is $BC > B_1C_1$, Q.E.D.

4.6 – 4.7 Questions and Problems.

1. Point M is located in the interior of $\triangle ABC$. The perimeter of which triangle is greater: ABC or AMC?

2. Is it possible to cut a scalene triangle into two congruent triangles?

3. *AM* is a median in △*ABC*. *AB>AC*. Which of the angles is greater: *AMB* or *AMC*? Substantiate.

4. *AM is a median in △*ABC*. *AB>AC*. Which of the angles is greater: *BAM* or *CAM*? Substantiate.

5. Show that the foot of the altitude drawn to the greatest side of a triangle falls in the interior of the greatest side (not on its extension).

6. In △*ABC* *BC>AB>AC*. *AM* is the median and *AH* is the altitude drawn from the vertex *A*. Show that *H* is located between *C* and *M*.

7. * *AM, AH,* and *AD* are, respectively, the median, the altitude, and the bisector drawn from the vertex *A* of △*ABC*. Show that if $AB \neq AC$, point *D* is located between *M* and *H*. (In other words: a bisector of a triangle lies between the median and the altitude drawn from the same vertex).

8. Does a triangle with the sides equal, respectively, 4cm, 1cm, and 6 cm exist? Give reasons.

9. Prove that the sum of the segments connecting an interior point of a polygon with its vertices is greater than half-perimeter of the polygon.

10. Two triangles have a pair of common vertices and the third vertex of the one triangle lies within the interior of the other triangle. Show that the perimeter of the *included triangle* (triangle that lies inside the other one) is less than the perimeter of the *including triangle* (triangle that contains the other one in its interior).

11. One triangle lies entirely within the interior of the other. Show that the perimeter of the included triangle is less than the perimeter of the including one.

12. Determine if the following statement it true or false: If a polygon is *included* in some other polygon (lies in its interior), the perimeter of the included polygon is the least of the two perimeters. – Support your answer.

13. Prove: The perimeter of a convex polygon included in some other polygon is less than the perimeter of the including polygon.

14. Fermat's principle of optics states: A ray of light requires less time along its actual path than it would along any other path having the same endpoints. (In other words: light travels in the fastest way possible). Based on that principle, prove that the ray of light traveling from *L* (light source) to *R* (receiver) in Figure1 is reflected from the mirror *m* at point *A* so that the *angle of incidence*, α, equals to the *angle of reflection*, β. α and β are the angles between the rays of light and the perpendicular to the mirror at the point of reflection *A*.

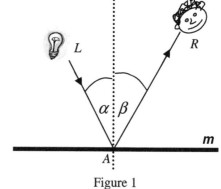

Figure 1

15. Prove that a median of a triangle is less than half-perimeter of the triangle.

16. Prove that a median of a triangle is less than half the sum of the sides that *include the median* (i.e. have a common vertex with the median) and greater than half the difference between the sum of the sides including the median and the third side.

17. It is known about the segments AB, AC, and BC: $AB < AC+BC$; $AB > AC$; $AB > BC$. Prove that A, B, and C are not collinear

4.8. Comparison of a perpendicular and the obliques

Theorem 4.8.1. *A perpendicular drawn from an external point to a given line is less than any oblique drawn to the same line from the same point.*

(Hereafter, wherever it is appropriate, we say for short "perpendicular" instead of "a segment of the perpendicular, bounded by a given point and the foot of the perpendicular". The same shortening is meant for "obliques").

Proof. In Fig. 4.8.1, $AB \perp MN$, and AC is an oblique drawn from A to MN. We shall prove that $AB < AC$.

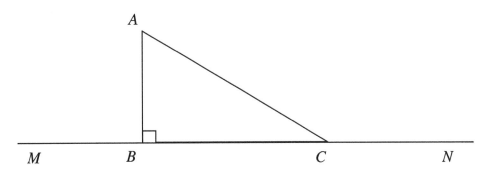

Figure 4.8.1

In $\triangle ABC$, $\angle ABC$ is right, then $\angle ACB$ is acute (why?), hence, by Th. 4.6.2, $AC > AB$, Q.E.D.

Remark. Whenever the words *the distance from a point to a line* are used, the *shortest* distance, i.e. the *length of the perpendicular* from the point to the line is suggested.

Theorem 4.8.2. *If a perpendicular and two obliques are drawn to a line from the same external point, then*

(i) if the feet of the obliques are equidistant from the foot of the perpendicular, then the obliques are congruent;

(ii) if the feet of the obliques are not equidistant from the foot of the perpendicular, the oblique whose foot is farther from the foot of the perpendicular, is greater than the other oblique.

Proof. (i) Let AC and AD (Fig. 4.8.2.) be two obliques drawn from A to MN, and let their feet, C and D, be equidistant from B, the foot of perpendicular AB: $CB = BD$. We have to prove that $AC = AD$.

In $\triangle ABC$ and $\triangle ABD$, AB is common, $\angle ABC = \angle ABD$, and $CB = BD$ by the hypothesis; then $\triangle ABC = \triangle ABD$, $\Rightarrow AC = AD$, Q.E.D.

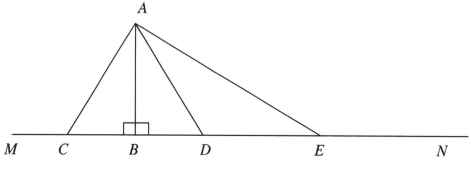

Figure 4.8.2

(ii) Let *AC* and *AE* be two obliques to *MN* from *A*, such that *BC* < *BE*. We have to prove that *AE* > *AC*.

Let us lay off segment *BD* on *BN*, from point *B* in the direction of point *N*, such that *BD* = *BC*. Then, since *BD* = *BC* < *BE*, *D* will be located between *B* and *E*.

It has already been shown in (i) that *AC* = *AD*. Since ∠*ADE* is exterior for Δ*ABD*, ⇒ ∠*ADE* is obtuse (why?), thus ∠*AED* is acute, and ∠*ADE* > ∠*AED*; ⇒ *AD* < *AE*. Then, *AC* = *AD* < *AE*, Q.E.D.

The converses to the statements of the latter theorem follow below. They can easily be proved by contradiction (RAA); students should prove them on their own.

Theorem 4.8.3 *If a perpendicular and two obliques are drawn from an external point to a line, then*

(i) if the obliques are congruent, then their feet are equidistant from the foot of the perpendicular;

(ii) if the obliques are not congruent, the foot of the greater one will be farther from the foot of the perpendicular.

4.8   Questions and Problems.

1. Prove that a segment joining a point on the greatest side of a triangle with the vertex opposite that side is less than the greatest of the two other sides.

2. Prove that a segment joining a point on a side of a triangle with the vertex opposite that side is less than the greatest of the other two sides.

3. Prove that a segment joining two points on the sides of a triangle is less than the greatest side of the triangle.

4. *Prove that the bisector of an acute angle of a right triangle divides the leg of the triangle into segments of which the greatest is the furthest from the vertex of the right angle.

4.9. Congruence of right triangles

Inasmuch as all right angles are congruent, the following two theorems of congruence do not require special proofs. They follow directly from, respectively, *SAS* and *ASA* conditions of congruence (see the Corollaries in Section 4.4):

Two right triangles are congruent if

(i)   two legs of the one are congruent, respectively, to the two legs in the other;

(ii)   an acute angle and its adjacent leg in the one are congruent, respectively, to an acute angle and its adjacent leg in the other.

The following two theorems may be called the *SSA and AAS conditions for right triangles*. Let us notice that the *SSA* condition for generic (non-right) triangles has not been proved or disproved so far. The *AAS* condition has been proved for any triangles, still we shall propose another proof for Th. 4.9.2.

Theorem 4.9.1 *Two right triangles are congruent if the hypotenuse and a leg of the one are, respectively, congruent to the hypotenuse and a leg of the other.*

Proof. Suppose in Fig. 4.9.1 $\angle C = \angle C_1 = d$, $AB = A_1B_1$ and $BC = B_1C_1$. We have to prove that $\triangle ABC = \triangle A_1B_1C_1$.

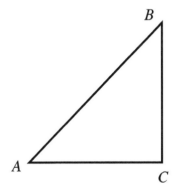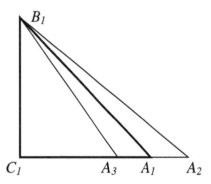

Figure 4.9.1

Let us superpose $\triangle ABC$ onto $\triangle A_1B_1C_1$ so as to make BC coincide with its congruent B_1C_1 and CA to fall along C_1A_2 (this is possible since $\angle C = \angle C_1$).

Suppose the hypotenuse AB will not coincide with A_1B_1 and will take position A_3B_1 or A_2B_1. That will create the situation where the feet of two equal obliques, A_1B_1 and A_3B_1 (or A_2B_1) are not equidistant from the foot of the perpendicular dropped from the same external point, B, onto the line lying along C_1A_1.

Thus, the suggestion of the non-coincidence of AB with A_1B_1 contradicts Th. 4.8.3 (i). Therefore, AB will coincide with A_1B_1 and the triangles will coincide in all their parts, i.e. $\triangle ABC = \triangle A_1B_1C_1$, Q.E.D.

Theorem 4.9.2 *Two right triangles are congruent if*
(i) *the hypotenuse and an acute angle of the one are, respectively, congruent to the hypotenuse and an acute angle of the other;*
(ii) *a leg and its opposite angle of the one are respectively congruent to a leg and its opposite angle of the other.*

Proof. Suppose in Fig. 4.9.2, $\angle C = \angle C_1 = d$, $AB = A_1B_1$, and $\angle A = \angle A_1$; Let us prove that $\triangle ABC = \triangle A_1B_1C_1$.

Let us superpose $\triangle ABC$ onto $\triangle A_1B_1C_1$ so as to make AB coincide with its congruent A_1B_1 and AC to fall along A_1C_1. The latter is possible because $\angle A = \angle A_1$.

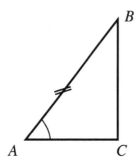 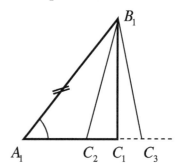

Figure 4.9.2

Then, if BC does not coincide with B_1C_1 and takes some other position, e.g. B_1C_2 or B_1C_3, that will lead to the existence of two or more perpendiculars, B_1C_1 and, e.g., B_1C_2 dropped from the same point, B onto the same line. Thus, BC and B_1C_1 will coincide, and so will do the triangles. Hence, $\triangle ABC = \triangle A_1B_1C_1$, Q.E.D.

4.9 Congruence of right triangles.

1. Prove that if in a triangle two altitudes are congruent, the triangle is isosceles.

2. Prove that two right triangles are congruent if a leg and the angle opposite to that leg of the one are respectively congruent to the leg and the opposite angle of the other.

3. Prove that two triangles are congruent if two sides and the altitude that emanates from their common vertex of one triangle are congruent, respectively, to two sides and the altitude that emanates from their common vertex of the other triangle.

4. Prove that two triangles are congruent if a side, the median and the altitude drawn to that side as a base of the one are respectively congruent to a side and the median and the altitude drawn to that side of the other.

5. Prove that two triangles are congruent if a median, the altitude that shares the vertex with that median, and the least of two angles between the altitude and a lateral side of the one are congruent, respectively, to a median, the altitude drawn from the same vertex, and the least of two angles between the altitude and the lateral side of the other.

6. Prove that two triangles are congruent if two angles and an altitude emanating from one of their vertices of the one are respectively congruent to two angles and the corresponding altitude of the other.

4.10 The perpendicular bisector of a segment and the bisector of an angle as loci.

The perpendicular bisector of a segment and the bisector of an angle possess similar symmetry properties, and we shall discuss them together.

Theorem 4.10.1a. *A point located on the perpendicular bisecting a segment is equidistant from the endpoints of the segment.*

Proof. Let *MN* be a line that passes through the midpoint *O* of some segment *AB* perpendicularly to the segment (Fig. 4.10.1a).

If *K* is a point on *MN*, then *KA* and *KB* are two obliques to *AB* from *K*, and their feet are equidistant from *O* since by the hypothesis *AO* = *OB*; therefore, *KA* = *KB*, Q.E.D.

Theorem 4.10.1b. *A point located on the bisector of an angle is equidistant from the sides of the angle.*

Proof. Let *K* be a point on the bisector *OM* of ∠*AOB* (Fig. 4.10.1b), and *C* and *D* be the feet of the perpendiculars dropped from *K* onto the sides of ∠*AOB*.

Then △*OCK* = △*ODK* since *OK* (the hypotenuse) is common and ∠1 = ∠2; therefore *KC* = *KD*, Q.E.D.

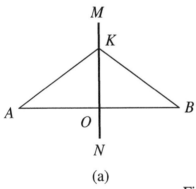

(a)

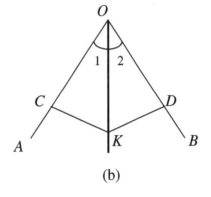

(b)

Figure 4.10.1

The converse theorems are also valid:

Theorem 4.10.2a. *If some point is equidistant from the endpoints of a segment, then this point is located on the perpendicular bisecting the segment.*

Proof. Given: in Fig. 4.10.1a, *AK* = *BK*. Let us drop the perpendicular, *KO*, onto *AB* from *K*. Then *A* and *B*, the feet of equal obliques, must be equidistant (Th. 4.8.3i) from the foot of the perpendicular, *O*. Thus *AO* = *OB*, Q.E.D.

Theorem 4.10.2b. *If some point is equidistant from the sides of the angle, then this point is located on the bisector of the angle.*

Proof. Given: in Fig. 4.10.1b, *KC* = *KD*. Let us draw the line through *K* and the vertex of the angle, *O*. Then right △s *OCK* and *ODK* are congruent by a *leg and hypotenuse condition*, therefore ∠1 = ∠2, Q.E.D.

Inasmuch as the theorems converse to Theorems 4.10.1(a,b) are true, the inverse theorems will be valid as well:

Theorem 4.10.3a. *If a point is not located on the perpendicular bisecting a segment, it is not equidistant from the endpoints of the segment.*

Theorem 4.10.3b. *If a point is not located on the bisector of an angle, it is not equidistant from the sides of the angle.*

It follows from the simultaneous validity of Th.4.10.1 and Th.4.10.2 (or 4.10.1 & 4.10.3) that perpendicular bisectors of segments and bisectors of angles are *loci* (recall the definition; we have already considered an example of a *locus*):

Corollary (a). *The perpendicular bisecting a segment is a locus of points equidistant from the endpoints of the segment.*

Corollary (b). *The bisector of an angle, is a locus of points equidistant from the sides of the angle.*

4.10 Questions and Problems.

1. In quadrilateral *ABCD* *AB=AD*, and the diagonals, *AC* and *BD*, are mutually perpendicular (such a quadrilateral is called a *kite*). Prove that *BC=CD*.

2. Prove as a direct theorem (not by RAA) that any point not located on the perpendicular bisector of a segment is not equidistant from the endpoints of the segment; particularly, it is closer to the endpoint with which it is located on the same side of the perpendicular bisector of the segment.

3. In △*ABC* in Figure 1 *AB=BC=18cm*. A perpendicular to *AB* drawn through its midpoint, *D*, cuts the base of the triangle, *AC*, at some point *E*. Determine the length of *AC*, if the perimeter of △*BEC* equals *45 cm.*.

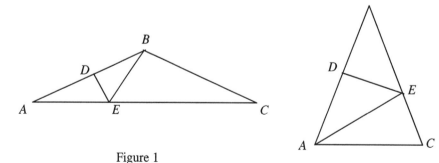

Figure 1

Figure 2

4. In △ABC in Figure 2 AB=BC=27cm. From D, the midpoint of AB, the perpendicular is erected; it cuts BC at some point E. The perimeter of △AEC equals 42 cm. Determine the length of AC.

5. A carpenter bisects ∠A (Figure 3) as follows: Lay off AB=AC. Place a steel square so that BM=MC, as shown in the figure. Mark point M and draw a line through A and M. Show that AM bisects ∠A. Would this method work if the legs of the steel plate did not form a right angle?

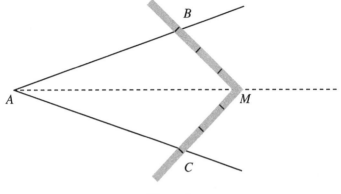

Figure 3

6. Prove that three bisectors of the interior angles of a triangle are concurrent.

4. REVIEW Questions and Problems

1. Prove that in an isosceles triangle two medians are congruent.

2. Prove that in an isosceles triangle two altitudes are congruent.

3. Prove that in an isosceles triangle two bisectors are congruent.

4. Prove that if two altitudes of a triangle are congruent, then the triangle is isosceles. (This is the converse to the proposition of Problem 2). #

5. If in an acute-angled isosceles triangle two perpendiculars are erected from the midpoints of the congruent sides, these perpendiculars cut off congruent segments on the congruent sides from the vertex of the isosceles triangle.

6. Reformulate the proposition of the preceding problem for the case of an obtuse-angled isosceles triangle. Substantiate your formulation.

7. Prove that the sum of the medians of a triangle is less than its perimeter but greater than half the perimeter.

8. Show that the sum of the diagonals of a convex quadrilateral is less than its perimeter but greater than half the perimeter.

9. Prove that two triangles are congruent if a median and the angles between the median and the including sides of the one are respectively congruent to the median and the angles between it and the including sides of the other.

10. In △ABC, AB>AC. AD is a bisector (D lies on BC). Which of the angles, ADB or ADC is greater? Which of the segments, BD or CD is greater?

11. Prove as a direct theorem that a point lying in the interior of an angle and exterior to the bisector of an angle is not equidistant from its sides.

12. Show that the median emanating from one of the vertices of a triangle is equidistant from the other two vertices.

13. Prove that if any two altutudes in a triangle are not congruent, then this triangle is not isosceles.

14. An acute angle *DOE* is given, and *A* is a point lying in the interior of that angle. Find points *B* and *C* on the sides *OD* and *OE,* respectively, such that the perimeter of $\triangle ABC$ would be minimum.

15. Prove that an isometry transforms three collinear points into three collinear points.

16. Prove that an isometry transforms three non-collinear points into three non-collinear points.

17. Prove that an isometry transforms a straight line into a straight line.

18. Prove Theorem 3.4.1(ii):
    In the same circle or in congruent circles:

    if two arcs are congruent, then the corresponding central angles are congruent.

19. *Prove that the sum of the interior angles of a triangle does not exceed two right angles.

5. CONSTRUCTION PROBLEMS

> "Well," said Owl, "the customary procedure in such cases is as follows."
> "What does Crustimoney Proseedcake mean?" said Pooh. "For I am a Bear of Very Little Brain, and long words Bother me."
> "It means the Thing to Do."
> A.A.Milne, *Winnie-The-Pooh*.

5.1 The axioms and tools for geometric constructions.

In plane geometry the use of only two tools for construction is allowed, since their existence has been *postulated*. These instruments are the *straightedge* and the *compass*. The existence of other drawing tools, such as, for example, drawing triangles, has not been postulated.

In accordance with AXIOM 1 (also called *the straightedge postulate*), one can use a straightedge for *drawing lines connecting given pairs of points* and for *extending given segments into straight lines*.

The use of a *compass* is postulated in AXIOM 3 or in its equivalent, the *compass postulate*. Thus, using a compass, one can *draw a circle of a given radius with the centre at a given point* or, given a ray emanating from a point, one can place the centre of the circle at this point and *lay off on the ray a segment equal to the radius of the circle (transfer the segment)*.

One may ask: can we also use some other tools, e.g. drawing triangles or rulers (Figure 5.1.1)?

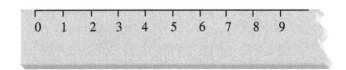

Figure 5.1.1

To answer this question let us suggest that we used a drawing triangle in some construction problem for drawing a perpendicular to a line. We took advantage of the fact that one of the angles of our drawing triangle was right. Then the following questions naturally spring up: How can we know that this angle is right? – We cannot be sure it is right unless we can prove this based on the axioms of Euclidean geometry! Maybe it is only approximately right? How was our triangle built? Suppose a mould was used, then how had *the very first* drawing triangle in history that served as a mould for all the others, been built? How did these ancient people draw the draft for constructing this triangle?

Evidently, they had been able to construct a right angle (or a perpendicular to a given line) before the first drawing triangle was built, and the only instruments they could use for their constructions were the instruments whose existence had been postulated by their axioms. (They believed that the whole Universe is built according to these axioms!). These instruments are a straightedge and a compass.

In future, after we have shown how to construct a right angle (erect or drop a perpendicular), a drawing triangle can be used to accelerate the process at some intermediate stages of constructions. A ruler and a protractor will also be used for the same purpose after we have established that we can do the basic constructions that are necessary for creating these tools (what would these constructions be?).

5.2 Basic construction problems.

In this section we shall consider a few basic constructions in Euclidean geometry. They are fundamental in the sense that one can use them as "building blocks" when performing more sophisticated constructions.

Also, these fundamental problems exemplify the basic elements involved in solving construction problems. There are usually four elements that are essential in lengthy and complicated constructions:
(i) preliminary analysis, (ii) construction, (iii) proof, and (iv) investigation.

They are summarized in the last section of this chapter.

In the first problem, which follows below, the bold font signifies the beginning of each new element.

a) To construct an angle congruent to a given angle, with the vertex at a given point and one side lying in a given line.

$\angle CAB$ is given (Figure 5.2.1a); also, point O and ray OM emanating form O (Figure 5.2.1b) are given. We have to construct an angle congruent to $\angle CAB$ with the vertex at O and one side coinciding with OM.

(i) **If the required** angle NOM **is constructed**, then any triangle, e.g., QOP, with two sides OQ and OP including $\angle NOM$, will be congruent to the $\triangle EAD$ with $AE = OQ$ and $AD = OP$, since these pairs of respectively equal sides include equal angles, NOM and CAB. Thus, we can construct the required angle by constructing a triangle with its base lying on OM with one of its extremities at O, congruent to a triangle containing CAB as an angle.

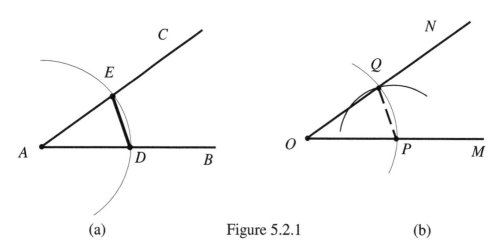

(a)   Figure 5.2.1   (b)

(ii) **Now let us construct.** From A as a centre, describe an arc of an arbitrary radius. This arc will cut *AB* at some point *D* and *AC* at some point *E*. Then describe an arc of the same radius from *O* as a centre. It will intersect *OM* at some point *P*. From *P* as a centre, describe an arc of radius equal *DE*. It will cut the former arc at some point *Q*. Draw a ray, *ON*, emanating from *O* and passing through *Q*. ∠*NOM* is a sought for angle.

(iii) **Let us prove it.** Δ*QOP* = Δ*EAD* by *SSS* condition (we have laid off *OP* by the radius congruent to *AD* and *AE*, hence *OP=OQ=AD=AE*, and similarly, by construction, *PQ=DE*). Therefore, the corresponding angles in these triangles are congruent: ∠*QOP* = ∠*EAD*. Also, *OP* lies along *OM*. Thus ∠*QOP* (or ∠*NOM*) is an angle that satisfies the given conditions, i.e. it is a *sought for* angle.

(iv) **How many solutions does the problem have?** The arcs described from *O* and *P* will intersect at two points, *Q* and Q_1 (Fig. 5.2.2.), and ∠Q_1OP also satisfies the condition of the problem.

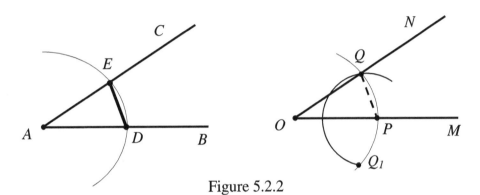

Figure 5.2.2

Thus, the problem has two solutions: there is an angle congruent to ∠*CAB* on either side of *OM*.

b) <u>To construct a triangle having given the three sides (SSS).</u>

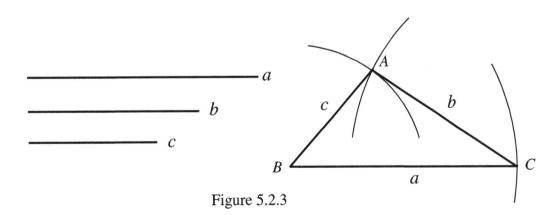

Figure 5.2.3

103

Three segments, - *a, b, c,* are given. It should be noticed that for each of them the triangle inequality is valid. Draw a ray emanating from some point *B* and describe from *B* an arc of a circle of radius *a*. The arc will intersect the ray at some point *C*. From *B* and *C* describe arcs of radii *c* and *b*, respectively. They will intersect at some point, *A*.

Now $BC = a$, $AC = b$, $AB = c$; thus, we have constructed the required triangle. Any other Δ with the same sides will be congruent to ΔABC.

c) To construct a triangle having given two sides and the included angle (*SAS*).

First, construct an angle congruent to the given one. Then from its vertex describe arcs with radii equal to the two given sides, making each arc to intersect one arm of the angle. Join the points of the intersections by a straight segment. The obtained Δ is uniquely defined by the *SAS* condition.

d) To construct a triangle having given two angles and the included side (*ASA*).
(To be solved by students as an exercise).

e) To construct a triangle having given two sides and an angle opposite to one of them (*SSA*).

Draw a ray, *AM*, emanating from some point *A*, and construct (as in problem [a]) another ray, *AN*, forming angle *NAM* congruent to the given angle α.

From *A* describe an arc of radius equal to one of the given sides, *b*. The arc intersects *AN* at some point *C*. From *C* describe a circle with radius equal to the other given side, *a*. This circle may intersect *AM* at two points, one point, or none points depending on whether *a* is greater, equal or less than the perpendicular, *h*, dropped from *C* onto *AM* (Figure 5.2.4.).

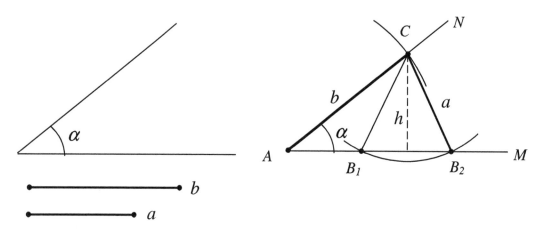

Figure 5.2.4

If $b>a>h$, both ΔAB_1C and ΔAB_2C satisfy the given condition, thus, the problem has two different solutions. This double solution is known as *the Ambiguous Case*.

If $a > b$ and $a > h$, there is a single solution (Figure 5.2.5.).

If $a = h$, the arc touches AM at one point only (this point is the foot of the perpendicular dropped from C onto AM). Then, there is a single solution in this case, and this solution is a right triangle. The latter is not surprising if we recall Th. 4.9.1 saying that given the hypotenuse (in our case, b) and a leg (in our case, a), the right triangle is determined uniquely.

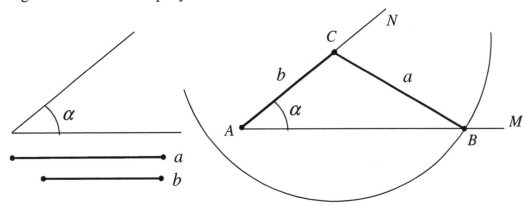

Figure 5.2.5

f) <u>To bisect an angle (draw the symmetry axis of an angle).</u>

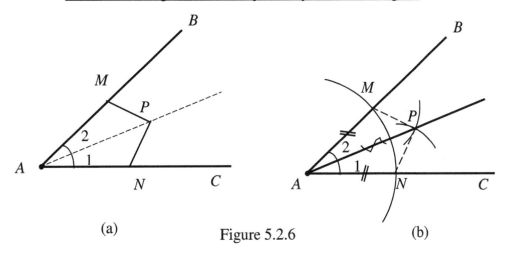

(a)　　　　　　Figure 5.2.6　　　　　　(b)

(i)　　　To draw a straight line we need two points lying in it. So far we know only one point through which the bisector passes, the vertex, A, of the given angle, BAC. Suppose, we know one more point, P, lying on the bisector. As the bisector of an angle is its symmetry axis, for any pair of points equidistant from A and lying on the arms of the angle, e.g., points M and N, triangles AMP and ANP will be congruent.

Thus, if we construct two congruent triangles with their common side in the interior of $\angle BAC$, this common side will be lying on the bisector.

(ii)   With point *A* as a centre and an arbitrary radius, describe an arc. It will cut the sides of ∠*BAC* at some points, *M* and *N*.

From *M* describe an arc with any radius greater than ½ *MN*. Then with the same radius describe an arc from *N*. The arcs will intersect at some point *P*. The line passing through *A* and *P* is the bisector of ∠*BAC*.

(iii)   Really, Δ*AMP* = Δ*ANP* by *SSS* condition; then ∠ 1 = ∠ 2, i.e. *AP* is the bisector.

(iv)   The problem has a unique solution: it has been proved (Th.4.4.6) that a bisector of an angle exists and (Th.4.4.7) that there is only one bisector for each angle.

g)   <u>To erect a perpendicular to a given line from a given point in the line.</u>

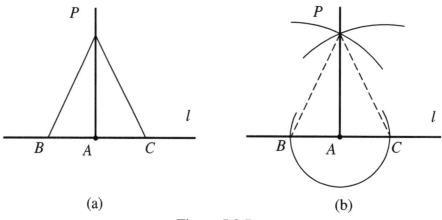

Figure 5.2.7

(i) Let *l* be a given line and *A* a point lying on this line (Fig.5.2.7a). Whenever we are going to use a straightedge for drawing a line, we need to know two points lying in the line. We know that *A* is the foot of the perpendicular we are looking for; therefore the problem of drawing the perpendicular can be reduced to finding another point, *P*, through which the perpendicular passes. *P* is, obviously, an exterior point of *l*.

Such a point, *P*, would be equidistant from any two points in the line, *B* and *C*, that are equidistant from the foot, *A*, of the perpendicular.

(ii) Use a circle of any radius with its centre at *A* to lay off equal segments *AB* and *AC* an on *l*. Describe an arc of any (greater than *AB*) radius from *B*, and then from *C*. The arcs intersect at some point *P*. *AP* is the required perpendicular.

(iii) Really, in Fig. 5.2.7b Δ*ABP* = Δ*ACP* (*SSS*), hence ∠*BAP* = ∠*CAP*, and as they are supplementary, each of them is right. That means, *AP* is the required perpendicular.

(iv) There is only one solution (why?).

h)   <u>To drop a perpendicular onto a line from a given exterior point.</u>

(i) Given a line, *l*, and an exterior point, A, we need another point through which the perpendicular to *l* will pass. Such a point will be equidistant from the same points B and C in the line *l* from which *A* is equidistant (Fig. 5.2.8a)

106

(ii) From *A* describe an arc of a radius, *r*, large enough to cut *l* at two points, *B* and *C*. Describe two arcs of equal radii from *B* and *C*. They (Fig. 5.2.8b) will intersect at some point *P*. Point *P* is equidistant from *B* and *C*. The same can be said about *A*. Thus, both of them (*A* and *P*) are located on the perpendicular bisecting *BC*; hence *AP* ⊥ *l*.

(iii) The solution is unique (why?) By the way, if we used in both steps of the construction the same radius, *r*, we would *construct point P symmetric to A about l*.

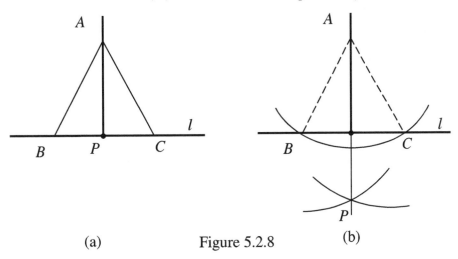

(a)　　　Figure 5.2.8　　　(b)

i) <u>To draw a perpendicular bisecting a given segment (the symmetry axis of a segment)</u>.

It is clear that each point, e.g. *C*, or *D*, of the required perpendicular will be equidistant from *A* and *B* (Fig. 5.2.9a). Thus, all we need is to construct two points that are equidstant from *A* and *B*. With an arbitrary radius *r* greater than ½ *AB*, describe two arcs, from *A* and from *B* as their centres. The arcs will intersect at some points, *C* and *D*. Draw a line passing through *C* and *D* (Fig. 5.2.9b). Both *C* and *D* are equidistant from *A* and *B*; then according to Th.4.10.2a, the line containing *CD* will be the required perpendicular.

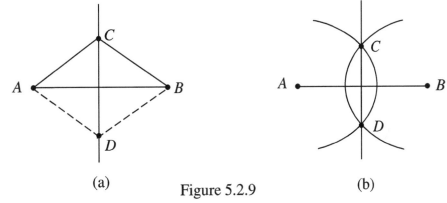

(a)　　　Figure 5.2.9　　　(b)

There is only one solution since there is only one perpendicular through a given point (in our case, the midpoint of *AB*) in a line.

5.3 Strategy for construction: analysis, construction, synthesis, investigation.

It may be observed from the above examples that the solution of any nontrivial construction problem consists of the following four parts.

1. **Analysis.** Draw the required figure as accurately as possible *without actually constructing it*, just suggest that the construction has been done. By the study of the completed figure, attempt to find how the construction of the required figure can be reduced to the known constructions. Create the plan of the construction.

2. **Construction.** Execute the construction in accordance with the plan.

3. **Synthesis (proof).** Prove that the constructed figure does satisfy the requirements imposed by the condition of the problem.

4. **Investigation.** This part of the solution of a construction problem contains the study of whether the problem is solvable for the given set of conditions, how many solutions it may possess depending on the given data, as well as the study of some particular cases that might be of special interest.

When a problem is simple, as, for instance, the construction of a triangle by the *SSS* condition, the **analysis** and **investigation** may be omitted or implicitly included in the other two steps of the solution, the **construction** and the **synthesis**. The latter two components of the solution are essential.

Remark. Do we need more axioms to complete some investigations?

Let us consider an example of a simple problem whose investigation cannot be completed unless another axiom is assumed.

Problem. Construct a triangle by *ASA* condition (having given two angles and their common side).

Solution. Let α and β be two angles of a triangle, and the side included between them is congruent to a given segment c.

CONSTRUCTION and SYNTHESIS.

Draw segment $AB = c$, and from its endpoints, A and B, draw two rays, m and n such that these rays form with AB angles α and β respectively (Figure 5.3.1). If the rays intersect at some point C, the obtained $\triangle ABC$ has a side $AB = c$ included between the angles α and β, and hence it is a sought for triangle.

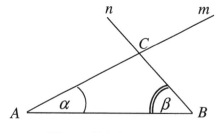

Figure 5.3.1

INVESTIGATION.

If the rays intersect, a triangle will be obtained, and any other triangle that satisfies the given conditions will be congruent to the triangle obtained by the above construction, according to *ASA* test.

There is however a question: what conditions should be imposed on the angles α and β to ensure that the rays *m* and *n* will intersect?

It is clear that for some angles the rays will not intersect. For example, they will not intersect if each of the angles (α and β) is right (Figure 5.3.2a), since the existence of a triangle with two right angles contradicts the exterior angle theorem.

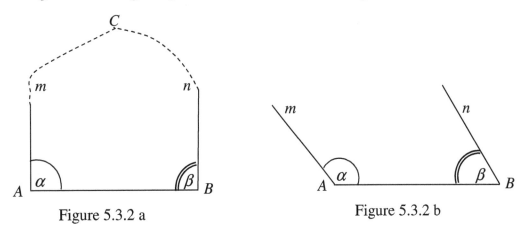

Figure 5.3.2 a   Figure 5.3.2 b

Similarly, it is easy to show (do this!) that the rays will not intersect if each of the angles α and β is obtuse.

It can be proved (it is not easy; see problem 19 from Review Problems to Chapter 4) that *the rays will not intersect if the sum of the angles α and β is equal to or greater than two right angles* (Figure 5.3.2 b).

Then, will they intersect if the sum of α and β is less than two right angles? (We know that the converse of a true statement is not necessarily true!) It seems natural to suggest that they will (it looks like in such a case the rays "converge" towards each other), however it does not follow from our axioms!

The statement claiming that *the rays will intersect if the sum of α and β is less than two right angles* has been assumed by Euclid in *The Elements* as the *fifth postulate*. It is also called the Euclidean *parallel postulate*. Without this postulate we cannot complete the investigation of our solution.

The story of the fifth Euclidean postulate is quite dramatic. Many scholars did not believe that this postulate is independent of the first four postulates of Euclidean geometry and since the time of Euclid until the late 19th century numerous "proofs" of the postulate had been proposed. Although all of these proofs turned out to be mistaken (usually they had been based on some hidden assumptions equivalent to the fifth postulate), many of these attempts contained interesting ideas and contributed greatly to the development of geometry. In the 19th century, N. Lobachevsky and, independently, J. Bolyai proved that the fifth postulate does not follow from the other axioms of Euclidean geometry: by assuming a different parallel postulate in addition to the first four Euclidean postulates, one can construct another non-contradictive geometry, different from Euclidean.

We shall consider the fifth postulate, also called the Euclidean parallel postulate, in the next chapter.

5. CONSTRUCTION PROBLEMS.

> …and all the time Pooh was saying to himself, "If only I could think of something!" For he felt sure that a Very Clever Brain could catch a Heffalump if only he knew the right way to go about it.
>
> A.A.Milne, *Winnie-The-Pooh*.

1. Construct the sum of two given angles.

2. Construct the difference of two given angles.

3. Bisect a given angle.

4. Given an acute angle, through its vertex draw a line that lies in its exterior and forms congruent angles with its sides.

5. *Euclid used for his constructions the so-called *collapsing compass*. Such a compass could be used to draw a circle with its centre at a given point, but when lifted would collapse. Show that a collapsing compass allows to complete the same constructions as a modern compass.

6. Given two points, use a compass only to find a point collinear with the given two. (Thus, in some sense can extend a segment into a straight line by means of a compass only).

7. Construct a triangle, having given :
   a) two sides and the included angle;
   b) two angles and the included side;
   c) two sides and the angle opposite to the least of them.

8. Construct an isosceles triangle having given:
   a) the base and a lateral side;
   b) a lateral side and the angle opposite the base;
   c) a lateral side and an angle adjacent to the base.

9. Construct a right triangle having given:
   a) two legs;
   b) a leg and the hypotenuse;
   c) a leg and the adjacent acute angle.

10. Construct an isosceles triangle having given:
    a) its altitude (drawn to the base) and a lateral side;
    b) its altitude (drawn to the base) and the angle opposed the base;
    c) its base and an altitude drawn to a lateral side.

11. Construct a right triangle having given its hypotenuse and an acute angle.

12. Through a point in the interior of a given angle draw a line that cuts off congruent segments on the arms of the angle.

13. Construct two segments having given their sum and difference.

14. In a line find a point equidistant from two given points exterior to the line.

15. Find a point equidistant from the three vertices of a triangle.

16. In a line that cuts the sides of an angle, find a point equidistant from the sides of the angle.

17. Find a point equidistant from the three sides of a triangle.

18. *Line *AB* divides the plane into two half-planes, and points *M* and *N* lie in the same half-plane. Find such a point *C* in *AB* that rays *CM* and *CN*, emanating from *C*, form congruent angles with rays *CA* and *CB*, respectively.

19. Construct a triangle having given :
    a) its base, an angle adjacent to the base (i.e. opposed to a lateral side), and the sum of the remaining sides;
    b) its base, an angle adjacent to the base, and the difference of the remaining sides;

20. Construct a right triangle having given its leg and the sum of the hypotenuse with the other leg.

21. Construct a right triangle having given its leg and the difference between the hypotenuse and the other leg.

22. Points *C* and *B* are located on the different sides of ∠ *A*. Find:
    a) point *M* equidistant from the sides of the angle and such that *MC=MB*.
    b) point *N* equidistant from the sides of the angle and such that *NC=CB*.

23. A railroad is virtually straight in the neighborhood of two villages, *A* and *B*. Find on the road a place for a station that would be equidistant from *A* and *B*.

24. Point *B* is located on a side of some angle, *A*. Find on the other side of the angle such a point *C*, that *CA+CB* = *l*, where *l* is a given segment.

25. Given a point in the interior of an acute angle, construct a triangle of minimum possible perimeter with one vertex at a given point and the other two vertices on the sides of the angle.

6. PARALLEL LINES

> The Piglet was sitting on the ground at the door of his house blowing happily at a dandelion, and wondering whether it would be this year, next year, sometime or never. He has just discovered that it would be never, and was trying to remember what "*it*" was…
>
> A.A.Milne, *Winnie-The-Pooh*

6.1 Definition. Theorem of the existence.

Lines that lie in the same plane but do not meet however far they may be extended are said to be *parallel*. If two lines do intersect, each of them is said to be a *transversal* to the other. Parallelism of lines (rays, segment lying on parallel lines) is denoted by // sign; e.g. in Figure 6.1.1, $l \,//\, m$; $AB \,//\, CD$. If l_1 is transversal to l_2 we shall denote it: $l_1 \not\!/\!/ \, l_2$.

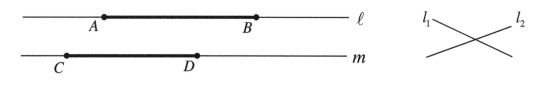

Figure 6.1.1

The fact of the existence of parallel lines can be proved.

Theorem 6.1.1. *Two perpendiculars to the same line are parallel.*

Proof. If two perpendiculars to some line l (Fig. 6.1.2.), e.g., the ones erected from points B and C did intersect at some point, A, there would be two perpendiculars drawn to the same line l from an external point, A. It would contradict Th. 3.2.4. Therefore, two perpendiculars to the same line do not intersect, i.e. they are parallel, Q.E.D.

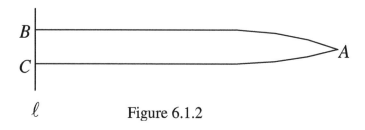

Figure 6.1.2

Problem 1. ***Through a given point draw a line parallel to a given line.***
Hint: Use the result of Th. 6.1.1.

6.2. Angles formed by a pair of lines and their transversal. Three tests for the parallelism. Neutral geometry. Parallel postulate (Playfair's formulation).

If two lines are cut by a transversal at two points (Fig. 6.2.1), eight angles are formed at those points, and certain sets of those angles have names as follows.

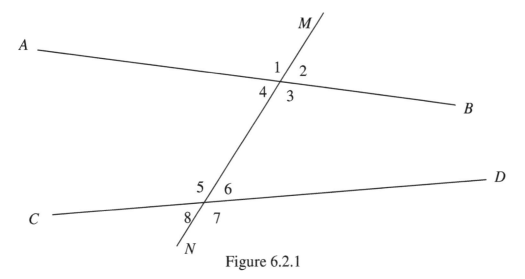

Figure 6.2.1

The angles 1, 2, 7, 8 are called *exterior* angles, and the angles 3, 4, 5, 6 are called *interior* angles.

The pairs of angles 1 and 7, 2 and 8, 3 and 5, 4 and 6 are called *alternate* pairs of angles (e.g., ∠1 and ∠7 are alternate to one another).

Of the angles 2 and 6, 2 is referred to as the *exterior* angle, and 6 as the *interior opposite* angle on the same side of *MN*. Such angles (as 2 and 6) are called *corresponding* angles. Similarly, 7 and 3, 1 and 5, 8 and 4 are pairs of corresponding angles.

Theorem 6.2.1. *If two lines are cut by a common transversal so as to make*
    (i)   *two corresponding angles equal; or*
    (i)   *two alternate angles equal; or*
    (ii)   *two interior (or two exterior) angles lying on the same side of the transversal supplementary,*
    then in each case the two lines are parallel.

Proof. (i) Let in Figure 6.2.2, ∠2 = ∠6. If we suggest that *AB* and *CD* intersect at some point *P* lying to the right of *MN*, we shall obtain a triangle with ∠2 as an exterior angle and ∠6 as an interior angle non-adjacent to ∠2. Therefore, ∠2 > ∠6. The latter contradicts to the hypothesis.

If we suggest that the intersection occurs to the left of *MN*, at some point *P´*, then for the triangle with a vertex at *P´* and sides lying on *AB* and *CD*, ∠6 is exterior, and therefore ∠6 > ∠4 = ∠2. Thus, *AB* and *CD* cannot intersect either to the left or to the right of *MN*. Then, *AB // CD*, Q.E.D.

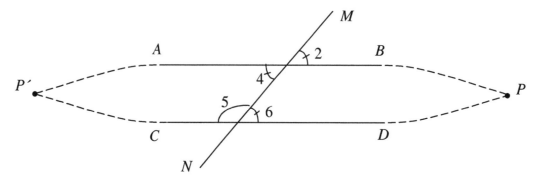

Figure 6.2.2

(ii) The proof copies the previous one if we take into account that, for example, ∠8 = ∠2 entails ∠4 = ∠2.

(iii) Suppose ∠4 + ∠5 = 2d. Inasmuch as ∠5 and ∠6 are supplementary, it follows that ∠4 = ∠6. The latter would be impossible if AB and CD did intersect at some point P´, since in this case ∠6 as an exterior angle of a triangle would be greater than ∠4. Then AB and CD do not intersect to the left of MN. Similarly it can be proved that the intersection cannot occur to the right of MN. Thus, AB // CD, Q.E.D.

Problem 1. *Through a given (external) point draw a line parallel to a given line.*
Solution. This problem has already been solved (section 6.1). Let us consider another solution, based on Th. 6.2.1.

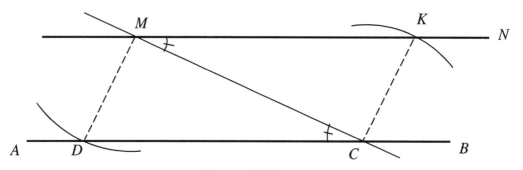

Figure 6.2.3

Suppose MN is a line drawn through M parallel to AB, and MC is a transversal of AB drawn through M. Then, by Th. 6.2.1, MN is parallel to AB if the alternate angles MCA and CMN are congruent.

Thus, in order to draw MN // AB we shall draw through M a transversal, cutting AB at some point C, and construct ∠CMN, congruent to ∠MCA. For instance, we can pick a triangle with its vertices at M, C and some other point, D, which lies on *AB*, and construct a congruent △*CMK* with *MC* as a common side and ∠*CMK* = ∠*MCD*. Then, according to Th. 6.2.1., *MK* // *AB*.

Now it seems natural to ask: are the statements converse to the results of Th.6.2.1 true? For example, can we assert that two parallel lines form congruent corresponding angles with their common transversal?

If we could prove such a statement, it would enable us to use it in some important constructions; for instance we would be able to draw through a given point *P* a line that makes a given angle α with a given line *l*, which does not contain *P* (Figure 6.2.4). Thus we would be able to *translate angles*.

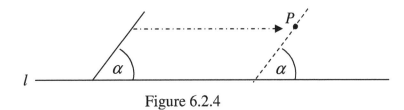

Figure 6.2.4

It turns out that in order to prove the converses of Th.6.2.1, we have to add to our system of axioms one more axiom, called the *parallel postulate*.

Let us notice that all the results we have obtained up to this point are true whether the parallel postulate is assumed or not. Moreover, they will be true even in case we assume a parallel postulate that is different from the Euclidean. That is why this results are often referred to as propositions (theorems) of *neutral geometry*.

Neutral geometry is based on AXIOMS 1 - 4 and AXIOM OF ISOMETRIES (or any other equivalent set of axioms) and does not assume any parallel postulate. It includes, in particular, all the results concerning the existence and uniqueness of perpendiculars, congruence of triangles (all four tests), the basic constructions, the existence of parallel lines, and the tests for parallelism based on relations between a pair of lines and their common transversal. If, in addition to the axioms of neutral geometry, the *Euclidean parallel postulate* is added, the geometry is called Euclidean. The geometry obtained from the neutral by adding the *Lobachevskian-Bolyai parallel postulate*, is called Lobachevskian-Bolyai or *hyperbolic geometry*. In this course we consider only Euclidean geometry, although, as we have already mentioned, quite a few results (and usually it is instructive to observe which of them) are valid in the hyperbolic geometry as well.

Now we are adding the Euclidean parallel postulate to our set of axioms for Euclidean plane geometry.

<u>AXIOM 5 (Parallel Postulate)</u> *Only one line can be drawn through a given point parallel to a given line*.
<u>Corollary 1.</u> *If a line is transversal to one of two parallel lines it will also be transversal to the other.*

Proof. In Figure 6.2.5, *CD // AB* and *CD₁* intersects *CD* (at *C*). If we suggest that *CD₁* is parallel to *AB*, we shall have two lines, *CD* and *CD₁* passing through *C* parallel to *AB*. That will contradict AXIOM 5.

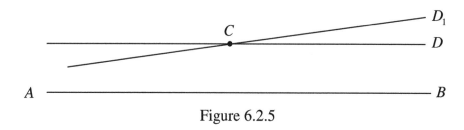

Figure 6.2.5

Corollary 2. **Lines parallel to the same line are parallel to each other.**
Proof. Suppose in Fig. 6.2.6 $l_1 // m$ and $l_2 // m$. Then, if l_1 and l_2 intersect, e.g., at some point *A*, it will lead to the existence of two lines passing through A parallel to *m*. The latter will contradict AXIOM 5. Therefore, l_1 and l_2 do not intersect.

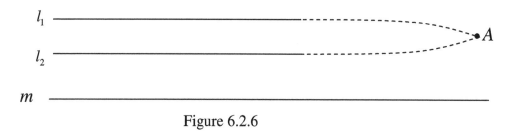

Figure 6.2.6

The parallel postulate is given here (in AXIOM 5) in Playfair's formulation. It will be shown shortly that this formulation is equivalent to the original one, called in **THE ELEMENTS** *the fifth postulate*.

Now we can prove the converse of Theorem 6.2.1:
Theorem 6.2.2. *If two parallel lines are cut by a common transversal, then*
    (i)   each exterior (interior) angle is congruent to its corresponding;
    (ii)   the alternate angles are congruent;
    (iii) the interior (exterior) angles on the same side of the transversal
are supplementary.
    Proof. (i) Let us prove that in Fig. 6.2.7, corresponding angles α and β are equal. Suggest the opposite, e.g., $\alpha > \beta$. Then we can draw a line, A_1B_1, forming
$\angle MEB_1 = \beta$ with our transversal. By Th.6.2.1(i), $A_1B_1 // CD$ that leads to the existence of two distinct lines, A_1B_1 and *AB*, passing through the same point *E*, that are parallel to *CD*. By Parallel Postulate (AXIOM 5) this is impossible. Similarly, α cannot be less than β. Then, $\alpha = \beta$, Q.E.D.

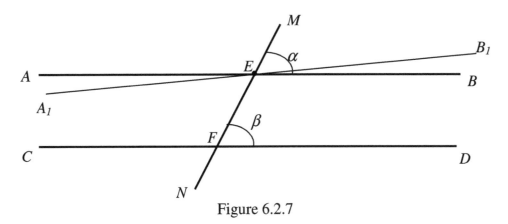

Figure 6.2.7

(ii) and (iii) can be proved in a similar manner (using RAA and Parallel Postulate).

Corollary 1 **Given two parallel lines, a perpendicular to one of them is also a perpendicular to the other.**

Proof. If in Fig. 6.2.8, $AB \parallel CD$ and $ME \perp AB$, then (why?) the extension of ME will intersect CD at some point F. It follows from Th.6.2.2 that $\angle EFD = \angle MEB$. Since $\angle MEB$ is right, then $\angle EFD = d$, i.e. $EF \perp CD$, Q.E.D.

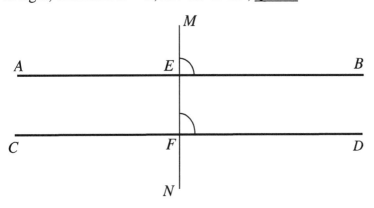

Figure 6.2.8

6.3 Tests for the nonparallelism of two lines. Parallel Postulate in Euclidean formulation.

Let us start with a reminder on logic. It is easy to establish (try to do that) that a theorem is logically equivalent to its contrapositive, and the converse and inverse theorems are logically equivalent. Then, based on the validity of Th.6.2.1 and Th.6.2.2, we can state the truth of two more theorems.

The following theorem follows from Th.6.2.1, which, as you remember, is true whether a parallel postulate is assumed or not; hence this theorem is also a result of the so-called *neutral geometry*.

Theorem 6.3.1 *If two nonparallel lines are cut by a common transversal, then*
- *(i)   the corresponding angles are not equal;*
- *(ii) the alternate angles are not equal;*
- *(iii) the interior [exterior] angles on the same side of the transversal are*
  not supplementary.

Now let us formulate the theorem that consists of the statements converse to the propositions of Th. 6.2.2. In order to prove Th.6.2.2, we had to assume the Euclidean parallel postulate (AXIOM 5), therefore each statement of the following theorem also follows from this postulate. Moreover, it can be proved (a good exercise!) that AXIOM 5 follows from each of them.

Theorem 6.3.2 *If two lines are cut by a common transversal, and*
- *(i)   the corresponding angles are not equal, or*
- *(ii) the alternate angles are not equal, or*
- *(iii) two interior (or exterior) angles lying on the same side of the transversal are not supplementary,*

then in each case the two lines are not parallel.

As we have already mentioned, each of the three propositions is true as the contrapositive of a respective proposition included in Th.6.2.2.

By historical reasons, part (iii) of Theorem 6.3.1 deserves our special attention: a similar statement has been utilized by Euclid as a parallel postulate in his original work *The Elements*. In addition to the statement that if two interior (exterior) angles are not the supplements of each other, then the lines are not parallel, the Euclidean formulation also specified on which side of the transversal the lines will intersect.

In order to recover the complete formulation, let us consider the proof of this test for the nonparallelism (even though we know that it is true as the contrapositive of Th.6.2.2 (iii)).

We shall consider the exterior angles; the case of interior angles can be given a similar consideration. Thus, it is given that the sum of two interior angles formed by a transversal to two lines is not equal to two right angles.

In Figure 6.3.1, the two possible cases, $\alpha + \beta < 2d$ and $\alpha + \beta > 2d$, are illustrated.

Now let us prove the following:

in the first case, $\alpha + \beta < 2d$, the extensions of *AB* and *CD* beyond points *B* and *D* will intersect;

in the second case, $\alpha + \beta > 2d$ the extensions of *AB* and *CD* beyond *B* and *D* will diverge and the extensions beyond *A* and *C* will intersect.

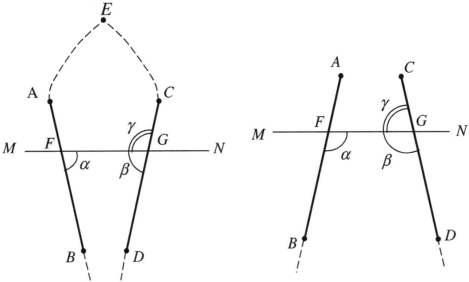

Figure 6.3.1

Proof. In the first case
$$\left.\begin{array}{r}\alpha+\beta<2d\\ \gamma+\beta=2d\end{array}\right\} \Rightarrow \alpha<\gamma.$$

Then the extensions of *AB* and *CD* beyond *A* and cannot intersect: if they do intersect they will form ΔAEC, in which the interior angle γ is greater than a non-adjacent exterior angle α. This would contradict to the exterior angle theorem.

Also, *AB* and *CD* cannot be parallel, since according to Theorem 6.2.2, in such a case we would have $\alpha+\beta=2d$, which contradicts to our hypothesis $\alpha+\beta<2d$.

Therefore, the extensions of *AB* and *CD* beyond *B* and *D* will intersect.

In the second case, $\alpha+\beta>2d$, the consideration is similar: if we suggest that the extensions of the lines beyond *A* and *C* intersect, we shall obtain a triangle *FEG* with an exterior angle, α, smaller than a non-adjacent interior angle, γ, which is impossible. Still, the lines must intersect, thus they will do it when being extended beyond *B* and *D*, Q.E.D.

Therefore, *if the sum of two interior angles is not equal to two right angles, the lines will intersect in the half-plane where their sum is less than 2d.*

This statement has been used by Euclid in his original formulation of the *parallel postulate*:

If a straight line falling on two straight lines makes the interior angles on the same side less than two right angles, the straight lines, if produced indefinitely, meet on that side on which are the angles less than the two right angles.

(The translation from Greek by sir Thomas Heath).

It is easy to prove (do it!) that the parallel postulate in Playfair's formulation (AXIOM 5) follows from the postulate in Euclidean formulation, and thus the two formulations are equivalent.

This axiom, known as *Euclid's fifth postulate*, is famous because of numerous attempts to prove that it follows from the other postulates of Euclidean geometry. Let us discuss the attempt of Proclus (c.a. V century A.D.).

Suppose in Fig. 6.3.2, $m \parallel l$ and m' is a line concurrent with m at some point A. The goal of the discussed "proof" was to show that m' must intersect l.

If m' is perpendicular to m, then by Corollary 1 of Th. 6.2.2, it will be also perpendicular to l. Suppose m' is not perpendicular to m.

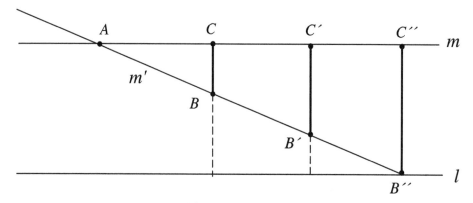

Figure 6.3.2

Choose some point, C, on m and construct a perpendicular to m at that point (C is chosen to the right of A so as to make the erected perpendicular intersect m' before reaching l). The perpendicular intersects m' somewhere at B (if, of course, m' is not perpendicular to m). If we imagine that C moves away from A (to the right), we see that the distance CB continually increases. Then there will eventually be a point C'' such that the corresponding B'' lies on l. Thus, m' will cut l somewhere at B'', and hence they (m' and l) are not parallel.

The above "proof" tacitly assumes that the distance between two parallel lines, in our case m and l, cannot increase. The assumption that two parallel lines are equidistant is an equivalent of Parallel Postulate (most apparently in Playfair's formulation), and therefore, it could not be used in the proof.

In conclusion of the section we shall adduce another two tests for nonparallelism, which we are going to use in the future. They can easily be proved, e.g. by contradiction (RAA method).

Theorem 6.3.3. *A perpendicular and an oblique drawn to the same line, are not parallel.*

Theorem 6.3.4. *If two lines intersect, their respective perpendiculars are not parallel.*

6.1–3. Questions and Problems.

1. How many lines parallel to a given line can be drawn in a plane?

2. Prove that if two lines in a plane are parallel to each other, then a line that is parallel to one of them is parallel to the other (*transitivity* of the parallelism).

3. Prove that if two lines in a plane are parallel, a transversal to one of them is a transversal to the other.

4. Use the theorem of existence of parallel lines to draw a line parallel to a given line and passing through a given point.

5. In Figure 1 two lines are cut by a transversal.
   Which of the obtained angles are: a) interior? b) exterior?
   List the pairs of the : c) corresponding angles; d) alternate angles.

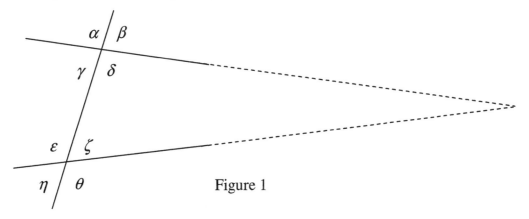

Figure 1

6. Which of the angles in Figure 1 is greater: ε or α? Substantiate.#

7. Given in Figure 1 $\beta = 72°$, and $\varepsilon = 116°$, find all the other angles.

8. In Figure 1, can the bisectors of angles η and γ be parallel? Explain.#

9. In Figure 1, can the bisectors of angles ε and γ be mutually perpendicular? Explain.

10. Lines l and m in Figure 2 are parallel. Given $\zeta = 66°$, find all the other angles.

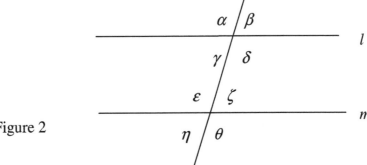

Figure 2

11. In Figure 2, l is parallel to m. Prove that
    a) the bisectors of θ and α are parallel;
    b) the bisectors of δ and η are mutually perpendicular.

12. In Figure 3, $l \parallel m$, and $\alpha = 1.5\delta$. Determine
    a) the value of γ,
    b) the angle between the bisector of β and line l.

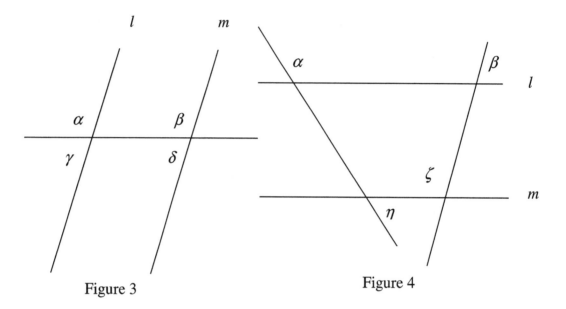

Figure 3

Figure 4

13. In Figure 4, $\alpha = 124°$, $\eta = 56°$, and $\zeta = 112°$. Determine the value of β.

14. Figure 5 illustrates how a drawing triangle may be used in practice for drawing a line parallel to a given line and passing through a given point. Explain and substantiate the method.

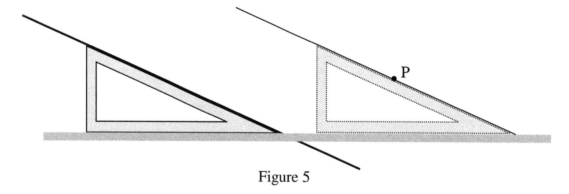

Figure 5

15. The diagonals of quadrilateral $ABCD$ meet at O, and $AO=OC$; $BO=OD$. Prove that the opposite sides of the quadrilateral are parallel.

16. Prove that a perpendicular and an oblique to the same line are not parallel.

17. Prove that if two lines intersect, their respective perpendiculars are not parallel.

18. Prove that a line drawn parallel to the base of an isosceles triangle through its vertex bisects its exterior angle.

19. Prove that a segment connecting the points, at which the bisectors of the base angles of an isosceles triangle cut the lateral sides, is parallel to the base.#

20. Propose a few methods of drawing a line parallel to a given line and passing through a given point. (Use a straightedge and a compass only).

21. Construct a triangle having given its side, the angle opposed that side and another angle (**SAA** condition).

22. Construct a right triangle having given its hypotenuse and an acute angle.

23. In an isosceles $\triangle ABC$ AC is the base, and $\angle C = 72°$. Determine the exterior angle with the vertex at B.

24. In $\triangle ABC$ $\angle A = 48°$, and $\angle B = 66°$. Determine $\angle C$.

25. A straight line, l, divides the plane into two *half-planes*. Segment AB lies entirely in one of the two half-planes formed by l. Prove that at least one of the rays, AB (that emanates from its vertex A and passes through B), or BA (that emanates from B and passes through A), lies entirely in the same half-plane. Use (i) the Euclidean formulation of the parallel postulate; (ii) Plaifair's formulation of the parallel postulate.

26. Point A' is a symmetric image of A in line l, and B' is a symmetric image of B in the same line. Prove that A, B, A', B' are located on a circle or on a straight line.

6.4 Angles with respectively parallel sides. Angles with respectively perpendicular sides.

Now, thanks to the parallel postulate, we can prove the propositions that will allow us to *translate angles* in the plane.

Theorem 6.4.1. *Two angles whose sides are respectively parallel are either equal or supplementary.*
Proof. The three cases shown in Fig. 6.4.1 are possible.

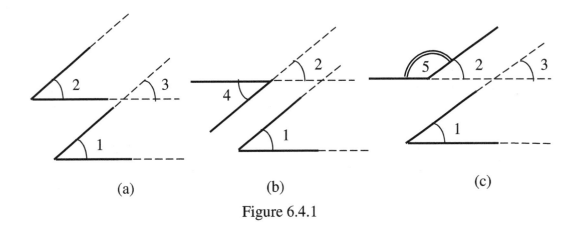

Figure 6.4.1

In case (a) the two sides of ∠ 1 are parallel to the respective sides of ∠ 2 and have the same direction. Extending the nonparallel sides till they intersect and form ∠ 3, we obtain ∠1 = ∠ 3 and ∠ 3 = ∠ 2 (by Theorem 6.2.2); hence, ∠1 = ∠2.

In case (b) the parallel sides of ∠ 1 and ∠ 4 have opposite directions. Thus,

∠ 4 = ∠ 2 as verticals, and the case reduces to the preceding one.

In case (c) one side of ∠ 1 is parallel to the respective side of ∠ 5, and the other two sides of these angles are *antiparallel*. Then ∠ 5 is supplementary for ∠ 2 that is congruent to ∠ 1, therefore ∠ 5 is supplementary to ∠ 1, Q.E.D.

Theorem 6.4.2. *Two angles whose sides are respectively perpendicular to each other are either equal or supplementary.*

Proof. Let ∠ 1, which is formed by *BC* and *BA*, be one of those angles; the second angle will be any of angles 2, 3, 4, or 5, formed by two intersecting lines, one of which, *QP*, is ⊥ to *AB*, and the other, *MN*, is ⊥ to *BC* (Fig. 6.4.2.).

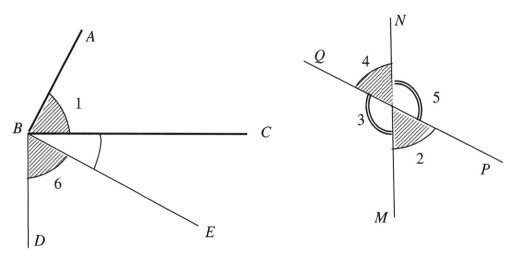

Figure 6.4.2

Draw from point *B* two auxiliary rays, *BD* ⊥ *BC* and *BE* ⊥ *BA*. These rays form ∠ 6 that equals to ∠ 1, because each of them is equal (*d* − ∠ EBC).

BD and *BE* are parallel respectively to *MN* and *PQ* as perpendiculars to the same lines, *BC* and *AB*. Then for ∠ 6 and ∠ 2, ∠ 3, ∠ 4, ∠ 5 the results of the previous theorem are applicable. Then ∠6 = ∠2 = ∠4, and ∠3 + ∠6 = ∠ 5 + ∠ 1 = 2*d*.

As we have shown above, ∠ 6 is congruent to ∠ 1, and therefore ∠ 1 = ∠ 2 = ∠ 4, and ∠ 3 + ∠ 1 = ∠ 5 + ∠ 1 = 2*d*, Q.E.D.

6.4 **Questions and Problems.**

1. Two angles have their sides respectively parallel. Evaluate the angles if:
   a) one of them is one fifth of the other;
   b) one of them is greater then the other by 49°.

2. The values of two angles with respectively parallel sides are in the ratio 4÷5 to each other. Evaluate the angles.

3. How would you find the value of an angle whose vertex is situated beyond the page on which it is drawn?

4. The sides of the two triangles in Figure 1 are respectively parallel. Find the angles of Δ *DEF* if ∠*A*=58°, ∠*B*=48°, ∠*C*=64°.

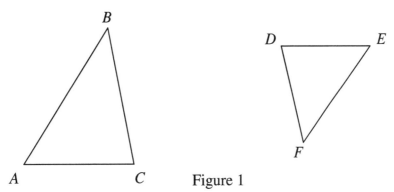

Figure 1

5. Given an angle and a point, construct an angle with the vertex at the given point and
   a) congruent to the given angle; b) supplementary to the given angle.

6. When the car shown in Figure 2 moves horizontally, its antenna is vertical. Find an angle between the antenna and the a) vertical; b) horizontal when the car climbs the slope shown in the figure.#

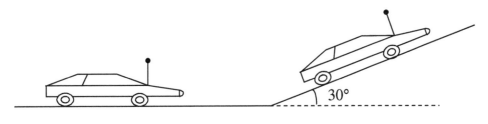

Figure 2

7. The values of two angles with mutually perpendicular sides are in the ratio 7:11. Evaluate the angles.

8. Through an interior point of ∠ *A* = 2/5 *d* two lines perpendicular to the arms of the angle are drawn. Find the degree measure of the least of the angles formed by these lines.

9. Prove that one of the angles formed by two altitudes of a triangle is congruent to the angle of the triangle formed by the sides to which the altitudes are drawn.

6.5 **The sum of the interior angles of a triangle. The sum of the interior angles of a polygon. The sum of the exterior angles of a convex polygon.**

Theorem 6.5.1. *The sum of the three interior angles of a triangle is equal to two right angles.*
Proof. Given Δ*ABC* (Fig. 6.5.1), let us prove that ∠*A* + ∠*B* + ∠*C* = 2*d*.

Extend AC beyond C and draw $CE \parallel AB$. Then $\angle 1 = \angle A$ as the corresponding; $\angle 2 = \angle B$ as the alternate; hence $\angle A + \angle B + \angle C = \angle 1 + \angle 2 + \angle C = 2d$, Q.E.D.

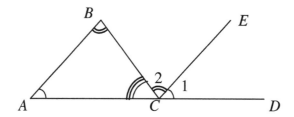

Figure 6.5.1

Corollary 1. **An exterior angle of a triangle is equal to the sum of its opposite (non-adjacent) interior angles.**

Corollary 2. **If two angles of a triangle are equal respectively to two angles of another triangle, then the third angles are also respectively equal.**

Corollary 3. **In any right triangle the two acute angles are complementary.**

Corollary 4. **In a right isosceles triangle each of the acute angles measures $45°$.**

Corollary 5. **Each angle of an equilateral triangle measures $60°$.**

Corollary 6. **In a right triangle with one of its acute angles equal $30°$, the leg opposite to this angle equals half of the hypotenuse.**

Corollary 7. **The sum of the angles of any quadrilateral is equal to four right angles.**

The above corollaries should be proved by students.

Theorem 6.5.2. **_The sum of the interior angles of a convex polygon with n sides is equal to two right angles repeated (n – 2) times._**

Proof. Any interior point, e.g. O in Fig. 6.5.2, of a convex polygon, $A_1 A_2 \ldots A_n$ can be connected with its vertices by segments lying completely inside the polygon.

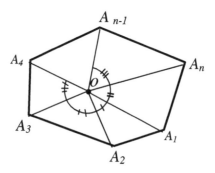

Figure 6.5.2

The sum of the angles of the polygon equals to the sum of angles of all the triangles with the vertices at O, i.e. $\triangle A_1 O A_3$, $\triangle A_2 O A_3$, …, $\triangle A_{n-1} O A_n$, minus the sum of angles formed by the segments emanating from *O*. The sum of the latter ones is the perigon (complete revolution), and thus, equals 4*d*. Hence, the sum of all interior angles of the polygon will be n·2*d* – 4*d* = 2*d* (*n*-2), Q.E.D.

It should be noticed that the same formula works for non-convex (concave) polygons as well. (The proof is left as an exercise.) Thus the general result is the following:

***Theorem 6.5.3.** The sum of the interior angles of a polygon with n sides is equal to two right angles repeated n – 2 times.*

This result can be used in order to find the sum of the exterior angles of a polygon.

***Theorem 6.5.4.** For any convex polygon the sum of the exterior angles (obtained by producing each side beyond one vertex, in order) is equal to four right angles.*

Proof. Extend the sides of a polygon as shown in Fig. 6.5.3. There is a straight angle associated with each vertex, so the sum of these angles will be 2*dn* where *n* is the number of sides of the polygon.

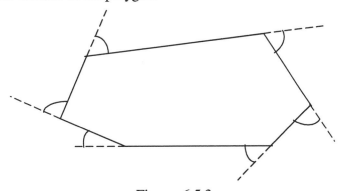

Figure 6.5.3

The sum of all interior angles equals 2*d* (*n* – 2), therefore, we shall obtain for the sum of all the exterior angles 2*dn* – 2*d* (*n* – 2) = 4*d*, Q.E.D.

6.5   Questions and Problems.

1. Two angles of a triangle are equal 45° and 69°. Evaluate the third angle.

2. Two angles of a triangle are equal 25°36′ and 78°15′. Evaluate the third angle.

3. In a triangle its bisectors drawn to the lateral sides form the angles of 36° and 15° with the base. Evaluate a) all interior angles of the triangle; b) an angle between the bisectors.

4. What can you tell about a triangle such that
    a)   one of its angles is congruent to the sum of the other two;
    b)   one of its angles is greater than the sum of the other two?

5. In Figure 1, segments *EC* and *BD* intersect at *A*, and ∠*D*=∠*C*. Prove that ∠*B*=∠*E*.

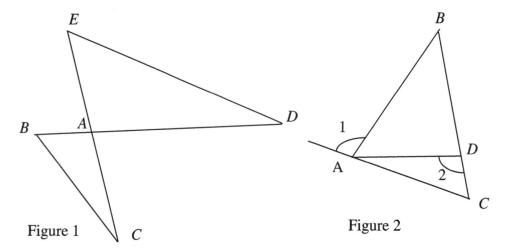

Figure 1

Figure 2

6. Side *AC* of △*ABC* (Fig.2) is extended beyond *A*. *AD* is a segment that forms an angle equal to ∠*C* with *AB*. Show that ∠1=∠2.

7. Find the angles of an isosceles triangle if the altitude drawn to the base equals half the base.

8. Find the angles of a right triangle if one of its legs is half the hypotenuse.

9. Prove that if in a triangle its median drawn to the greatest side equals half that side, then the triangle is right.

10. In a figure consisting of a square and an equilateral triangle attached to each other so that one of their sides is common (Fig.3), two greatest diagonals are drawn. Find angles 1, 2, and 3.

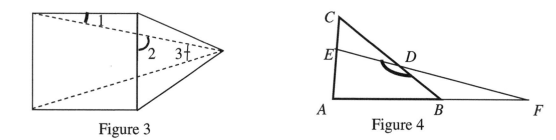

Figure 3

Figure 4

11. In △*ABC* in Figure 4, *AF* extends *AB*. *FE* is a straight segment that forms ∠*BDE*=140°. Given ∠*A*=84°, and ∠*ABC*=55°, evaluate ∠*CED* and ∠*DFB*.

12. In a triangle two angles are equal respectively 65° and 42°. Find an angle between their bisectors.

13. An acute angle of a right triangle equals 24°. Find the angle between the bisectors of
    a) this angle and the right angle;
    b) this angle and the other acute angle of the triangle.

14. In a triangle, the *vertical* angle (the angle opposed the base) equals 28°. Find the angle between the bisectors of the *base angles* (the angles that include the base).

15. In a triangle, two angles are equal respectively 66° and 78°. Find the greatest angle between the bisectors of these angles.

16. Construct a right triangle having given its
    a) leg and the opposite angle;
    b) hypotenuse and an acute angle.

17. Construct an isosceles right triangle, having given its hypotenuse.

18. Prove that two isosceles triangles are congruent if they have their bases and vertical angles respectively congruent.

19. *CD* is a median in $\triangle ABC$. Prove that the altitudes drawn in $\triangle DBC$ and $\triangle DAC$ through their vertices at *B* and *A*, respectively, are congruent.

20. *A*, *B*, and *C* are three given non-collinear points in the plane. Through *B* draw a line that is equidistant from *A* and *C*.

21. Prove that the bisector of an exterior angle at the vertex of an isosceles triangle is parallel to the base.

22. Find the angles of a right triangle if one of its exterior angles equals 123°.

23. In $\triangle ABC$ in Figure 5, $\angle A= 50°$, and $\angle B=72°$. *AC* is extended in both directions by the segments *CE=BC* and *AD=BA*, respectively. Find the angles of $\triangle BDE$.

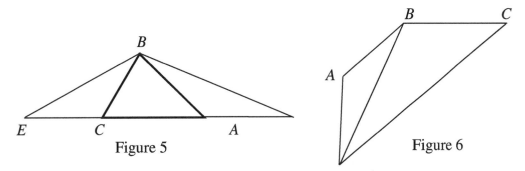

Figure 5   Figure 6

24. Two lines, *m* and *n* form acute angles α and β, respectively, with line *l*. Determine the least angle between *m* and *n*. How many solutions does the problem have?#

25. In Figure 6, *DC // AB*, $\angle A=120°$, and $\angle DBC = 2\angle C=100°$. Prove that *AD=AB*.

26. In right $\triangle ABC$ *AB = c*, $\angle C$ is right, and $\angle A = 2\angle B$. Find the projection of *AC* onto the hypotenuse.#

27. Construct a right triangle with one of its angles equal 30° having given its hypotenuse.

28. Construct a right triangle with one of its angles equal 60°, if it is known that its hypotenuse is by 4 cm greater than the least of its legs.#

29. Construct an equalateral triangle having given its altitude.

30. *The least of the acute angles of a right triangle equals α. Determine the angle between the altitude and the bisector drawn to the hypotenuse.

31. *Prove that in a right non-isosceles triangle, the bisector of the right angle bisects the angle between the median and the altitude drawn from the vertex of the right angle.

32. Evaluate the forth angle of a quadrilateral if the other three angles are equal
    a) 27°, 36°, 155°.
    b) 66°15', 91°36', 96°17'.

33. Can a quadrilateral have
    a) two right angles?
    b) three acute angles?
    c) three obtuse angles?

34. Prove that the sum of the interior angles of a general (not necessarily convex) polygon equals *2d(n-2)*, where *n* is the number of its vertices.#

35. How many sides has the convex polygon each of whose exterior angles equals 30°?

36. Three exterior angles of a quadrilateral are equal respectively 40°, 66°, and 158°. Find the shortest way to evaluate the sum of its exterior angles.#

37. Can a quadrilateral be non-convex if two of its angles are right?#

38. Two opposite angles of a quadrilateral are right. Prove that bisectors of the other two angles either are parallel or they coincide.

39. Show that the entire plane can be paved by tiles in the shape of congruent regular hexagons, i.e. these hexagons can cover the whole plane in such a way that none of them overlap.

40. Check if the plane can be paved with regular
    a) triangles; b) quadrilaterals; c) pentagons.

6.6. Central Symmetry

If two points, A and A', are collinear with some point, O, that is located between them and they are equidistant from this point, they are said to be *symmetric with respect to (about) this point*. The point O is called the *centre of symmetry*. This type of symmetry is called *central*.

Figure 6.6.1

In Figure 6.6.1, points A, O, and A' are collinear and $AO = A'O$; then A' is *symmetric* to A about the centre of symmetry O. A' is also called a *symmetric image* of A and A is a *symmetric pre-image* of A'.

Theorem 6.6.1. *Given a centre of symmetry, each point in the plane has exactly one symmetric image about the centre.*

Proof. This proof is *constructive*, i.e. we shall prove the statement by means of the appropriate construction of the symmetric image of a given point. In Fig. 6.6.1, O is the centre of symmetry and A is a point in the plane.

Draw a line passing through *A* and *O* and extend it beyond *O*. On this extension lay off a segment *OA´ = AO*. Then *A´* is a symmetric image of *A* about *O*. By AXIOM 1, there is only one line containing *A* and *O* and, by AXIOM 4, there is only one point *A´* on the ray extending the segment *AO* beyond *O* such that *OA´ = AO*. Then, *A´* is defined uniquely, Q.E.D.

Theorem 6.6.2. *If for two points (A and B) their symmetric images (A´ and B´) about a given centre (O) are constructed, then*

    (i) *the segment joining the points (A and B) will be parallel and equal to the segment joining their images (A´ and B´);*

    (ii) *for each point of the segment connecting the given points (A and B), its symmetric image lies on the segment (A´B´) connecting the images of the given points.*

Proof. (i) In Figure 6.6.2 △*AOB* = △*A´OB´* by *SAS* condition. Then *A´B´ = AB* and ∠*OAB* = ∠*OA´B´*, and the latter angles are alternate for AB, *A´B´* and their transversal *AA´*. Then, *A´B´ // AB*, Q.E.D.

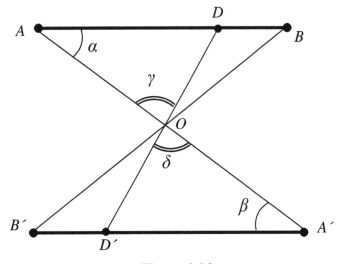

Figure 6.6.2

(ii) Pick a point, *D*, on *AB* and draw a line through *D* and *O* till it cuts *A´B´* at some point *D´*. *AO = OA´*; α = β as alternate, γ = δ as vertical; hence △*AOD* = △*A´OD´*. Therefore, *OD´ = DO*, and *D´* is the symmetric image of *D* about *O*, Q.E.D.

Two figures (or parts of the same figure) are said to be *symmetric about a given point* (O) if each point of the one figure is a symmetric image of some point of the second figure about this point (O) and vice versa. Each figure is called a *symmetric image* of the other *about a given centre*. (See the illustration in Figure 6.6.3).

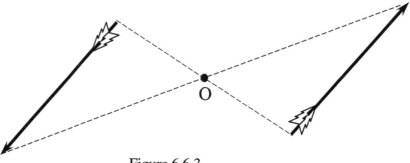

Figure 6.6.3

It is also common to say that a figure is subjected to a symmetry (transformation) about a point (*O*). Each point, *A´*, that is a symmetric image of some point, *A*, about a given centre, *O*, may be considered as obtained by rotating its preimage, *A*, about the centre, *O*, through a straight angle (Fig. 6.6.4).

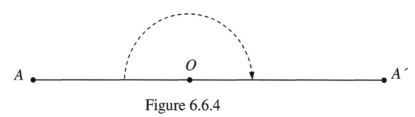

Figure 6.6.4

The same will be true for segments (Fig. 6.6.5): *A´B´* may be seen as *AB* rotated about *O* through 180°. (That should be proved by students). Geometers prefer to say that *A´B´* is the image of *AB* if **the plane is subjected to a rotation** through *180° about O*.

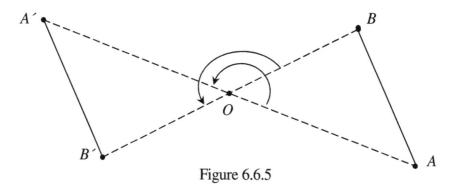

Figure 6.6.5

The described *transformation* of the plane is called a *symmetry about point O*. As we have seen, it can be realized as a rotation about *O* through 180°.

The results of Theorem 6.6.2 particularly assert that a symmetry about a point transforms a segment into a congruent segment. Then the following result ensues.

Theorem 6.6.3. ***Symmetry in a point is an isometry.***

Corollary 1 **If two figures are symmetric in a point, they are congruent.**

Remark. It should be noticed that in contrast with reflection in a line (that is also an isometry), symmetry about a point transforms a figure into its symmetric image without taking the figure out of the plane. To make the figure coincide with its symmetric about point *O* image, it would be enough to rotate the plane about *O* through 180°.

Central symmetry is very common in nature and in everyday life (see, for instance, Fig. 6.6.6).

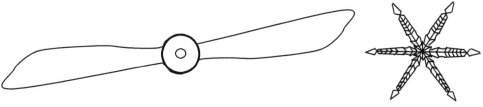

Figure 6.6.6

6.6 Questions and Problems.

1. Prove that given a segment and an exterior point, the segment has exactly one symmetric about this point image.#

2. Segments *AB* and *CD* are symmetric images of each other about point *O*. Are *AC* and *BD* symmetric images of each other about *O*?

3. Given a triangle and an exterior point, construct the symmetric image of the triangle about the point.

4. Prove that if a segment is being rotated through a straight angle about a point, it will totally coincide with its symmetric image about that point.

5. Adduce examples of figures that possess axial symmetry but do not possess central symmetry.

6. Prove the congruence of triangles by two sides and the included median (the median drawn from the common vertex of these sides).

7. Given a point in the interior of an angle, draw through this point a segment with the endpoints on the arms of the angle and such that the segment is bisected by the given point. (In other words: draw a *chord of the angle* that is bisected by the given point.)

7. PARALLELOGRAM AND TRAPEZOID

> …and Owl was telling Kanga an Interesting Anecdote full of long words like Encyclopædia and Rhododendron…
>
> A.A.Milne, *Winnie-The-Pooh*

7.1 Parallelogram.

If opposite sides of a quadrilateral are parallel, it is called a *parallelogram*. Such a quadrilateral may be obtained, for example, by intersecting two parallel lines by a pair of parallel transversals (Fig. 7.1.1).

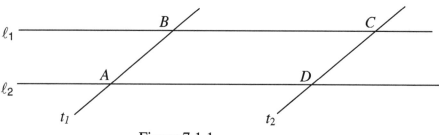

Figure 7.1.1

The following theorem describes properties of sides and angles of a parallelogram.

Theorem 7.1.1 *In a parallelogram opposite sides are congruent, opposite angles are congruent, and two angles with a common side are supplementary.*

Proof. Given a parallelogram, ABCD in Fig. 7.1.2, draw diagonal BD.

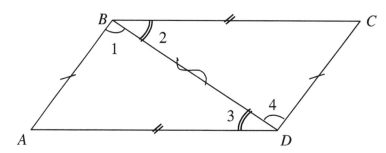

Figure 7.1.2

$\angle 1 = \angle 4$, and $\angle 2 = \angle 3$ as pairs of alternate angles, then $\triangle ABD = \triangle CDB$ by **ASA** (side BD is common). Then AD = BC, AB = CD, and $\angle A = \angle C$, Q.E.D.

$\angle ABC = \angle 1 + \angle 2 = \angle 4 + \angle 3 = \angle ADC$, Q.E.D.

$\angle A$ and $\angle ABC$ (as well as $\angle C$ and $\angle ADC$, $\angle A$ and $\angle ADC$, $\angle C$ and $\angle ABC$) are supplementary as interior angles lying on the same side of a transversal cutting two parallel lines, Q.E.D.

Remark. The congruence of the opposite sides of a parallelogram is often expressed in the following form: *the segments of parallel lines included between parallel lines are congruent.*

Corollary 1. *If two lines are parallel, the distance from any point of one line to the other line will be the same for all points of the first line.*

Proof. Lines *l* and *m* in Fig. 7.1.3 are parallel. Drop perpendiculars, *MP* and *NQ* from two points, *M* and *N* lying on *l*, onto *m*. Then *MNPQ* is a parallelogram (why?), therefore *MP = NQ*, Q.E.D.

Figure 7.1.3

The following theorem provides two *sufficient* conditions for a quadrilateral to be a parallelogram. (As follows from Th.7.1.1, each of these conditions is also *necessary*).

Theorem 7.1.2. *If a convex quadrilateral has*
    (i)   both pairs of opposite sides equal, or
    (ii)   one pair of sides equal and parallel,
then the quadrilateral is a parallelogram.

Proof.

(i) Given *ABCD* (Fig. 7.1.2) with *AB = CD* and *BC = AD*, prove that *AB // CD* and *BC // AD*.
Draw diagonal *BD*; then △ *ABD* = △*CDB* by **SSS** condition, whence it follows that ∠ 1 = ∠ 4 ⇒ *AB // CD*, and ∠ 2 = ∠ 3 ⇒ *BC // AD* (if the alternate angles are equal ⇒ the lines are parallel). Then *ABCD* is a parallelogram, by definition, Q.E.D.

(ii) If *AB = CD* and *AB // CD*, then ∠ 1 = ∠ 4 (as alternate) and, hence, △*ABD* = △*CDB* by **SAS** condition (*BD* is common). Then ∠ 2 = ∠ 3, and therefore *BC // AD*. So, *ABCD* is a parallelogram, Q.E.D.

The following results express the *properties of diagonals of a parallelogram.*

Theorem 7.1.3. *The diagonals of a parallelogram bisect each other.*
Proof. In Fig. 7.1.4, *ABCD* is a parallelogram, and its diagonals *AC* and *BD* intersect at some point *O*. Then △*AOD* = △*BOC* by **ASA**: *AD = BC* as a pair of opposite sides, and ∠ 1 = ∠ 4; ∠ 3 = ∠ 2 as alternate.

BO = DO and *AO = CO* as pairs of corresponding sides (opposed to equal angles) in congruent triangles, Q.E.D.

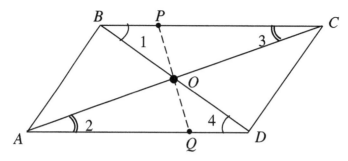

Figure 7.1.4

The conversion of Th.7.1.3. is also valid:

Theorem 7.1.4. *If in a quadrilateral diagonals bisect each other, the quadrilateral is a parallelogram.*

The proof is left for students.

Theorem 7.1.5. *In a parallelogram the point of intersection of the diagonals is the centre of symmetry of the parallelogram.*

Proof. Since ABCD in Fig. 7.1.4 is a parallelogram, it follows from Th.7.1.3 that $BO=OD$ and $AO=OC$. Also, $\angle BOC = \angle DOA$ as verticals; hence $\triangle BOC = \triangle DOA$ by **SAS**. Therefore if the plane is rotated about point O through two right angles (180°), OD will be transformed into OB, AO – into OC, and $\triangle BOC$ into $\triangle DOA$. Similarly, it will transform $\triangle AOB$ into $\triangle COD$. Thus a symmetry about O will transform the parallelogram into itself, which means it is the symmetry centre of the parallelogram, Q.E.D.

Remark. In order to establish the symmetry of ABCD, it would be enough to prove that for an arbitrary point P lying on a side of the parallelogram, its symmetric about point O image Q (Fig. 7.1.4) is located on the opposite side of the parallelogram.

7.1 Questions and Problems.

1. The sum of two angles of a parallelogram equals 152°. Find the angles of the parallelogram.

2. Evaluate all angles of a parallelogram if one of its angles is: a) greater than the other by 56°; b) one third of the other.

3. The perimeter of a parallelogram equals 144cm, and one of its sides is by 16 cm greater than the other. Evaluate all sides.

4. The perimeter of a parallelogram equals P, and the sides of the parallelogram are in a ratio of 3÷5. Find all sides.

5. Figure 1 illustrates a method of finding the distance between two points (palms in the figure) in case when it cannot be done directly. Explain the method.

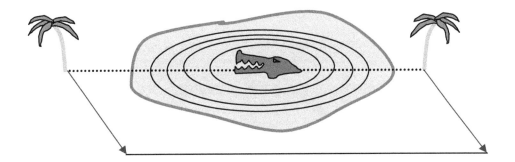

Figure 1

6. In parallelogram *ABCD* in Figure 2, point *O* of intersection of the diagonals lies in segment *MN* that connects the opposite sides. Show that *AM=CN*.

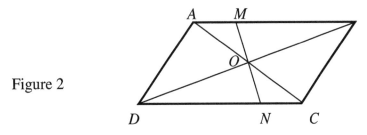

Figure 2

7. The bisector of an obtuse angle of a parallelogram cuts the side in a ratio 2÷1 from the vertex of the acute angle. Find the sides of the parallelogram if its perimeter equals *P*.

8. The bisector of an obtuse angle of a parallelogram cuts the side in a ratio 2÷1 from the vertex of the obtuse angle. Find the sides of the parallelogram if its perimeter equals *P*.

9. An acute angle of a parallelogram equals 60°. The altitude dropped from the vertex of an obtuse angle bisects the side of the parallelogram. Determine the least diagonal if the perimeter equals *P*.

10. The obtuse angle of a parallelogram is three times its acute angle. Evaluate the altitude dropped from the vertex of an obtuse angle if it cuts the side into two pieces equal 3cm and 4cm.

11. Can such a parallelogram exist that:
    a) its sides are 20cm and 34cm and one of the diagonals is 52cm?
    b) one of its sides equals 8cm, and its diagonals are 6cm and 10 cm?
    c) one of its diagonals is 6cm, and the sides are 20 cm and 38 cm?

12. In Figure 3 *ABCD* is a parallelogram, and *AM=CN*. Show that *AMCN* is a parallelogram.

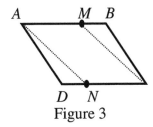

Figure 3

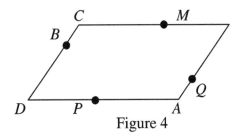

Figure 4

13. In Figure 4, *ABCD* is a parallelogram, and *BM=DP; BQ=DN*. Show that *M, Q,, P, N* are the vertices of a parallelogram.

14. Explain why the device shown in Figure 5 can be used for drawing parallel lines.

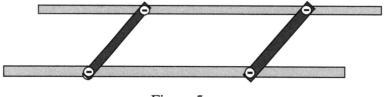

Figure 5

15. Explain why the axis of the lamp in Figure 6 is always vertical.

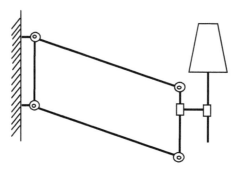

Figure 6

16. Construct a parallelogram, having given:
    a) two sides and the included angle;
    b) two (neighbouring) sides and a diagonal;
    c) a side, a diagonal and their included angle;
    d) a side and two diagonals;
    e) two diagonals and an angle between them;
    f) a side, a diagonal and an angle opposed that diagonal.

17. How many different parallelograms can constitute two congruent triangles if these triangles are:
    a) scalene? b) isosceles? c) equilateral?

18. Prove that in a parallelogram opposite vertices are equidistant from the diagonal that joins the other two opposite vertices.

19. Prove that in a parallelogram the bisectors of the opposite angles either are parallel or coincide.

20. Prove that in a parallelogram bisectors of two angles that have a common side are perpendicular.

21. In parallelogram *ABCD* bisectors of angles *B* and *D* cut diagonal *AC* at points *M* and *N*, respectively. Prove that points *B, N, D,* and *M* are the vertices of a parallelogram.

22. Bisectors of the angles of a parallelogram cut the sides at points *M, N, P,* and *Q* (see Figure 7). Determine what kind of a quadrilateral is *MNPQ*.

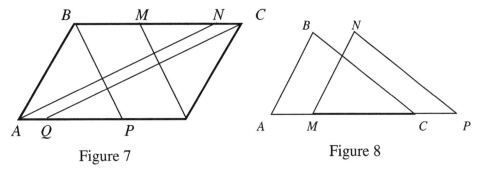

Figure 7 Figure 8

23. Prove that if in a quadrilateral two neighbouring angles are supplementary, and two opposite angles are congruent, then the quadrilateral is a parallelogram.

24. In Figure 8 triangles *ABC* and *MNP* are congruent. *AP=20cm; MC=12cm*. Evaluate *BN*.

25. Prove that if in a quadrilateral its diagonals bisect each other, the quadrilateral is a parallelogram.

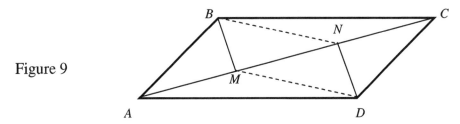

Figure 9

26. In Figure 9, *BM* and *DN* are perpendiculars dropped from the vertices *B* and *D*, respectively, of parallelogram *ABCD* onto its diagonal *AC*. Show that *BN=DM*.

27. *Prove that in a parallelogram the greatest diagonal is opposed to the greatest angle.

28. Prove that in a parallelogram the angle between the altitudes drawn from the vertex of an obtuse angle is congruent to the acute angle of the parallelogram.

29. Prove that in a parallelogram the angle between the altitudes drawn from the vertex of an acute angle is congruent to the obtuse angle of the parallelogram.

30. Two angles of a parallelogram are in a ratio 1÷3. Find the angle between the altitudes drawn from the vertex of an a) acute angle; b) obtuse angle.

31. Prove that two triangles are congruent if two sides and the included median of the one are respectively congruent to two sides and the included median of the other.

32. Construct a triangle, having given two sides and the included median.

33. In quadrilateral *ABCD*, *AB //DC*, *AB=DC*, and *FD=BG* (Figure 10). Prove that *FG* is bisected by diagonal *AC*.

Figure 10

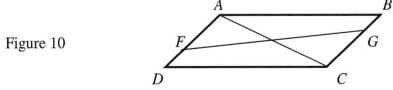

34. Given two intersecting lines and an exterior (for both of them) point, construct a parallelogram with two of his sides lying in the given lines and one of its vertices located at the given point.

35. Given a segment and an exterior point, construct a parallelogram with that segment as a side and the given point as a vertex.

36. Given a segment and an exterior point, construct a parallelogram with that segment as a side and the given point as a midpoint of one of the other sides.

37. In Figure 11, *E* and *F* are the midpoints of the opposite sides of parallelogram *ABCD*. Prove that *BE* and *DF* trisect diagonal *AC*.

Figure 11

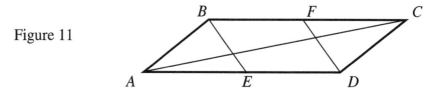

7.2 Particular cases of parallelograms: rectangle, rhombus (diamond), square. Symmetry properties of parallelograms.

a) <u>Rectangle.</u> If one of the angles of a parallelogram is right, so will be the other three (why?). A parallelogram with right angles is called a *rectangle*. In addition to properties common for all parallelograms, a rectangle possesses its special features expressed in the following theorem.

b)

Theorem 7.2.1. *In a rectangle*
  (i)   the diagonals are congruent;
  (ii) each of two lines passing through the centre of symmetry of a rectangle parallel to its sides, is an axis of symmetry of the rectangle.

Proof. (i) In rectangle *ABCD* (Fig. 7.2.1.), $\triangle ABD = \triangle DCA$ since they are right with equal legs $AB = CD$ and a common leg *AD*.
Then $AC = BD$, Q.E.D.

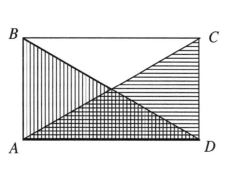
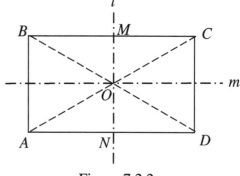

Figure 7.2.1   Figure 7.2.2

(ii) In Fig. 7.2.2, *l* and *m* are lines passing through the centre of symmetry of rectangle *ABCD*, point *O*. With that $l \parallel AB \parallel CD$; $m \parallel BC \parallel AD$.

Then $ON \perp AD$, $OM \perp BC$, and by the theorem about an isosceles triangle, *ON* is a symmetry axis for $\triangle AOD$ and *OM* for $\triangle BOC$. Then *A* is the reflection image of *D*, and *B* is the reflection image of *C* (in line *MN* as a symmetry axis). Hence, *NABM* will coincide with *NDCM* as a result of the reflection in *MN*. Then *MN* (i.e. *l*) is a symmetry axis of *ABCD*, Q.E.D.

Similarly, *m* is another symmetry axis.

c)   Rhombus (Diamond)

If all sides of a quadrilateral are equal, it is called a *rhombus*. It is easy to prove (do it!) that such a quadrilateral is necessarily convex. Then it follows immediately from Th.7.1.2. that any ***rhombus is a parallelogram.***

Theorem 7.2.2 *In a rhombus,*

i) the diagonals are mutually perpendicular and bisect the angles of the rhombus;

ii) each diagonal is an axis of symmetry of the rhombus.

Proof. (i) In rhombus *ABCD* (Fig. 7.2.3) *AC* and *BD* are diagonals; they intersect at some point *O*. Then, by Th.7.1.3, $AO = OC$, hence $\triangle ABO = \triangle CBO$ by **SSS** condition. Therefore, $\angle 1 = \angle 2 \Rightarrow BD \perp AC$, and $\angle 3 = \angle 4$, Q.E.D. Similar consideration applied to $\triangle BOC$ and $\triangle DOC$, completes the proof.

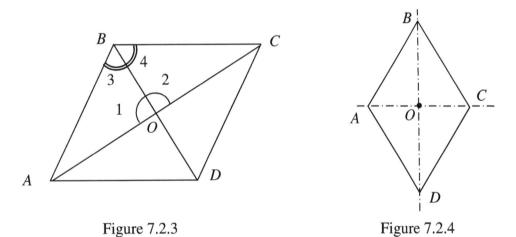

Figure 7.2.3   Figure 7.2.4

(ii) For rhombus *ABCD* in Fig. 7.2.4, $\triangle ABD$ if being reflected in *BD*, will coincide in all its parts with $\triangle CBD$ (why?), therefore *BD* is an axis of symmetry of *ABCD*, Q.E.D. Similar consideration applies to *AC*.

d)   Square. If in a rhombus one of its angles is right, then the others will also be right (why?). Such a rhombus is called a *square*.

Apparently, a square combines all the properties of a rhombus with all the properties of a rectangle. For instance (Fig.7.2.5.), it has four symmetry axes.

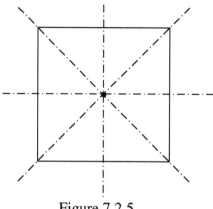

Figure 7.2.5

7.2 Questions and problems.

1. In a rectangle its diagonals form an angle of 50°. Evaluate the angles between the diagonals and sides of the rectangle.

2. A perpendicular dropped from a vertex of a rectangle upon its diagonal divides the right angle in a ratio of 2÷3. Determine:
   a) the angles between the diagonals and sides of the rectangle;
   b) an angle between the perpendicular and the second diagonal.

3. Construct a rectangle, having given:
   a) two adjacent sides;
   b) a diagonal and an angle that it forms with a side;
   c) a side and a diagonal;
   d) a diagonal and an angle between the diagonals.

4. Prove that in a quadrilateral with three right angles the opposite sides are parallel.

5. Prove that if in a quadrilateral the diagonals are congruent and bisect each other, the quadrilateral is a rectangle.

6. Prove that the bisectors of the interior angles of a parallelogram, when cutting each other, form a rectangle.

7. A carpenter made a wooden window frame, and the customer who had ordered it has doubts whether the frame has a rectangular shape or not. How can he check that having given only a ruler?

8. Prove that in a right triangle, the median drawn to the hypotenuse is half the hypotenuse.

9. In right $\triangle ABC$ $AC=BC=a$. Through a point lying on the hypotenuse, two lines parallel to the legs of the triangle are drawn, thus forming a quadrilateral. Find the perimeter of that quadrilateral.

10. A parallelogram is inscribed in a quartercircle of radius R as shown in Figure1: one of its vertices is located on the arc, the opposite vertex – at the centre of the circle, and the other two – on the radii that bound a sector equal one quarter of the disk. Determine segment MN.

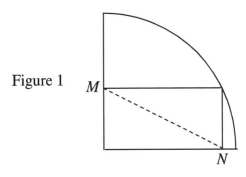

Figure 1

11. Given a segment and an acute angle, describe the segment in an angle in such a way that the segment is perpendicular to one of the arms.

12. Find a point that is located at a given distance a from a given line as well as from a given point. How many solutions does the problem have?

13. Find a point which is equidistant from the sides of an angle and is located at a given distance a from a given line.

14. Construct a triangle, having given two sides and an altitude drawn to one of these sides.

15. Construct a triangle, having given a side, an altitude drawn to that side, and an angle between that side and another altitude.

16. Construct a triangle, having given its angle and two altitudes dropped upon the sides that include that angle.

17. Construct a parallelogram, having given its side, an altitude, and a diagonal.

18. Construct a triangle, having given its side, an altitude drawn to that side and a median drawn to some other side.

19. Construct a parallelogram, having given the two sides and an altitude.

20. Two triangles constitute a rhombus; what can you tell about these triangles?

21. Four congruent triangles constitute a rhombus; what can you tell about these triangles?

22. In a rhombus a diagonal is congruent to a side. Evaluate angles between the diagonals and the sides.

23. In a rhombus, the angles between its sides and diagonals are in a ratio 2÷3. Evaluate the angles of the rhombus.

24. Prove that any rhombus is necessarily convex.

25. The altitudes drawn through a vertex of a rhombus form an angle of 30°. Evaluate a) the angles of the rhombus; b) The angles between the sides and diagonals of the rhombus.

26. In a rhombus, an altitude drawn from the vertex of an obtuse angle, bisects the side. Determine: a) the angles of the rhombus; b) the perimeter of the rhombus, given its least diagonal equals 2cm.

27. Prove that if in a parallelogram one of its diagonals bisects an angle, then the parallelogram is a rhombus.

28. Prove that any parallelogram in which a diagonal bisects an angle is a rhombus.

29. Prove that if in a quadrilateral its diagonals bisects all angles, the quadrilateral is a rhombus.

30. Construct a rhombus, having given:
    a) a side and a diagonal;
    b) two diagonals;
    c) a side and an adjacent angle;
    d) an altitude and a diagonal;
    e) an angle and a diagonal that passes through the vertex of that angle;
    f) a diagonal and an opposite angle.

31. Use only a straightedge with two parallel sides to bisect a given segment. (The width of the straightedge is less than the length of the segment).

32. Use only a straightedge with two parallel sides to erect a perpendicular to a line from a given interior point.

33. Explain, why the hinges located at the points A, B, C, D, E in Figure 2 are necessarily collinear.

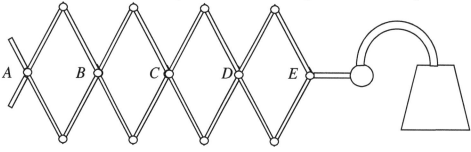

Figure 2

34. Two triangles constitute a square. What can you say about them?

35. Given a square, split it into four congruent triangles. In how many ways can you do that?

36. Is a quadrilateral necessarily a square, if its diagonals are congruent and perpendicular?

37. Prove that a rhombus with congruent diagonals is a square.

38. Construct a square, having given a) its side; b) its diagonal.

39. In Figure 3, $ABCD$ is a square, and $AA'=BB'=CC'=DD'$. Prove that $A'B'C'D'$ is a square.

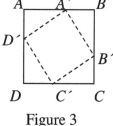

Figure 3

40. Prove that a quadrilateral is a square if its diagonals are congruent, perpendicular, and bisect the angles. Which of the conditions in this hypothesis is superfluous?

7.3.1. Some theorems based on properties of parallelograms.

Theorem 7.3.1. *If three or more parallel lines cut off equal segments on one transversal, they cut off equal segments on any other transversal.*

Proof. Given $DM \parallel EN \parallel FP$ and $DE = EF$ in Fig. 7.3.1, draw auxiliary lines DK and EL parallel to MA. Then $DMNK$ and $ENPL$ are parallelograms (as both pairs of their opposite sides are formed by parallel segments).

$\Delta DEK = \Delta EFL$ by **ASA** condition (explain why), therefore, $DK = EL$. On the other hand, $DK = MN$ and $EL = NP$ as opposite sides of parallelograms. Then, $MN = NP$, Q.E.D.

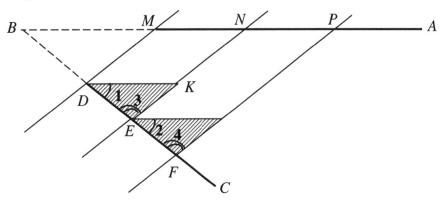

Figure 7.3.1

<u>Remark 1.</u> The above proof suggested that MA is not $\parallel DC$, i.e. they intersect at some point, B. If, in addition, $BD = DE$, then $BM = MN = NP$ (why?) If the transversals, MA and DC, are parallel, the conclusion of the theorem follows directly from the properties of parallelograms.

<u>Corollary 1.</u> *The line drawn through the midpoint of one side of a triangle parallel to another side, bisects the third side of the triangle.*

Proof. (Follows directly from the Remark to Th.7.3.1). If in Figure 7.3.2 $AD = DB$ and $DE \parallel AC$ then $BE = EC$.

Hereafter we shall call a segment connecting midpoints of two sides of a triangle a *midline (or a midjoin) of a triangle*. (Note: these terms are not common).

Theorem 7.3.2. *(Properties of a midline of a triangle). The segment joining the midpoints of two sides of a triangle is parallel to the third side and equal to half the third side.*

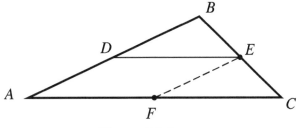

Figure 7.3.2

Proof. Let in Fig. 7.3.2 *D* and *E* be the midpoints of *AB* and *BC*, respectively. Suppose, *DE* is not // *AC*, then there must exist some other straight segment passing through *D* parallel to *AC*. According to Corollary 1 to Th.7.3.1, this segment will cut *BC* at the midpoint, and the latter is point *E*. Then the endpoints of this segment coincide with D and E, and by AXIOM 1, it will be *DE*. Thus, *DE // AC*.

Now, draw *EF // AB*. By Corollary 1, *AF = FC*. On the other hand, *ADEF* is a parallelogram, therefore *DE = AF*. Hence, *DE* = ½ AC, Q.E.D.

The solution of the following important construction problem is based on the property of parallel transversals.

<u>Construction problem.</u> *Divide a given segment into a given number of equal segments*

<u>Solution 1</u>. Given a segment *AB* in Fig.7.3.3, draw a ray emanating from one of its endpoints, *A*. Lay of on this ray *n* sequential congruent segments ($n = 3$ in Fig. 7.3.3).

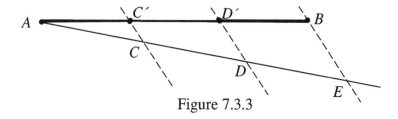

Figure 7.3.3

Through the farthermost from *A* endpoint of the farthest from *A* segment (point *E* in Fig.7.3.3), draw a straight line passing through *B*, the other endpoint of *AB*.

Then through the other endpoints of the *n* congruent segments (*C* and *D*) draw straight lines parallel to *EB* till they intersect AB (at points *C´* and *D´* respectively). Then, according to Th. 7.3.1, *AB* is divided into *n* equal parts (*AC´ = C´D´ = D´B*).

<u>Solution 2.</u> The idea of this solution is illustrated by Fig. 7.3.4 where *AE // BE´* and *AC = CD = DE = BE´ = E´D´ = D´C´*. *AB* is divided into four equal parts. Explain the solution in detail.

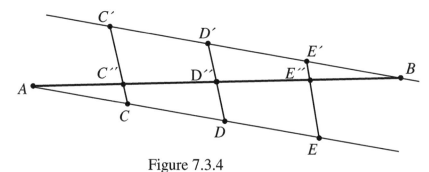

Figure 7.3.4

7.3 Some theorems based on properties of parallelograms.

1. Trisect a given segment (divide a segment into three equal parts).

2. Is the converse of Theorem 7.3.1 true?

3. Divide a given segment into five congruent parts. (Can you invent something that will allow you to avoid the necessity to construct each of five parallel lines?)

4. Use only a straightedge with two parallel edges and one mark on it (see Figure 1) to divide a given segment, *AB,* into five congruent parts.

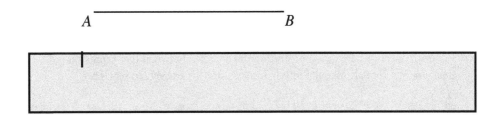

Figure 1

5. Prove that projections of congruent and parallel segments onto the same line are congruent.

6. The sides of a triangle are in a ratio of 7:8:9. The length of the shortest midline of the triangle is 3.5 cm. Find the perimeter of the triangle.

7. Prove that if in a triangle two medians are congruent, then the triangle is isosceles. #

8. Prove that the four segments that connect consecutively the midpoints of the sides of a quadrilateral form a parallelogram.

9. Triangle $A'B'C'$ is formed by three lines drawn through the vertices of $\triangle ABC$ parallel to the opposite sides: A' is opposed to A, B' to B, and C' to C, respectively.
   a) Find the perimeter of $\triangle A'B'C'$ given the perimeter of $\triangle ABC$ equals P;
   b) Find the altitude dropped from A' upon $B'C'$ if the altitude of $\triangle ABC$ dropped from A equals h.

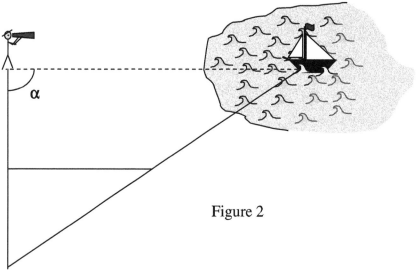

Figure 2

10. Figure 2 illustrates a method of finding the distance between two points one of which is inaccessible. Explain the method. Is it necessary for angle α to be right?

11. In Figure 3, $MC=12cm$ and $NC=8cm$ are projections of AC and BC onto a line drawn through C. Find the projections of all medians of $\triangle ABC$ onto that line.

12. In Figure 4, BD is an altitude of the triangle; $BK=KD$, $AL=LC$, $MN \parallel AC$, and $PQ \parallel BD$. Prove that PQ is bisected by KL. Which of the conditions is superfluous?

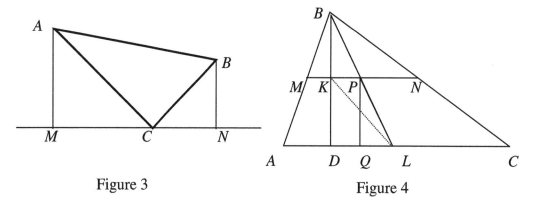

Figure 3   Figure 4

7.4 Trapezoid.

If two sides of a quadrilateral are parallel, it is called a *trapezoid*. (It should be noticed that in older texts the term *trapezium* was used, whereas "trapezoid" sometimes denoted a general quadrilateral).

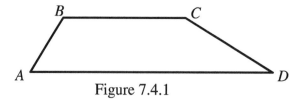

Figure 7.4.1

The parallel sides of a trapezoid are called its *bases*, and its nonparallel sides are called *lateral sides* or just *sides*. In Fig.7.4.1, *AD* and *BC* are the bases and *AB* and *CD* are the sides (lateral sides) of trapezoid *ABCD*. If the lateral sides of a trapezoid are congruent, it is called *isosceles*.

The segment joining the midpoints of the (lateral) sides of a trapezoid is called its *midline*. It possesses the following properties.

Theorem 7.4.1. *The midline of a trapezoid is parallel to its bases and equal half of the sum of the bases.*

Proof. In Fig. 7.4.2, *E* and *F* are the midpoints of the sides *AB* and *CD*, respectively.

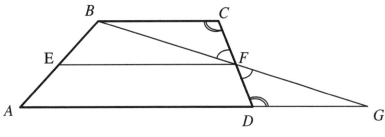

Figure 7.4.2

Draw *BF* and extend it till its intersection with the extension of *AD*, at some point *G*.

$\Delta BCF = \Delta GDF$ by ASA condition (explain and substantiate!), hence $BF = FG$, and *EF* is a midline of ΔABG. Thus, *EF // AG*, whence *EF // BC*, and $EF = \frac{1}{2} AG = \frac{1}{2}(AD + DG) = \frac{1}{2}(AD + BC)$. Q.E.D.

Remark. It is easy to prove based on Th.7.4.1 and 7.3.2 that the converse of Th.7.3.1 is true. Formulate and prove! The converse of Th.7.3.1 can also be proved by contradiction.

7.4 Questions and Problems.

1. Two angles that include a base of a trapezoid are equal respectively 112° and 144°. Find all the other angles of the trapezoid.

2. May one of the angles which include the greater base of a trapezoid be obtuse? May both angles which include the greater base of a trapezoid be obtuse? Explain. #

3. Can the values of the angles of a trapezoid, when taken in a consecutive order, be in a ratio of
   a) 6:5:4:3 ?   b) 7:8:5:11 ?

4. A diagonal of a trapezoid forms an angle of 100° with one of its lateral sides, and the acute angle opposed the diagonal is 50°. The least base of the trapezoid is congruent to the other lateral side which equals *a*. The other base equals *b*. Find: a) all the other angles of the trapezoid;
b) the distance between the lines drawn parallel to the above diagonal through the vertices opposed that diagonal.

5. In a trapezoid with the least base equal 6cm, a line is drawn through a vertex common to that base and a lateral side parallel to the other lateral side. The perimeter of the obtained triangle equals 15 cm. Determine the perimeter of the trapezoid.

6. Find the perimeter of an isosceles trapezoid if one of its angles equals 60°, and the bases are equal a and b, respectively.

7. Prove that if a diagonal of a trapezoid bisects its angle, then one of its lateral sides is congruent to one of the bases.

8. Formulate and prove a converse of the statement of the preceding problem. (Is there only one converse?)

9. The endpoints of a segment are H and h distant from a line that does not cut the segment. Determine the distance from the midpoint of the segment to the line.

10. *The endpoints of a segment are at the distance H and h respectively from a line that intersects the segment. Determine the distance from the midpoint of the segment to the line.#

11. Evaluate the midline of an isosceles trapezoid of perimeter P if its diagonal bisects an acute angle, and the bases are in a ratio of 3:8.

12. In a trapezoid the midline is perpendicular to a lateral side, three consecutive angles are in a ratio of 3/2:1:1, and the bases are to each other as $\frac{8}{15} : 0.4$. Find the least lateral side if the midline equals m.

13. Four lines parallel to the bases of the trapezoid are drawn through the points that partition a lateral side into four equal segments. Find the segments of these lines located within the trapezoid if the bases are a and b.

14. The midline of a trapezoid equals m. A diagonal partitions off the midline in a ratio of $k:1$. Find the bases of the trapezoid.#

15. Show that the midline of a trapezoid bisects any segment with the endpoints on the bases.

16. Show that the midpoints of the sides of an isosceles trapezoid are the vertices of a rhombus.

17. Prove that the midline of an isosceles trapezoid with mutually perpendicular diagonals equals to its *altitude* (the distance between the bases).

18. Formulate and prove the converse to the statement of Problem 16.

19. A *right trapezoid* (i.e. such a trapezoid that its lateral side is perpendicular to the bases) is divided by its diagonal into two triangles: one of them is right, and the other is equilateral with the side a. Find the midline of the trapezoid.

20. The greatest base of a trapezoid equals a, one of the lateral sides equals l, and their included angle equals 120°. One diagonal of the trapezoid is perpendicular to the bases. Determine the midline of the trapezoid.

21. Construct a trapezoid, having given its base, an adjacent (acute) angle and two lateral sides.

22. Construct a trapezoid, having given the two bases, a lateral side, and an angle between that side and the greatest base.

23. Construct a trapezoid, having given a lateral side, a base and two acute angles including the given base.

24. Construct a trapezoid, having given two lateral sides, a base, and a diagonal.

25. *Construct a trapezoid, having given the difference between the bases, two lateral sides and a diagonal.

26. Construct an isosceles trapezoid, having given a base, an altitude and a lateral side.

27. *Construct a trapezoid, having given its two bases and two diagonals.

7.5 a) Translations
b) Translations and symmetry in construction problems.

> ...he felt that Heffalump was as good as caught already, but there was just one other thing which had to be thought about, and it was this. *Where should they dig the very Deep Pit?*
>
> Piglet said that the best place would be somewhere where a Heffalump was, just before he fell into it, only about a foot farther on.
>
> A.A.Milne, *Winnie-The-Pooh.*

a) <u>Translation.</u>

Transformation of a figure is called a *parallel translation* or just a *translation* if it shifts each point of the figure through the same segment and in the same direction parallel to a fixed straight line.

For example, in Figure 7.5.1, the Heffalump (see the epigraph to this section) is being *translated* into the position just above the Deep Pit: each point of his figure is moved through the same distance and along the same direction. A segment connecting the original position of some point of the figure with the new position of this point is congruent and parallel to any other segment that joins the original and new positions of any other point of the figure (some of these segments are shown as arrows in Fig.7.5.1).

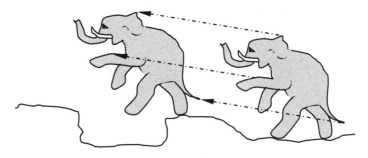

Figure 7.5.1

Mathematicians prefer to say that *translation* is a transformation of a plane. We say that a plane is subjected to a *parallel translation* if every point of the plane is translated through the same distance and in the same direction. In other words, for any two points, *A* and *B*, in the plane and their *images*, *A´* and *B´*, obtained as a result of a translation, *AA* = *BB´* and *AA´* // *BB´* (Fig. 7.5.2).

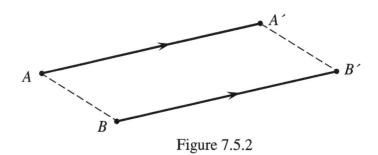

Figure 7.5.2

According to Th.7.1.2 (ii), *AA´B´B* is a parallelogram, whence it follows that *A´B´* = *AB*. Thus, a translation transforms segments into congruent segments, hence it is an isometry!

Theorem 7.5.1. *A translation is an isometry.*

Corollary 1. *A translation transforms a figure into a congruent figure.*

Two successive transformations (of a plane) result in a transformation called a *composition* of the two.

Theorem 7.5.2. *A composition of (two) translations is a translation.*
Proof. Suppose *A* and *B* are two arbitrary points in a plane, which is subjected to successive translations: the first one carries *A* into *A´* and *B* into *B´*; the second translation transforms *A´* into *A´´* and *B´* into *B´´* (Fig. 7.5.3).

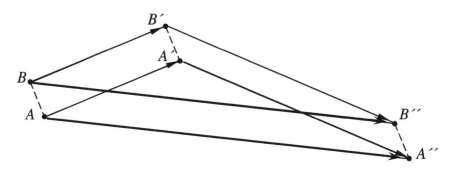

Figure 7.5.3

The resulting transformation is a composition of the two translations: it carries *A* into *A´´* and *B* into *B´´*. In order to prove that it is a translation, we have to show that *AA´´* = *B B´´* and *A A´´*// *B B´´*.

By the definition of translation, *AA´* = *BB´* and *AA´* // *BB´*; also *A´A´´*=*B´B´´* and *A´A´´* // *B´B´´*. Therefore, according to Th.7.1.2, both *ABB´A´*

and $A'B'B''A''$ are parallelograms. Then, since the opposite sides of parallelograms are congruent and parallel, $AB = A'B'$, and $AB // A'B'$; also $A'B' = A''B''$, and $A'B' // A''B''$.

The latter implies that $AB = A''B''$ as each of them is congruent to $A'B'$; also $AB // A''B''$ since each of them is parallel to $A'B'$. Thus the opposite sides AB and $A''B''$ of $ABB''A''$ are congruent and parallel, hence, by Th.7.1.2 it is a parallelogram, and its opposite sides AA'' and BB'' are parallel (by definition) and congruent (by Th.7.1.1). Therefore the resulting transformation, which is a composition of two translations, translates all the points of the plane (recall that A and B are arbitrary points) through the same distances and in the same direction. Thus it is a translation, Q.E.D.

Remark: Vectors. As we have seen, each translation is determined by a single *directed segment* that joins the original position of a point (in a plane) with its position after the translation. Such a directed segment is called a *vector*.

All vectors that determine the same translation are identified as *equal* (*equivalent*) to each other. That means a vector does not change as a result of a parallel translation.

The law of composition of translations proved in Th.7.5.2, determines the addition of vectors: a vector representing the composition of two translations is said to be the sum of two vectors representing these translations. For example, in the notations of Th.7.5.2, the composition is determined by the vector that translates A into A'' (or any other vector equal to it, e.g. the one that carries B into B''). We shall denote such a vector $\overrightarrow{AA''}$. The law of composition can be written as $\overrightarrow{AA''} = \overrightarrow{AA'} + \overrightarrow{A'A''}$.

Vectors are often used in physics to represent such physical entities that are characterized by both the magnitude and direction. For example, force, velocity, acceleration, strength of a field, are vectors. It is amazing that the law of composition of vectors, which follows only from the axioms of Euclidean geometry, provides us with the right rule for adding, for example, forces exerted by physical fields (gravitational, electromagnetic, etc.)!

b) <u>Translations and symmetry in construction problems.</u>

<u>Problem 7.5.3.</u> *Two towns are located on opposite sides of a river, which is virtually straight and has a constant width in some neighbourhood of these towns (see Fig. 7.5.4a). Where should a bridge that is perpendicular to the river banks be constructed so as to minimize the walking distance between the towns?*
<u>Solution.</u>
Analysis. The total path from A to B will consist of the three segments: AC, CD, and DB. CD is equal to the width of the river, and therefore this part will be the same for all possible routes. Let us temporarily remove it from consideration. We can achieve this by performing a parallel translation of DB so that it starts a C. As a result of the translation, DB is transformed into a segment CE such that $CE = DB$, and $CE//DB$, i.e. $BDCE$ is a parallelogram.

Now the problem has been reduced to minimizing the distance between A and E. This distance is minimal along a straight segment joining the two points; it is shown in Fig.7.5.4b in dotted line. The point where this segment intersects the "upper" bank of the river should be chosen as an endpoint of the bridge.

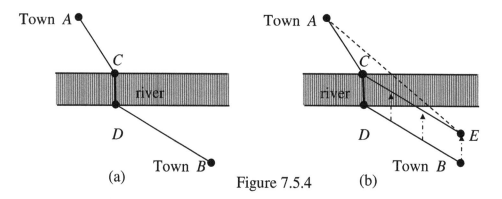

Figure 7.5.4

Construction. Through point B, draw a line l perpendicular to the banks of the river. On this line from point B, lay off a segment BE congruent to the width of the river (i.e. congruent to a segment cut off by the banks of the river from l). Join E with A by a straight segment; it will intersect the "upper" bank at some point M.

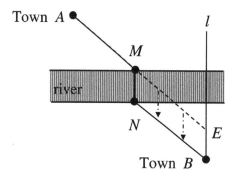

Figure 7.5.5

Translate ME parallel to itself along EB. Then M will be moved into some point N that lies on the "lower" bank, since EB is equal to the perpendicular distance between the banks. MN is the sought for location of the bridge.

Proof. By construction, $BEMN$ is a parallelogram, hence $AM+ME+MN = AM+NB+MN$, which is the total walking distance between A and B. $AM+ME$ is the shortest possible route between A and E, since it is a straight segment. Since MN component is the same for all routes, that makes $AM+NB+MN$ the shortest path, Q.E.D.

Investigation. There exists exactly one straight line through B that is perpendicular to the banks of the river (AXIOM 5). One can lay off on this line from point B in the direction of the river exactly one straight segment BE congruent to the distance between the banks (AXIOM 4). Also, the segment AE will intersect the "upper" bank at exactly one point (AXIOM 1). Therefore, the location of point M is defined uniquely, and so is the solution of the problem.

Problem 7.5.1. *Construct a quadrilateral given its sides and a segment joining the midpoints of two opposite sides.*

Analysis. Suppose the required quadrilateral, $ABCD$ in Fig. 7.5.5, is constructed. E and F are the midpoints of AB and DC, respectively.

Let us subject AD and BC to translations that map points A and B into E. Thus, the first translation transforms AD into ED', and the second translation transforms BC into EC'. $AED'D$ and $BEC'C$ are parallelograms, hence $DD' \parallel AB$ and $CC' \parallel AB$, $\Rightarrow$ $\Rightarrow DD' \parallel CC' \Rightarrow \angle D'DF = \angle C'CF$ (as alternate).

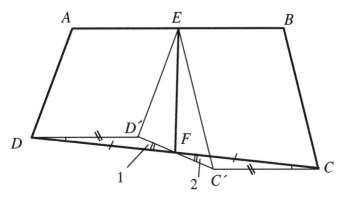

Figure 7.5.5

In addition,
$$\left. \begin{array}{l} DD' = AE = \tfrac{1}{2}AB \\ CC' = BE = \tfrac{1}{2}AB \end{array} \right\} \Rightarrow DD' = CC'.$$

Taking into account $DF = FC$, we conclude that $\triangle DD'F = \triangle CC'F$ by *SAS* condition. The latter means $\angle 1 = \angle 2$, hence $D'FC'$ is a straight segment (D', F, and C' are collinear – why?); then $D'FC'F$ is a triangle with EF as a median.

Construction. We can construct this $\triangle$ given its two sides, $ED' = AD$ and $EC' = BC$ (AD and BC are known) and the included median EF (this segment is also known). The construction is shown in Figures 7.5.6 and 7.5.7.

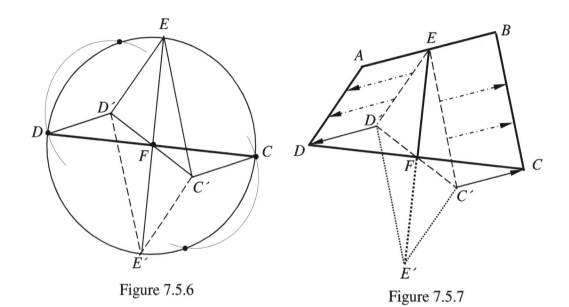

Figure 7.5.6   Figure 7.5.7

(i) construct a triangle with the sides $EE' = 2EF$; $EC = BC$; $C'E' = ED' = AD$;
(ii) complete it into a parallelogram $EC'E'D'$;
(iii) cut the parallelogram into two triangles by the second diagonal, $C'D'$. $\Delta ED'C'$ is constructed.

Now we shall complete the construction of the required quadrilateral, ABCD.

(iv) From C' and F describe arcs of radii $\tfrac{1}{2}AB$ and $\tfrac{1}{2}CD$, respectively; the point of their intersection will be C. Point D can be obtained similarly (Fig. 7.5.6).
(v) Then translate ED' through $D'D$ and EC' through $C'C$: $ED' \to AD$; $EC' \to BC$ (Fig. 7.5.7).
(vi) Connect vertices A and B, C and D. ABCD is constructed.

Remark. It was not necessary to use translations in order to find A and B. For instance, in order to find A, we could describe arcs of radii AD and $\tfrac{1}{2}AB$ from D and E, respectively.

Proof. It just repeats the analysis in the opposite direction.

Investigation. All steps of the construction except (iv) are defined uniquely. Since two circles intersect, in general, at two points, two pairs of points, C and D, and C'' and D'' may be obtained as a result of step (iv).

Triangles $D''D'F$ and $C''C'F$ are the reflection images of $\Delta DD'F$ and $\Delta CC'F$, respectively, in line $D'C'$. Thus D'', F, and C'' are collinear and the whole construction (steps (v) and (vi)) can be completed for $C''D''$ as a base of the required quadrilateral (Fig. 7.5.6).

Thus, the problem has, in general, *two solutions*. Both solutions, quadrilaterals $ABCD$ and $A''B''C''D''$, are shown in Figure 7.5.8. The first quadrilateral is constructed by translating ED' along $D'D$ and EC' along $C'C$. The sides $A''D''$ and $B''C''$ of the second quadrilateral are obtained as a result of the translations of ED' and EC' along $D'D''$ and $C'C''$ respectively (the translations are illustrated by arrows).

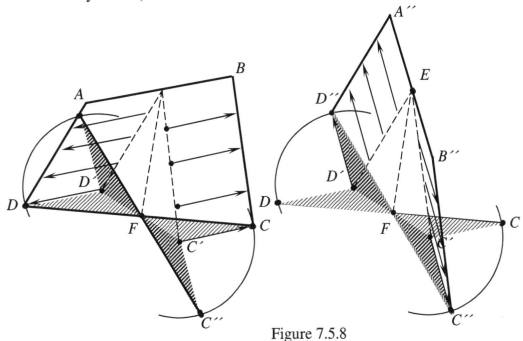

Figure 7.5.8

There are also cases when the problem has (a) no solutions; (b) only one solution; (c) infinitely many solutions. Examine when (a), (b), and (c) may occur.

<u>Problem 7.5.2.</u> *Given a line and a pair of exterior points lying on the same side of the line, find a point in the line for which the sum of the distances to the given two points is minimal.*

<u>Solution.</u> In Fig. 7.5.9, we are looking for such a point P on AB that the sum $MP + PN$ is minimal, that is for any other point Q lying on AB, $MQ + QN > MP + PN$.

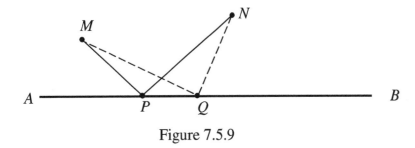

Figure 7.5.9

Let us reflect M in AB: $M \to M'$ (Fig. 7.5.10). Then for any point Q lying on AB, $MQ = M'Q$; hence $MQ + QN = M'Q + QN$. The latter sum is minimal if MQN is a straight segment.

Thus, to find the required point *P*, we reflect *M* in *AB* and connect *M´*, its reflection image, with *N* by a straight segment. *M´N* will cut *AB* at *P*.

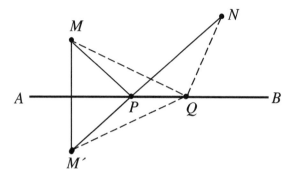

Figure 7.5.10

The problem has a unique solution (why?).

7.5 Questions and Problems.

1. Construct a trapezoid, having given an angle, two diagonals and the midline.

2. Construct a quadrilateral, having given three sides and two angles that include the unknown side.

3. Construct a trapezoid, having given the four sides.

4. Construct a quadrilateral, having given the four sides if it is known that one of its diagonals bisects an interior angle.

5. Given a line and a pair of exterior points, find an interior point of the line such that the segments connecting that point with the given points form equal angles with the line.

6. *Two balls, A and B are located on the surface of a rectangular billiard-table (Figure 1). In which direction A to be pushed so that after being reflected consecutively from all the four walls it would hit B?

Figure 1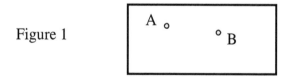

7. Construct a parallelogram having given the location of one of its vertices and the midpoints of
    (i) two opposite sides;
    (ii) two sides intersecting at the given vertex;
    (iii) two sides that do not pass through the given vertex.

7. Review.

1. Prove that if a median equals half the side to which it is drawn, then the triangle is right.

2. Prove that in a right triangle the angle between the median and the altitude drawn to the hypotenuse is congruent to the difference between the acute angles of the triangle.

3. In $\triangle ABC$ the bisector of $\angle A$ meets BC at D; the line drawn through D parallel to CA meets AB at E; the line drawn through E parallel to BC meets AC at F. Prove that $EA=FC$.

4. An angle is constructed in the interior of another angle in such a way that the sides of the constructed angle are respectively parallel and equidistant from the sides of the other angle. Prove that the bisector of the constructed angle lies on the bisector of the other angle.

5. A line is drawn parallel to the base of a triangle through the point of the intersection of the bisectors of the angles that include the base. Prove that the segment of the line that lies in the interior of the triangle equals to the sum of the segments cut by the line on the lateral sides counting from the base.

6. Prove that in an isosceles triangle the sum of the distances of any point in the base from the lateral sides is a constant equal to an altitude dropped onto a lateral side.

7. How should the statement of the previous problem be changed if a point is chosen on the extension of the base?

8. Prove that in an equilateral triangle the sum of the distances from an interior point of the triangle to the bases is a constant equal to the altitude.

9. Show that a parallelogram with congruent diagonals is a rectangle.

10. Prove that a parallelogram with mutually perpendicular diagonals is a rhombus.

11. Prove that a parallelogram in which a diagonal bisects an angle is a rhombus.

12. From the centre of symmetry of a rhombus perpendiculars are dropped onto the sides. Prove that the feet of these perpendiculars are the vertices of a rectangle.

13. Prove that the bisectors of all interior angles of a rectangle, when cutting each other, form a square.

14. Prove that if the midpoints of the sides of a quadrilateral are consecutively connected, the obtained quadrilateral is parallelogram. Find the conditions under which that parallelogram is a a) rectangle; b) rhombus; c) square.

15. Find the locus of the midpoints of segments connecting a point to all points lying in a given line.

16. Find the locus of points equidistant from two parallel lines.

17. Find the locus of the vertices of triangles that have a common base and congruent altitudes.

18. Given two angles of a triangle, construct the third angle (propose at least two different methods).

19. Draw a straight line at a given distance from a given line.

20. Through a given point draw a line that forms a given angle with a given line.

21. Through a given point draw a transversal to two parallel lines in such a way that the segment of the transversal included between the parallels is congruent to a given segment.

22. Given an angle and a segment, draw a straight line that is perpendicular to one arm of the angle and the segment of the line included between the arms equals to the given segment.

23. Given an angle and a segment, locate the segment in the interior of the angle with the endpoints in the arms of the angle in such a way that the segments included between the each endpoint and the vertex of the angle are congruent.

24. Construct a right triangle, having given an acute angle and the opposite leg.

25. Construct a triangle, having given two angles and the side opposite to one of them (**SAA** condition).

26. Construct an isosceles triangle, having given the base and the vertical (i.e. opposite the base) angle.

27. Construct an isosceles triangle, having given an angle adjacent to the base and an altitude to a lateral side.

28. Construct an isosceles triangle, having given a lateral side and the altitude to that side.

29. Construct an equilateral triangle, having given its altitude.

30. Trisect (partition into three equal parts) the right angle.

31. Construct a triangle, having given the base, the altitude to the base and a lateral side.

32. Construct a triangle, having given the base, the altitude to the base and an angle adjacent to the base.

33. Construct a triangle, having given its angle and two altitudes to the sides that include the angle.

34. Construct a triangle, having given a side , the sum of the other two sides , and the altitude dropped onto one of the latter sides.

35. Construct a triangle, having given its altitude, an angle adjacent to the base, and the perimeter.

36. Given a triangle, draw a line that is parallel to its base and cuts on the lateral sides counting from the base the segments whose sum equals to the segment of that line that lies in the interior of the triangle.

37. Construct a a) quadrilateral; b) pentagon congruent to a given one.

38. Construct a quadrilateral, having given three angles and two sides that include the fourth angle.

39. Construct a quadrilateral, having given three sides and two diagonals.

40. Construct a parallelogram, having given two unequal sides and one diagonal.

41. Construct a parallelogram, having given a side and two diagonals.

42. Construct a parallelogram, having given two diagonals and an angle between them.

43. Construct a parallelogram, having given a base, the altitude and a diagonal.

44. Construct a rectangle, having given its diagonal and an angle between the diagonals.

45. Construct a rhombus, having given a side and a diagonal.

46. Construct a rhombus, having given the two diagonals.

47. Construct a rhombus, having given an altitude and a diagonal.

48. Construct a rhombus, having given an angle and the diagonal passing through the vertex of that angle.

49. Construct a rhombus, having given its diagonal and an opposite angle.

50. Construct a rhombus, having given the sum of the diagonals and an angle between a side and a diagonal.

51. Construct a square, having given its diagonal.

52. Construct a trapezoid, having given a base, an adjacent angle, and two lateral sides.

53. Construct a trapezoid, having given a base, the altitude, and two diagonals.

54. Construct a trapezoid, having given two bases and two diagonals.

55. Construct a square, having given the sum of the side and the diagonal.

56. Construct a square, having given the difference between the diagonal and the side.

57. Construct a parallelogram, having given two diagonals and an altitude.

58. Construct a parallelogram, having given a side, the sum of the diagonals, and an angle between the diagonals.

59. Construct an isosceles triangle, having given its base and a median drawn to a lateral side.

60. Construct a triangle, having given a base, the altitude, and the median drawn to a lateral side.

61. Construct a right triangle, having given the hypotenuse and the sum of the legs.

62. Construct a right triangle, having given the hypotenuse and the difference between the legs.

63. Two given points, *A* and *B*, are located on the same side of a given straight line. Locate on that line a segment of a given length in such a way that if it takes position *CD*, the sum *AC+CD+DB* would be minimal.

8. CIRCLES.

> "He's going *round* and *round*," said Roo, much impressed.
>
> "And why not?" said Eeyore coldly.
>
> "I can swim too," said Roo proudly.
>
> "Not round and round," said Eeyore. "It's much more difficult."
>
> A.A.Milne, *The House At Pooh Corner.*

8.1 Shapes and location.

Given a point in the plane, one can draw as many circles with various centres and radii passing through this point as one wishes (Figure 8.1.1).   (Substantiate this statement based on AXIOMS 1-4).

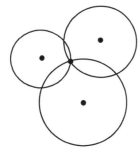
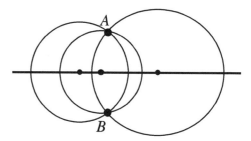

Figure 8.1.1   Figure 8.1.2

Given a pair of points (e.g., *A* and *B* in Fig. 8.1.2), there are still infinitely many circles of different radii passing through these two points, however all their centres will be located on one line only (what is that line?)

If three points are given, are there any circles passing through these points?

Theorem 8.1.1. ***There is one and only one circle passing through three given noncollinear points.***

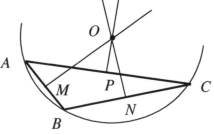

Figure 8.1.3

Proof. Let points *A*, *B*, and *C* in Fig. 8.1.3 be noncollinear. Then the perpendiculars bisecting *AB* and *BC* will intersect at some point *O* (why?). $AO = OB$ and $OB = OC$ (why?); therefore $OA = OB = OC$. Thus, a circle with its centre at *O* and radius *OA* will pass through *A*, *B*, and *C*.

This circle is unique, since two nonparallel straight lines (in our case – the perpendicular bisectors of AB and BC) intersect at one and only one point. The proof is complete.

Remark 1. *If A, B and C were collinear, the perpendicular bisectors to AB and BC would be parallel, and so point O would not exist as well as a circle passing through A, B, and C.*

In *projective geometry* a straight line is regarded as a circle of infinitely long radius (and infinitely distant centre). From this point of view, given three collinear points, their common line will be the circle passing through these three points.

Remark 2. It follows from Th.8.1.1 that four points will not, in general, lie on one circle (why?). If, by an accident, four or more points lie on a circle, these points are called *concyclic*. One may reformulate the contents of the above theorem and Remark 2 by saying that a triangle is *inscribable* in a circle, and a quadrilateral (in general) is not.

Corollary 1. *Three perpendicular bisectors drawn to the three sides of a triangle, are concurrent.*

Proof. Let in Figure 8.1.3 MO and NO are the perpendicular bisectors of AB and BC. Since O is located on MO, it is equidistant from A and B: AO = BO. Also, it is located on NO, hence it is equidistant from B and C: BO = CO. Thus AO = BO = CO; therefore O is equidistant from A and C. At the same time the perpendicular bisector of AC is the locus of points equidistant from A and C, therefore O is lying on this bisector. Q.E.D.

Theorem 8.1.2. *A diameter that is perpendicular to a chord, bisects the chord and its subtended arcs.*

Proof. Suppose in Fig. 8.1.4 diameter AB is ⊥ to chord CD; O is the centre of the circle. CO = DO as two radii of the same circle; hence $\triangle COD$ is isosceles, and its altitude OE is also a median and a bisector; i.e. CE = ED (Q.E.D.) and $\angle COB = \angle DOB$. According to Th. 3.4.1, the latter leads to $\cup DB = \cup CB$, Q.E.D.

$\angle COA = \angle DOA$ as supplements of equal angles, therefore $\cup CA = \cup DA$, Q.E.D. This completes the proof.

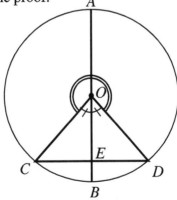

Figure 8.1.4

Statements converse to Th.8.1.2 are also valid. They are represented by the next theorem.

Theorem 8.1.3 *(i) A diameter that bisects a chord is perpendicular to the chord and bisects the arc subtended by the chord.*

(ii) A diameter that bisects an arc is a perpendicular bisector of the chord that subtends the arc.

<u>Proof.</u> (Students should prove by RAA).

Theorem 8.1.4. *Two arcs included between a pair of two parallel chords, are congruent.*

<u>Proof.</u> In Fig. 8.1.5 $CD \parallel AB$. Draw a diameter $EF \perp CD$. It will be perpendicular to AB as well (why?). According to Th. 8.1.2, EF is the symmetry axis of CD and AB; then D is the reflection image of C, and B is the reflection image of A.

Therefore, $AC = BD$; $\Rightarrow \cup AC = \cup BD$, <u>Q.E.D.</u>

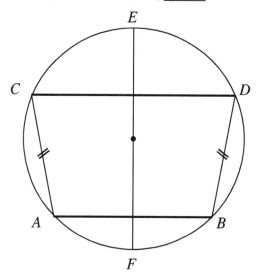

Figure 8.1.5

Solutions of the following construction problems are based on the above theorems.

<u>Problem 1.</u> *Bisect a given arc of a circle.*

<u>Solution.</u> Bisect the corresponding chord by a perpendicular (or draw a diameter bisecting the chord), then it will bisect each of the subtended arcs.

<u>Problem 2.</u> *Find the centre of a circle.*

<u>Solution.</u> Draw two nonparallel chords and bisect them by the corresponding perpendiculars. These perpendiculars will be diameters, $\Rightarrow$ their point of intersection will be the centre.

8.2. Relations between arcs and chords and their distances from the centre

Theorem 8.2.1. *In the same circle, or in congruent circles,*

>   *(i) congruent arcs are subtended by congruent chords that are equidistant from the centre;*
>
>   *(ii) of two unequal arcs, each less than semicircle, the greater is subtended by a greater chord that is nearer to the centre than the shorter chord.*

Proof. (i) In Fig. 8.2.1, $\cup AB = \cup CD$; O is the centre of the circle; $OE \perp AB$; $OF \perp CD$.

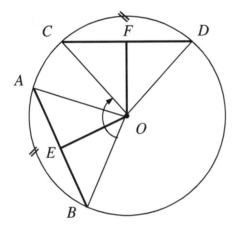

Figure 8.2.1   Figure 8.2.2

Rotate ray OB clockwise through $\angle BOC$. Then OB will fall onto OC, and due to the congruence of $\angle AOB$ and $\angle COD$ (central angles subtending equal arcs), OA will coincide with OD (thus the whole sector BOA has been rotated to be superimposed onto sector COD).

Then, AB will coincide with CD (why?) and OE – with OF, as there is only one perpendicular from O onto CD. Thus, $AB = CD$ and $OE = OF$, Q.E.D.

(ii) Suppose in Fig. 8.2.2 $\cup AB < \cup CD$, and each of these arcs is less than the semicircle. Let OF and OE be the perpendiculars dropped from the centre O of the circle onto the arcs CD and AB respectively. We have to prove that $AB < CD$, and $OE > OF$.

Lay off, starting from point C, $\cup CK = \cup AB$. Then chord CK is equal to AB and equidistant with AB from the centre, O: if $OL \perp CK, \Rightarrow OL = OE$.

In $\triangle$s COK and COD, two sides of the one, CO and OK, are congruent respectively to the two sides, CO and OD, of the other. $\cup CD > \cup CK$; therefore $\angle COD > \angle COK$, whence $CD > CK$ (why?), Q.E.D.

In right $\triangle OMF$, its leg OF is less than its hypotenuse OM and the latter is only part of OL. Hence, $OL > OM > OF$, and since $OL = OE$, it follows that $OE > OF$, Q.E.D.

Obviously, the theorem is also valid for congruent circles, as they can be made to coincide by means of an isometry.

The propositions converse to the ones of Th.8.2.1 are formulated below. They can be easily proved by RAA.

Theorem 8.2.2. *In the same circles, or in congruent circles,*

(i) congruent chords are equidistant from the centre and subtend congruent arcs;

(ii) chords that are equidistant from the centre are congruent and subtend congruent arcs;

(iii) of two unequal chords, the longer is nearer the centre and subtends the greater arc;

(iv) of two chords that are unequal distances from the centre, the one nearer the centre is greater and subtends the greater arc.

Theorem 8.2.3. *A diameter is the greatest chord in a circle.*

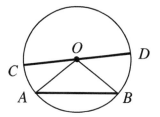

Figure 8.2.3

Proof. In Fig. 8.2.3, CD is a diameter, and AB is some other chord of a circle described from O by the radius r. According to the triangle inequality,

$$\left. \begin{array}{l} AB < AO + OB = 2r \\ CD = CO + OD = 2r \end{array} \right\} \Rightarrow AB < CD, \underline{\text{Q.E.D.}}$$

8.1-2   Questions and Problems.

1. Explain, why the following "definition" is not correct: *a circle is a set of points equidistant from a given point called the centre.* Correct the above statement. #

2. Construct the locus of the centres of circles passing through a given pair of points. Which of these circles has the least radius? Explain.

3. Given a line and a pair of points, find the centre of a circle that is situated in the line and passes through a given pair of points.

4. Given a circle and a pair of points, find the centre of a circle that lies on the given circle and passes through a given pair of points.

5. An equilateral triangle with altitude *H* is given. *Circumscribe a circle about* that triangle (i.e. draw a circle through the three vertices) and determine its radius.

6. Circumscribe a circle about a given right triangle.

7. The endpoints of two mutually perpendicular diameters are consecutively connected. What kind of a quadrilateral is obtained?

8. Two lines intersect at the centre of a circle, and the points where they cut the circle are consecutively connected. What kind of a quadrilateral is obtained?

9. Prove the converse to Theorem 8.1.2, i.e. show that
    (i) a diameter that bisects a chord is perpendicular to the chord and also bisects the arc subtended by that chord;
    (ii) a diameter that bisects an arc is a perpendicular bisector of the chord that subtends the arc.

10. Prove that two parallel chords drawn through the endpoints of the same diameter are congruent.

11. Prove that if a chord is a perpendicular bisector of another chord, then it is a diameter.

12. Prove that a chord that bisects a radius perpendicular to that chord subtends an arc of 240°.

13. Through a given point in the interior of a circle, draw a chord that is bisected by the given point.

14. A circle passes through the four vertices of an isosceles trapezoid. Show that the angles formed by the radii drawn to the endpoints of each of the lateral sides are congruent.

15. In Figure 1, two chords, *CC'* and *DD'*, are perpendicular to diameter *AB*. Prove that the segment that connects the midpoints of the chords *CD* and *C'D'* is also perpendicular to that diameter.

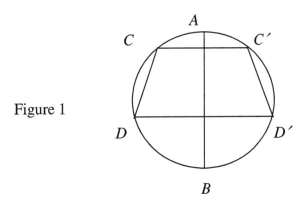

Figure 1

16. In Figure 1, two chords, *CC'* and *DD'*, are perpendicular to diameter *AB*. Prove that
    a) ∠*D*=∠*D'*; b) *BC*=*BC'*.

17. Bisect a given arc of a circle.

18. Find the centre of a circle.

19. Given a chord in a circle, construct another chord which is congruent and parallel to the given one.

20. A chord subtends an arc of 120° in a circle of radius R. Determine the segments into which the chord partitions the perpendicular diameter.

21. A circle is *circumscribed about* (passes through the four vertices of) a parallelogram. What can you tell about such a parallelogram?

22. A circle is circumscribed about a trapezoid. What can you tell about that trapezoid?

23. Prove Theorem 8.2.2.

24. A circle is circumscribed about a trapezoid. Prove that the centre of the circle is located between the greatest base and the midline, closer to the midline.

25. If a central angle is doubled, is the subtended chord doubled?

26. Show that the least chord among all chords passing through a given interior point of a circle is bisected by that point.

27. In a circle of radius R there is a chord equal $0.9R$. Can that chord be subtended by a central angle of 56°?

8.3 Respective positions of a straight line and a circle

Only the following three situations may take place:

1. *The distance from the centre of a circle to a line is greater than the radius of the circle* (Fig. 8.3.1).

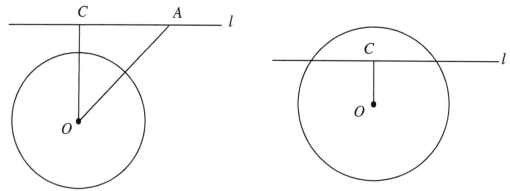

Figure 8.3.1   Figure 8.3.2

Point C, the foot of the perpendicular dropped from the centre, O, upon line l, is the closest to O point on l (since a segment connecting O to any other point A on l will be an oblique, therefore it will be greater than the perpendicular).

Thus, if point C is exterior for the circle, so will be all other points of l. *The line will not have any common points with the circle.*

2. *The distance from the centre of a circle to a line is less than the radius* (Fig. 8.3.2).

Then the foot of the perpendicular dropped from the centre onto the line will be located in the interior of the circle; thus *the line will intersect the circle*. It can be proved (students are encouraged to try) that no straight line can be drawn to meet the circle in three points.

3. *The distance from the centre is equal to the radius* (Fig. 8.3.3).
Point C, the foot of the perpendicular dropped onto the line from the centre, belongs both to the circle and to the straight line. This point is the radius distant from the centre, and the other points of *l* will be further from the centre (why?). Therefore, *there is one and only one point that is common for the line and for the circle*. This point is the foot of the perpendicular drawn to the line from the centre of the circle.

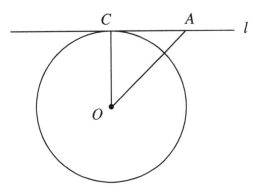

Figure 8.3.3

A line that has exactly one common point with a circle is called a *tangent line* or just a *tangent*.

A point common to a tangent and a circle is called a *point of tangency (or a point of contact)*. One may also say that this is a point where a line and a circle *touch* each other.

The above consideration of the third case, shown in Fig. 8.3.3 is summarized in the following theorems.

Theorem 8.3.1. *If a line is perpendicular to a radius of a circle at the endpoint of the radius lying on the circle, then the line is tangent to the circle*, and conversely:

Theorem 8.3.2. *If a line is tangent to a circle, the radius drawn to the point of tangency is perpendicular to the line.*

The hints to the proofs are given above, in the comments to Fig. 8.3.3; students should complete the proofs.

Theorem 8.3.3. *If a tangent is parallel to a chord, then the point of tangency bisects the arc subtended by the chord.*

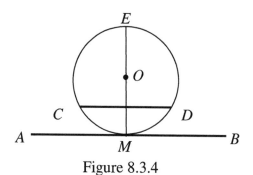

Figure 8.3.4

Proof. Suppose AB touches a circle (O, r) at some point M, and AB // CD (Fig. 8.3.4). We have to prove that $\cup CM = \cup MD$. Let us draw a diameter, ME, through the point of tangency M. Then $EM \perp AB \Rightarrow EM \perp CD$, hence EM bisects CD and the subtended arc CMD, i.e. $\cup CM = \cup MD$, Q.E.D.

<u>Problem.</u> *Given a circle, draw a tangent that will be parallel to a given line.*

<u>Solution.</u> Through the centre O of a given circle draw a perpendicular to the given line *l*.

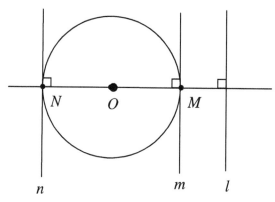

Figure 8.3.5

Through the points M and N, at which this diameter cuts the circle (Fig.8.3.5), draw perpendiculars, *m* and *n* to the diameter. These perpendiculars are the required tangents. Really: *m* and *n* are parallel to *l* as perpendiculars to the same line MN; *m* and *n* are tangents to the circle as perpendiculars to radii OM and ON passing through their endpoints located on the circle (Th.8.3.1.)

The problem has exactly two solutions, since each diameter has exactly two endpoints.

Properties of tangents are applied to solve some practical problems, e.g., in road construction for planning smooth transitions from one road to the other.

An arc, *AB* in Fig. 8.3.6, and a line, *BC*, are said to be *conjugate* at some point, *B*, if they extend each other and the line is tangent to the arc at the common endpoint.

The same term, *conjugate,* is applied to two arcs, *AB* and *BC* in Fig. 8.3.6b, which extend each other and have a common tangent at their common endpoint (*B*).

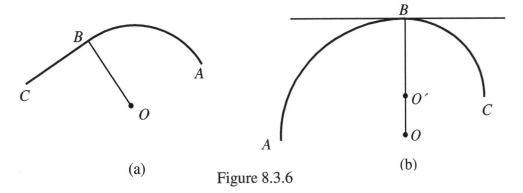

(a) Figure 8.3.6 (b)

For the conjugation of a line and an arc, the centre from which the arc is described, must lie on the perpendicular drawn to the line at the point of the contact.

In case of conjugated arcs, the centres from which they are described, lie on the perpendicular drawn to the common tangent at the common endpoint.

8.3 Questions and Problems.

1. Given a point on an arc of a circle with inaccessible centre, draw a tangent touching the circle at the given point.

2. A circle of radius *R* is inscribed in an isosceles trapezoid with an interior angle of 30°. Determine the lateral side of the trapezoid.

3. Given a pair of concentric circles, prove that the chords of the greater circle which are tangent to the least circle, are congruent.

4. Draw a tangent to a given circle parallel to a given line in each of the following cases: a)) the centre is known and accessible; b) the centre is inaccessible.

5. Given a circle, draw two tangents that form a given angle.

6. Given a circle, draw a tangent that forms a given angle with a given line.

7. Draw a circle of a given radius that touches a given line at a given point.

8. In a circle with the centre at *O*, diameter *AB* and chord *AC* form an angle of 30°.The tangent that passes through *C* cuts *AB* at *D*. Show that *OC=0.5OD.*

9. Given an angle, inscribe in it a circle of a given radius.

10. In a circle with the centre at *O, AB* is a chord that subtends an arc of less than 90°. The tangent that touches the circle at *B* is drawn, and the chord *AB* is extended. The extension of the

diameter perpendicular to *OA* cuts the tangent and the extension of the chord at points *C* and *D*, respectively. Show that *BC=CD*.

11. Prove that the segments of two tangents drawn to a circle from one point are congruent.

12. Two tangents form an angle of 60°. Show that:
    a) the segment joining their point of intersection with the centre of the circle equals the diameter;
    b) the segment joining the points of contact equals to the segments of the tangents.

13. Prove that two tangents to the same circle are symmetric to each other in a line. What is that line if a) the tangents intersect; b) the tangents do not intersect.

14. In Figure 1, *MA, MB,* and *KL* are tangent to the circle. Show that the perimeter of △*MKL* does not depend upon the location of point *C* on ∪*AB* (the least of two arcs is suggested).

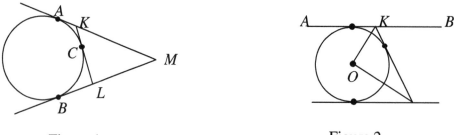

Figure 1   Figure 2

15. In Figure 2, *AB, CD,* and *KL* are tangents to a circle with its centre at *O*, and *AB//CD*. Show that ∠*KOL=d*.

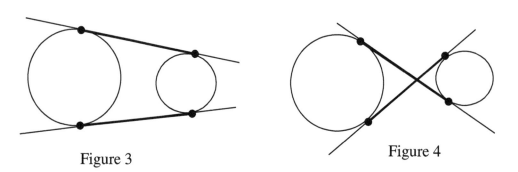

Figure 3   Figure 4

16. Prove that the segments of common *external tangents* (Figure 3) to two circles between the points of contact are congruent.

17. Prove that the segments of common *internal tangents* (Figure 4) to two circles between the points of contact are congruent.

8.4   Respective position of two circles

If two circles have exactly one common point they are said to *touch, or to be tangent.* Their common point is called *the point of tangency*, or *the point of contact.* If two circles have two common points, they are said to *intersect* each other.

According to Th.8.1.1, two distinct circles cannot have three or more common points.

The line drawn through the centres of two circles we shall call the *centre line*.

Theorem 8.4.1. *If two circles have a common point that is exterior to their centre line, then they have one more common point, symmetric in their centre line with the first common point.*

Proof. The centre line contains the diameters of both circles, and thus, it is a symmetry axis of the whole figure (Fig. 8.4.1). Therefore, for a common point A, there will be a corresponding common point A', a reflection image of A in OO'.

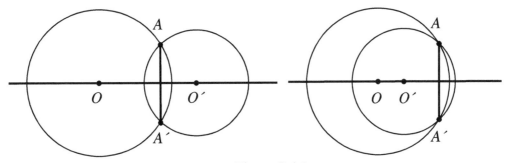

Figure 8.4.1

Corollary 1. *If two circles have a common point exterior to their centre line, the circles intersect.*

Corollary 2. *If two circles intersect, their common chord is perpendicular to the centre line and is bisected by that line.*

Theorem 8.4.2. *If a common point of two distinct circles lies in their centre line, the circles are tangent.*

Proof. The circles cannot have another common point: If they had a common point exterior to their centre line, then, by Th.8.4.1, they would have one more common point, symmetric to the conjectural common point in a centre line. Thus, the circles would have three common points, and therefore, they would coincide.

If the circles had another common point lying in their centre line, then they would have a common chord connecting their common points. But this chord would lie in the centre line, therefore it would be a diameter of each of the two circles. Then, the circles would coincide.

Hence, the circles can have only one common point, Fig. 8.4.2 and they touch each other at this point, Q.E.D.

Remark. Circles are said to touch *internally* (Fig. 8.4.2a) or *externally* (Fig. 8.4.2b), according as they lie on the same side or on the opposite sides of the point of tangency.

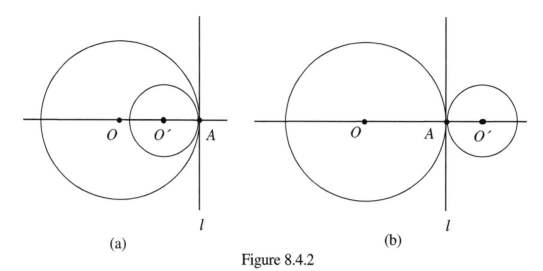

(a)   (b)

Figure 8.4.2

Theorem 8.4.3. (converse to 8.4.2) *If two circles touch each other, the point of tangency lies in their centre line.*

Proof. If the point of tangency (A in Fig.8.4.2) lied beyond the centre line, the circles would have one more common point (why?). The latter would contradict to the hypothesis of the theorem.

Corollary 1. *If two circles touch each other, they have a common tangent at the point of contact.*

Proof. If line l in Fig.8.4.2 is perpendicular to radius OA, it will also be perpendicular to OA', since both of these radii lie in the centre line. Then, by Th.8.3.1 line l is tangent to both circles, Q.E.D.

Let us consider different cases of relative position of two circles. We denote the radii of the circles r and R, and d will stand for the distance between their centres.

The following five situations may occur:

1. *One of the circles lies outside of the other and they do not touch* (Fig. 8.4.3); in this case $d > R + r$.

2. *The circles touch externally* (Fig. 8.4.4); then $d = R + r$, since the point of contact lies in the centre line.

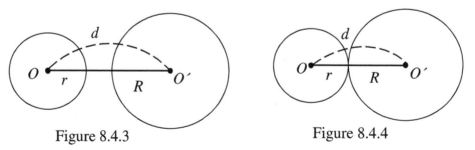

Figure 8.4.3   Figure 8.4.4

3. *The circles intersect* (Fig. 8.4.5); then $d < R + r$, and at the same time $d > R - r$ (since in $\triangle OAO'$ side $OO' = d$ is less than the sum but greater than the difference of the other sides, equal respectively R and r).

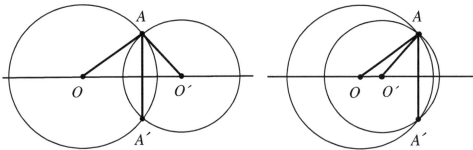

Figure 8.4.5

4. *If the circles touch internally* (Fig. 8.4.2a), then $d = R - r$, since the point of contact is in the centre line.

5. *If one circle lies in the interior of the other* (Fig.8.4.6), then, evidently, $d < R - r$. In a particular case when $d = 0$, the centres of the circles coincide; such circles are called *concentric*.

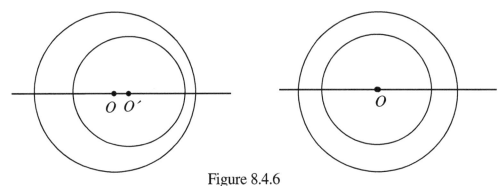

Figure 8.4.6

Students should check the validity of the converse statements, particularly:

1. *If $d > R + r$, then one circle lies in the exterior of the other and they do not touch.*

2. *If $d = R + r$, the circles touch externally.*

3. *If $R - r < d < R + r$, the circles intersect.*

4. *If $d = R - r$, the circles touch internally.*

5. *If $d < R - r$, one circle lies in the interior of the other, and they do not touch.*

(These statements can be proved by RAA).

8.4 Questions and problems.

1. Describe the respective position of two circles in each of the following cases:
    a) the distance between their centres is 20 cm, whereas their radii are 4 cm and 8 cm, respectively;
    b) the distance between their centres is 10 cm, whereas their radii are 4 cm and 8 cm, respectively;
    c) the distance between their centres is 2 cm, whereas their radii are 4 cm and 8 cm, respectively;
    d) the distance between their centres is 20 cm, whereas their radii are 12 cm and 8 cm, respectively;
    e) the distance between their centres is 4 cm, whereas their radii are 4 cm and 8 cm, respectively;
    f) their centres coincide and their radii are 4 cm and 8 cm, respectively;

2. How many axes of symmetry each of the following figures possesses? A figure formed by :
    a) two intersecting circles of different radii;
    b) two congruent intersecting circles;
    c) two congruent circles with the distance between the centres congruent to their diameter;
    d) two congruent circles with the distance between their centres greater than their diameter;
    e) two concentric circles.

3. The radii of two concentric circles are in a ratio of 3:4. Find their radii, if the circles form a ring 3cm wide.

4. Find the centres of two intersecting circles, having given their points of intersection and the angles formed by each of their radii drawn to a point of intersection with their centres line.

5. Two circles touch each other internally. Their radii are to each other as 2÷3, and the distance between their centres is l. Determine their radii.

6. Two circles touch each other internally. Their radii are to each other as 4÷3, and the distance between their centres is L. Determine their radii.

7. Find the greatest and the least distances between the points of two circles with radii R and r, respectively and the distance d between their centres. (Consider all possible cases).

8. Given a circle, how many circles of the same radius can be located around that circle in such a way that each of them touches that circle and two neighbouring circles?

8.5 Inscribed angles and some other angles related to circles. Construction of tangents.

An angle formed by two chords emanating from the same point lying on a circle is called *inscribed*. For example, ∠*ABC* in Fig.8.5.1 (a, b, or c) is an inscribed angle *intercepting* (minor) ∪*AC*. It is also common to say that ∠*ABC stands* on ∪*AC*, ∠*ABC* is *inscribed* in ∪*AC*, or ∠*ABC is subtended* by ∪*AC*.

Theorem 8.5.1. *An inscribed angle is half the central angle subtended by the same arc.*

Proof. We shall consider three possible cases represented in Fig.8.5.1.

a)   *One of the arms of the angle, BC, is a diameter.*

Draw radius OA. Then ∠AOC = ∠OAB + ∠ABC as an exterior angle of ΔAOB. ∠OAB = ∠ABC, hence ∠AOC = 2 ∠ABC, Q.E.D.

One also can say that ∠ABC is measured by ½ ∪AC.

b)   *The centre of the circle lies in the interior of the inscribed angle.*

Draw diameter BD. Then ∠ABD = ½ ∠AOD; ∠CBD = ½ ∠COD. Adding these equalities, we obtain

∠ABC = ∠ABD + ∠CBD = ½ ∠AOD + ½ ∠COD = ½ ∠AOC, Q.E.D.

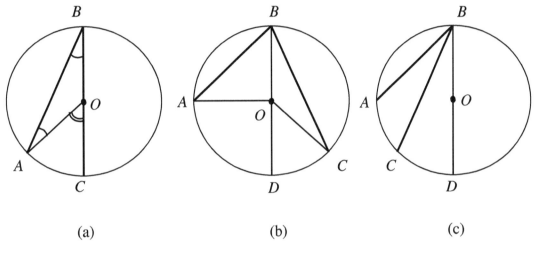

(a)   (b)   (c)

Figure 8.5.1

c)   *The centre of the circle lies outside the inscribed angle.* The proof is similar to the case (b).

Corollary 1. (Another formulation of the theorem.) ***An inscribed angle is measured by one half of its intercepted arc.***

Corollary 2. ***All angles inscribed in the same arc or in congruent arcs are congruent.***

Proof. Any of these angles is measured by one half of its intercepted arc, e.g. ∪AnB in Fig.8.5.2a.

Corollary 3. ***An angle inscribed in a semicircle (an angle standing on a diameter) is a right angle.***

Proof. Each angle of this kind, e.g., ∠ACB, ∠ADB, ∠AEB, or any other angle standing on diameter *AB* in Fig. 8.5.2b, equals ½ ∠AOB = ½ 2 d = d, since ∠AOB is a straight angle. Q.E.D.

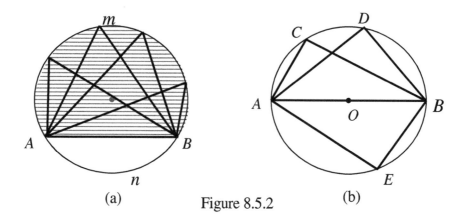

(a)   Figure 8.5.2   (b)

Construction problems.

1. *Construct a right triangle, given its hypotenuse, c, and one leg, b.*

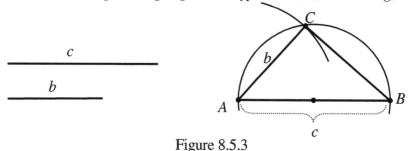

Figure 8.5.3

Solution. On a line lay off a segment $AB = c$. Draw a circle with AB as a diameter. (How would you do this?) From A describe an arc with radius b. The arc will intersect the circle at some point C. $\angle ACB$ is right since it stands on a diameter; $AB = c$; $AC = b$; therefore, $\triangle ABC$ is the sought for triangle.

2. *Erect a perpendicular to a ray at its vertex, without extending the ray.*

Solution. Let some point A be the vertex of the given ray (Fig.8.5.4a). From any exterior point O that lies, for example, above the ray, describe a circle with radius OA. It will intersect the ray at one more point, C. Draw diameter CD with C as an endpoint. Then DA is the sought for perpendicular, since $\angle DAC$ stands on a diameter, and therefore it is right.

Question1. In this solution we used for the construction an arbitrary point O, lying above the given ray. Finding such a point does not seem to be a difficult task in practical terms; yet from a theoretical viewpoint it is a substantial problem. First of all: what do we mean by *above the ray*? We can say that O is located above ray AB if $\angle OAB$ is acute: then a perpendicular dropped from O would fall on AB (Fig.8.5.4b). Then how can we know that $\angle OAB$ is acute if there is no right angle with its vertex at A to which we could compare ∠OAB? Propose and substantiate a procedure for finding point O.

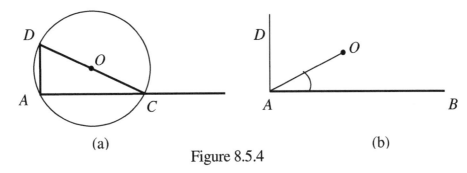

(a)   (b)

Figure 8.5.4

Question 2. Is the above solution valid in neutral geometry? If not, – can the problem be solved within neutral geometry?

3. *Through a given point, draw a tangent to a given circle.*

Solution. There are two possibilities.

(i) The given point, C (Fig. 8.5.5), lies on the circle.

Draw radius CO and use the procedure of Problem 1 to draw a perpendicular to CO passing through C. By Th.8.3.1 this perpendicular is tangent to the circle at C.

(ii) The given point, A in Fig. 8.5.6, is exterior to the circle.

Find the midpoint, O_1, of segment AO. Draw a circle with OA as a diameter. Draw lines AB and AB_1 joining A with the intercepts of the two circles. ∠OBA = ∠OB_1A = d as angles standing on diameter OA. Therefore, AB and AB_1 are tangent to the circle.

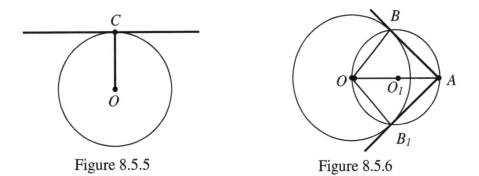

Figure 8.5.5   Figure 8.5.6

Question: Is the above solution for drawing a tangent to a circle through an exterior point of the circle valid in neutral geometry? Can you solve this problem in neutral geometry? (Hint: Is it possible to construct a right triangle by its hypotenuse and a leg in neutral geometry?)

<u>Corollary</u>. *Two tangents drawn from an exterior point to a circle form congruent angles with the line connecting this point with the centre of the circle.*

Proof. It follows from the congruence of right triangles ABO and AB_1O.

Problem 4. *Given two circles, draw their common tangents.*

Analysis. Suppose the problem has been solved: AB in Fig. 8.5.7, either (a) or (b), is a common tangent, A and B are the points of contact (the given circles are shown in solid lines as well as the sought for tangents, whereas auxiliary circles and tangents are shown as dotted lines).

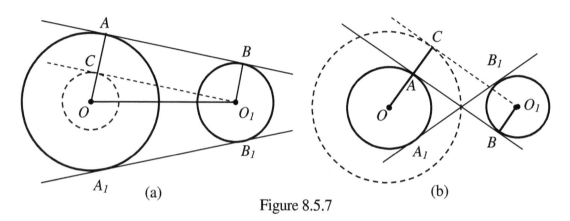

Figure 8.5.7

The tangents shown in figures (a) and (b) are called, respectively, *exterior* and *interior*.

First, consider the drawing of the *exterior* tangents (case (a)). Draw radii OA and O_1B to the points of contact. These radii are perpendicular to their common tangent AB, so they are parallel. Therefore, if we draw $O_1C \parallel AB$, then $\angle O_1CO$ is right. Then from O describe a circle with radius OC. It will touch O_1C at C. The radius OC of this circle is congruent to $OA - O_1B = R - r$ (the difference of the radii of the given circles).

Construction. From O as a centre describe a circle of radius $R - r$. Use the method of Problem 3 to draw a tangent, O_1C, to this circle from O_1. Draw radius OC and extend it till it meets the circle (at A). Through A draw AB parallel to O_1C. AB is tangent to both circles.

Another exterior tangent, A_1B_1, can be drawn using the same procedure or obtained as a reflection image of AB in OO_1 (prove the latter!)

Now consider Fig.8.5.7 b, where AB and A_1B_1 are interior tangents. From O_1 draw a ray parallel to AB and extend radius OA till it meets this ray at some point C. $OA \perp AB$ and $O_1C \parallel AB$; then $OC \perp O_1C$. Therefore, O_1C is a tangent to the circle described from O with radius $OC = R + r$.

Construction, therefore, includes the following:

from O, describe a circle of radius $R + r$;

from O, draw a tangent to this circle

connect the point of tangency, C, with O;

point A, where OC intersects the given circle, is the point of contact of the circle and a common tangent of the two circles;

draw $AB \parallel O_1C$.

AB is tangent to the given circles. A_1B_1, another interior tangent, can be constructed in a similar way or obtained by means of a reflection of AB in OO_1.

Theorem 8.5.2. *An angle with the vertex lying inside a circle is measured by one half of the sum of the arcs intercepted by its arms and extensions of its arms.*

Proof. In Fig. 8.5.8 $\angle ABC = \angle A + \angle D$ as an exterior angle of $\triangle ABD$, therefore it is measured by $½ \cup DE + ½ \cup AC$. The latter proves the theorem (just factor out ½).

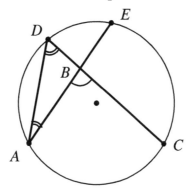

Figure 8.5.8

Lemma 8.5.3. *An angle formed by a chord and a tangent touching the circle at one of the endpoints of the chord, is measured by one half the arc intercepted by the chord.*

Proof. First, we consider a case when a chord is a diameter (Fig.8.5.9a). According to Th.8.3.2, $\angle DCA$ is right, hence it is measured by $½ \cup CmD$ that is a semicircle. (Also, $\angle DCB$ is measured by $½ \cup CnD$ that is also a semicircle).

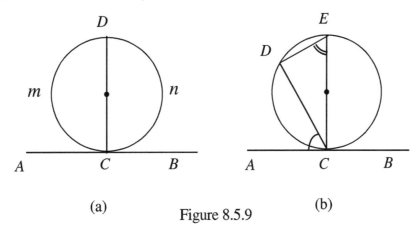

(a)　　Figure 8.5.9　　(b)

If CD is not a diameter, as in Fig. 8.5.9b, where CE is a diameter, then

∠DCA = ∠DEC (why?). The r.h.s. (right-hand side) of this equality, ∠DEC, is measured by $\frac{1}{2}\cup DC$, whence the Lemma follows.

Similar consideration is applicable to ∠DCB = ∠ECB + ∠DCE.

Theorem 8.5.4. *An angle formed by two secants, two tangents, or a secant and a tangent, that intersect outside the circle, is measured by one half the difference of the intercepted arcs.*

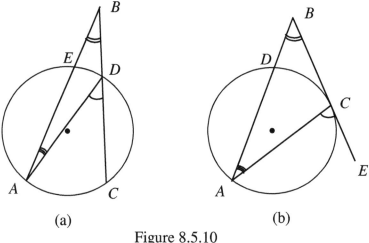

Figure 8.5.10

Proof. In case of two secants (Fig. 8.5.10a), ∠B = ∠ADC − ∠A (why?). The latter difference is measured by ½ ∪AC − ½ ∪ED = ½ (∪AC − ∪ED), Q.E.D.

When one of the arms is tangent to the circle (Fig.8.5.10b), ∠B = ∠ACE − ∠A. According to Lemma 8.5.3, mes(∠ACE) = ½ ∪AC, and thus mes(∠B) = ½ ∪AC - ½ ∪CD, whence the statement follows.

The case of two tangents should be considered by students.

Problem 5. *Given a straight segment and an angle, draw a circle in which the given segment is a chord and an angle inscribed in an arc intercepted by this chord is congruent to the given angle.*

Analysis. Suppose the problem is solved: the given segment is a chord *AB* in Fig. 8.5.11, and an inscribed angle, for instance ∠ACB, subtended by ∪AnB, is congruent to the given angle α.

Draw an auxiliary straight ray, *AE* tangent to the circle at *A*. Then ∠BAE = α, since it is measured by ½ ∪AnB, as well as ∠ACB.

The centre of the circle, *O*, lies on the perpendicular, *DO*, drawn to the chord *AB* through its midpoint, and, at the same time, *O* lies on the perpendicular to the tangent *AE*, which is drawn through *A*.

Draw an auxiliary straight ray, *AE* tangent to the circle at *A*. Then ∠*BAE* = α, since it is measured by ½ ∪*AnB*, as well as ∠*ACB*.

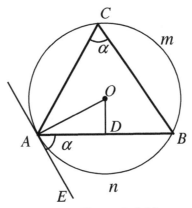

Figure 8.5.11

The centre of the circle, *O*, lies on the perpendicular, *DO*, drawn to the chord *AB* through its midpoint, and, at the same time, *O* lies on the perpendicular to the tangent *AE*, which is drawn through *A*.

<u>Construction.</u> Construct ∠*BAE* = α, with *A* as a vertex and *AB* as a side. From *A* erect a perpendicular to *AE* and draw a perpendicular through the midpoint, *D*, of *AB*. From the point of intersection of the two perpendiculars, *O*, describe a circle with radius *OA* (or *OB* = *OA*).

<u>Proof.</u> The described circle is the required one, since *AB* is a chord and any angle inscribed in ∪*AnB* is measured by ½ ∪*AnB*, and the latter is the measure of ∠*BAE* = α.

<u>Remark.</u> The reflection image of the circle in *AB* also satisfies the condition of the problem. Thus, the *locus of points from which AB is seen at angle α, consists of two congruent arcs that include α as an inscribed angle, and each of these arcs is a reflection image of the other in AB.*

8.5 Questions and Problems.

1. Find the angles of an isosceles triangle inscribed in a circle, if a lateral side of the triangle intercepts an arc
   of a) 27°30′; b) 48°12′.

2. Find the angles of a triangle whose vertices divide the circumscribed circle into arcs that are to each other as 2:5:11.

3. Can a non-convex quadrilateral be inscribed in a circle?

4. Prove that the opposite angles of any inscribed quadrilateral are supplementary.

5. Find the necessary and sufficient condition for a parallelogram to be inscribable in a circle.

6. Find the necessary and sufficient condition for a trapezoid to be inscribable in a circle.

7. Find the locus of the vertices of right triangles with a common hypotenuse as a base.

8. Construct a right triangle, having given its leg and the projection of that leg onto the hypotenuse.

9. The vertical angle of an isosceles triangle equals α=55°. A semicircle standing on a lateral side of the triangle as on a diameter is cut by the sides of the triangle into three arcs. Evaluate the degree measures of these arcs.

10. In Figure 1, *AB* and *CD* are straight segments drawn through the points of intersection of two circles, and *AB // CD*. Show that *ABDC* is a parallelogram.

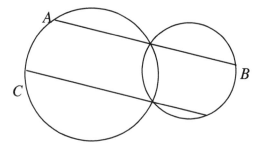

Figure 1

11. A chord subtends an arc of 68°. Find the angles between the chord and the tangents drawn through its endpoints.

12. A common secant is drawn through the point of contact of two circles that touch externally. Show that the arcs cut on the two circles by the secant and located on the different sides of the secant subtend congruent central angles in the respective circles.

13. A circle inscribed in an angle is divided by the points of contact into arcs that are in a ratio of 5÷7. Evaluate the angle.

14. If a triangle is inscribed in a circle, any line parallel to the tangent drawn through the vertex of the triangle cuts from the triangle an inscribable quadrilateral if that line intersects the lateral sides.

15. Given the angles of an inscribed triangle, find the angles of the triangle formed by the tangents that touch the circle at the vertices of the inscribed triangle.

16. An angle between the diagonals of an inscribed trapezoid equals 50°. The least base of the trapezoid equals to a lateral side. Evaluate the angles of the trapezoid.

17. Two secants are drawn through the point of contact of two circles that touch externally. Show that the chords which connect the endpoints of the secants in the circles are parallel.

18. *BC* is a common tangent (*B* & *C* are the points of contact) of two circles touching at point *A*. Show that ∠ *BAC* is right. *

19. Quadrilateral *ABCD* is inscribed in a circle. ∠ *DAB*=α=48°, ∠ *ABC*=β=124°. Evaluate an angle between the extensions of *DA* and *CB*.

20. Draw a tangent to a given circle through a given point.*

21. Given a point inside a circle, draw through that point a chord congruent to a given segment.

22. Draw common tangents to two given circles.*

23. Given a straight segment as a chord, draw a circle such that an inscribed angle standing on that chord equals to a given angle.

8.6 Constructions using loci

Many construction problems can be solved by exploiting the notion of loci. This method is known to be used since Plato's time (c.a. IV Century B.C.)

Suppose, there is a problem of finding a point that satisfies a certain set of conditions. Ignore one of these conditions, then the problem becomes *underdetermined*, i.e. there are many (usually – infinitely many) points that satisfy the remaining conditions. These points form some locus. Construct this locus if it is possible.

Then take the ignored condition into account and exclude from the original set some other condition. Another loci will be obtained as a solution of this new underdetermined problem. The sought for point satisfies all conditions of the original set, therefore it must belong to both of the constructed loci. Hence, this point may be found as a point of intersection of these loci.

The number of the points of intersection will determine the number of solutions of the problem. In particular, if there are no points of intersection of the two loci, then the problem has no solutions; in this case the set of conditions is called *inconsistent*.

We shall apply the method to solve the following problem.

<u>Problem 1.</u> *Construct a triangle, having given its base, a, the opposite angle, $\angle A = \alpha$, and the sum, s, of the two lateral sides.*

<u>Solution.</u> Suppose $\triangle ABC$ in Fig. 8.6.1a is a required triangle. To visualize the sum of the lateral sides, extend BA till point M such that $BM = s$. Then $AM = AC$ since $AC = s - AB$.

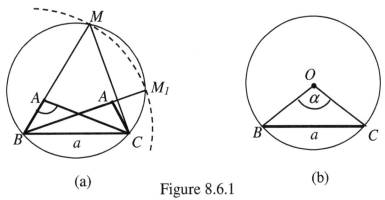

(a)   Figure 8.6.1   (b)

As $\triangle CAM$ is isosceles, $\angle BAC = 2 \angle AMC$ (why?), i.e. $\angle M = \frac{1}{2}\alpha$, where α is the given angle.

Thus, point M satisfies the following two conditions:

1. its distance from B is congruent to s;

2. $\angle BMC = \frac{1}{2}\alpha$.

Condition (1) without condition (2) determines a locus consisting of all points that are equidistant from B by distance s, so this is a circle described from B as a centre with radius s.

If we ignore condition (1), condition (2) will determine a locus of the vertices M of all angles BMC such that $\angle BMC = \frac{1}{2}\alpha$. This locus is an ("upper") arc of such a circle that contains BC as a chord, and the angles inscribed in $\cup BC$ of this circle are equal $\frac{1}{2}\alpha$. It is easy to draw such a circle; a hint for such a construction is presented in Fig. 8.6.1b.

Point M is a point of intersection of the circle obtained from condition (1) and the arc obtained from condition (2). They may intersect at two points (M and M_1 in Fig. 8.6.1), one point if they are tangent (M and M_1 merge in this case), or they may not intersect at all. The problem will have, respectively, two, one, or no solutions.

Sometimes a construction problem may be reduced to finding a line that satisfies certain conditions. If one of the conditions is excluded, there will be a set of lines that satisfy the remaining conditions. These lines may have a common property that defines a certain curve (For example, all the lines may turn out to be tangent to the same circle).

Excluding another condition of the set and accepting the one that had been ignored before may lead to obtaining another set of lines, which determines another common curve. If the two curves are found, the sought for line may be obtained. An example of a problem where such a method is applied, follows below.

Problem 2. *Draw such a secant of two given circles, that the parts of the secant lying inside the circles are equal to two given segments, a and a_1, respectively.*

Solution. Let O and O_1 be the centres of the given circles. Suppose, only one condition, concerning the chord(s) congruent to a in the circle centred at O, is imposed. There are infinitely many chords equal a in this circle, and all of them are equidistant from the centre (why?). Denote the distance from the centre to any of these chords h. Then all these chords are tangent to the concentric circle with radius h (Fig. 8.6.2).

Draw this *auxiliary* circle. In order to do this, describe an arc of radius a from an arbitrary point A on the given circle. Since a is less than the diameter of the circle (otherwise the problem cannot be solved), the arc will cut the circle at some other point B. Erect a perpendicular to AB through its midpoint M. Then $OM = h$ is a radius of the auxiliary circle.

Subject the second circle, centred at O_1, with chords equal a_1, to an analogous consideration. Another auxiliary circle, concentric to O_1, will be constructed. All the chords tangent to this circle are congruent to a_1.

If we draw a required secant, whose segments lying in the interior of the circles are congruent to a and a_1 respectively, these segments, and therefore the secant itself, will be tangent to each of the auxiliary circles.

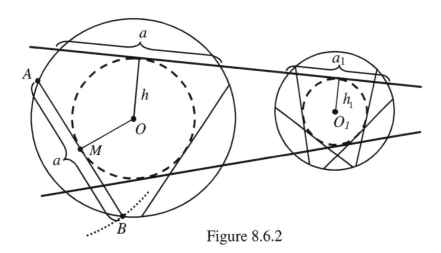

Figure 8.6.2

Hence the sought for secants will be common tangents of the two auxiliary circles, shown in dotted lines in Fig.8.6.2. The problem is reduced to the construction of common tangents of two circles (Problem 4, section 8.5).

8.6 Questions and Problems.

1. Divide a given arc into 4, 8, 16, ... congruent arcs.

2. Given the sum and the difference of two arcs, find the arcs.

3. Describe a circle with the centre at the given point such that it cuts a given circle into two semicircles.

4. In a line find the point closest to a given circle.

5. Given a chord in a circle, draw another chord that is bisected by the given chord and forms with it a given angle (is the solution possible for any angle?).

6. Given a point in the interior of the circle, draw a chord that is bisected by the given point.

7. Given a point on a side of an angle, describe from this point as a centre a circle that cuts off on the other side of the angle a chord equal to a given segment.

8. Given a radius and an angle, describe a circle with its centre on one side of the angle such that it cuts on the other side of the angle a chord equal to a given segment.

9. Given a radius, describe a circle which touches a given line at a given point.

10. Given a circle and a line, draw a tangent to the circle parallel to the given line.

11. Draw a circle that passes through a given point and touches a given line at a given point.

12. Given an angle, inscribe a circle that touches one of the sides at a given point.

13. A point located between two parallel lines is given. Describe a circle that passes through that point and touches the lines.

14. Given a circle, draw a tangent that forms a given angle with a given line (how many solutions may exist?).

15. Given a point in the exterior of a circle, draw a secant whose interior part equals to a given segment.

16. Given a radius, describe a circle that passes through a given point and touches a given line.

17. Given a line, find a point in it such that the segments of tangents drawn from that point to a given circle, are congruent to a given segment.

18. Construct a triangle, having given an angle and two altitudes one of which is drawn from the vertex of the given angle.

19. Given two points and two segments, draw a line such that the distances from these points to the line would be congruent to the given segments.

20. Describe a circle that passes through a given point and touches a given circle at a given point.

21. Describe a circle that touches two given parallel lines and a circle located between them.

22. Given a radius, describe a circle that touches a given circle and passes through a given point a) in the exterior; b) in the interior; c) on the circle.

23. Describe a circle of a given radius such that it touches a given line and a given circle.

24. Given a radius, describe a circle that cuts off chords equal to given segments on the sides of a given angle.

25. Describe a circle tangent to a given circle at a given point and to a given line (find all solutions).

26. Describe a circle tangent to two given circles, one of the points of contact is given.

27. Describe a circle tangent to three given circles a) internally; b) externally.

28. Inscribe a circle in a given sector of a circle (i.e. describe a circle tangent to the radii and the arc of a given sector).

29. Given a circle, inscribe in it three congruent circles tangent to each other and to the given circle.

30. Given a point in the interior of a circle, draw a chord that is divided by the given point into segments whose difference is given.*

31. Through a point of intersection of two given circles, draw a secant with the sum of its segments lying in the interiors of the circles equal to a given segment.

32. Given a point in the exterior of a circle, draw a secant whose exterior and interior segments are congruent.

33. Draw an arc passing through a given point and conjugate to a given straight line at a given point.

34. Connect two given non-parallel lines by a conjugate arc or arcs. (Use two or more arcs only if necessary). Consider the following cases:
    a) the points of conjugation and the radius of the arc are not specified;
    b) only radius is given;
    c) only one radius and one point of conjugation are given (this case of conjugation is often used for rounding turns of railroads; see Figure 1).

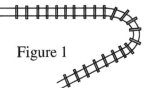

Figure 1

8.7 Inscribed and circumscribed polygons

If all vertices of a polygon (*ABCD* in Fig. 8.7.1) lie on a circle, it is said to be *inscribed in the circle*. One can also say that the circle is *circumscribed about the polygon*.

If all sides of a polygon are tangent to a circle (MNPQ in Fig. 8.7.1) the polygon is said to be *circumscribed about the circle* or a circle is said to be *inscribed in the polygon*.

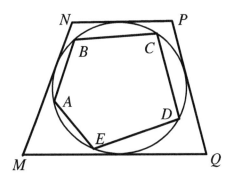
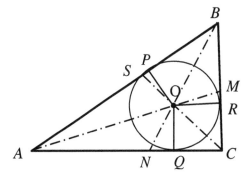

Figure 8.7.1   Figure 8.7.2

Theorem 8.7.1 *One and only one circle can be circumscribed about a triangle.*

Proof. Inasmuch as vertices of a triangle are three noncollinear points, this theorem follows immediately from Th.8.1.1.

Theorem 8.7.2 *One and only one circle can be inscribed in a triangle.*

Proof. If such a circle exists, its centre must be equidistant from the sides of the triangle (Fig. 8.7.2). Bisector *AM* of ∠*A* is a locus of points equidistant from *AB* and *AC*; bisector *BN* of ∠*B* is a locus of points equidistant from *BC* and *AB*.

Therefore, the point, O, of intersection of these bisectors is equidistant from all three sides of $\triangle ABC$: if $OP \perp AB$, $OQ \perp AC$, and $OR \perp BC$, then $OP = OQ = OR$. Any of these perpendiculars is a radius of the inscribed circle.

Therefore, a perpendicular dropped from O onto any side of the triangle is a radius of the inscribed circle:.

There is only one circle inscribed in $\triangle ABC$, since two bisectors intersect only at one point (thus, the centre is defined uniquely) and there is only one perpendicular dropped from a point onto a line (thus, the radius is defined uniquely).

<u>Corollary.</u> ***Three bisectors of a triangle meet at one point (are concurrent).***

<u>Proof.</u> Point O in Fig. 8.7.2 is equidistant from all three sides of $\triangle ABC$, therefore it must belong to the locus of points equidistant from AC and BC. The latter is the bisector of $\angle C$, hence O lies on this bisector. In other words, the third bisector will also pass through O, Q.E.D.

<u>Remark.</u> Students can check that

(i) the centre of an inscribed circle (*the incentre*) is always located within the triangle;

(ii) the centre of a circumscribed circle (*the circumcentre*) lies within the triangle if the triangle is acute-angled, in the exterior of the triangle if it is obtuse-angled, and at the midpoint of the hypotenuse if the triangle is right.

A circle is called *escribed* to a triangle if it is tangent to one side and the extensions of the other two sides of the triangle (Fig.8.7.3). Apparently, there are three circles escribed to each triangle. Students should be able to specify the locations of their centres (called *excentres*) and to construct the circles.

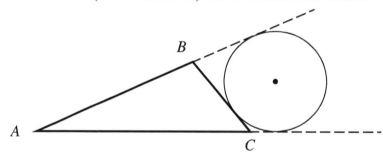

Figure 8.7.3

It has already been mentioned (see section 8.1) that four points do not necessarily lie on one circle, i.e., in general, a quadrilateral is not inscribable. The following theorem provides a criterion of the inscribability.

Theorem 8.7.3. ***A convex quadrilateral can be inscribed in a circle iff (if and only if) its opposite angles are supplementary.***

<u>Proof.</u> Suppose $ABCD$ (Fig. 8.7.4) is an inscribed quadrilateral. Let us prove that

$\angle B + \angle D = \angle A + \angle C = 2d$. As the sum of all angles of a quadrilateral equals $4d$, it is enough to prove that $\angle B + \angle D = 2d$.

$\angle B$ as an inscribed angle is measured by $½\cup ADC$; similarly, mes($\angle D$)= $½\cup ABC$. Therefore mes($\angle B + \angle D$) = $½\cup ADC + ½\cup ABC$ = $½ (\cup ADC + \cup ABC)$ = half revolution, whence the statement follows.

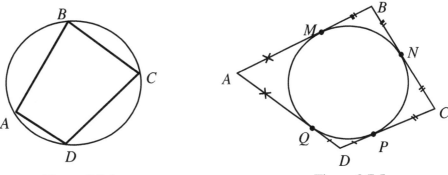

Figure 8.7.4   Figure 8.7.5

Conversely, if in ABCD $\angle B + \angle D = 2d$, and, thus, $\angle A + \angle C = 2d$, point D must lie on a circle circumscribed about A, B, and C, because otherwise $\angle D + \angle B \neq 2d$.

Corollary 1. *A parallelogram is inscribable if and only if it is a rectangle.*

Corollary 2. *A trapezoid is inscribable if and only if it is isosceles.*

Students should be able to prove the above corollaries.

Theorem 8.7.4. If a circle can be inscribed into a quadrilateral, then the sums of the opposite sides of this quadrilateral are equal to each other.

Proof. ABCD in Fig. 8.7.5 is circumscribed about a circle, i.e. its sides touch the circle at some points M, N, P, and Q. It is required to prove that $AB + CD = BC + AD$.

Since two tangents drawn to a circle from one point are equal, $AM = AQ$, $BM = BN$, $CN = CP$, and $DP = DQ$. Then, $AM + MB + CP + PD = AQ + QD + BN + NC$, $\Rightarrow$ $\Rightarrow AB + CD = AD + BC$, Q.E.D.

8.7 Questions and Problems.

1. Prove that in any triangle a bisector lies between the median and the altitude drawn from the same vertex.

2. Show that the incentre is always located in the interior of a triangle.

3. Show that the circumcentre is located in the interior of a triangle if it is acute-angled, in the exterior if it is obtuse-angled, and at the midpoint of the hypotenuse if it is right.

4. Construct a right triangle, having given the radii of the inscribed and circumscribed circles.*
Think of possible generalizations of this problem.

5. The segments that connect the vertex of the right angle of a triangle with the centres of the inscribed and circumscribed circles form an angle equal α=10°. Determine the angles of the triangle.

6. In acute triangle *ABC*, its vertex *A* is joined by a straight segment to the centre, *O*, of the circumscribed circle. The altitude, *AD*, is drawn from *A* to the opposite side. Prove that ∠ *BAD*=∠ *OAC*. *

7. Prove that in any triangle the bisectors of two exterior angles are concurrent with the bisector of the interior non-adjacent to them angle.

8. Given a triangle, escribe a circle.

9. Two arbitrary secants are drawn through the midpoint of a given arc, ∪*AB*, of a circle. These secants cut the chord *AB* at some points, *M* and *N*, and they cut the circle at some points, *P* and *Q*. Prove that quadrilateral *MNPQ* is inscribable.

10. The bisectors of the four angles of a convex quadrilateral form (when cutting each other) an inscribable quadrilateral.

11. What can you tell about a parallelogram in which a circle can be inscribed?

12. Construct a rhombus, having given its acute angle and the radius of the inscribed circle.

13. What kind of a quadrilateral is both inscribable and circumscribable (a sufficient condition)?

14. Construct a regular hexagon and evaluate its perimeter, having given the radius of the inscribed circle.

8.8   Four remarkable points of a triangle

We have seen that

1.   The perpendicular bisectors of the sides of a triangle meet in a point (the circumcentre of the triangle);

2.   The bisectors of the angles of a triangle meet in a point (the incentre of the triangle).

The following two theorems concern the other two remarkable points of a triangle.

Theorem 8.8.1. The altitudes of a triangle meet in a point.

Proof. Through each vertex of $\triangle ABC$ draw a line parallel to the opposite side (Figure 8.8.1). That creates an *auxiliary* $\triangle A_1B_1C_1$. The altitudes of $\triangle ABC$, *AD*, *BE*, *CF* are perpendicular to the respective sides of $\triangle A_1B_1C_1$.

AC_1BC and ABA_1C are parallelograms, therefore $C_1B = AC = BA_1$ (why?). Hence, *B* is the midpoint of A_1C_1. Similarly, *A* and *C* are the midpoints of B_1C_1 and B_1A_1 respectively.

Thus, *AD*, *BE*, and *CF* are the bisecting perpendiculars of the sides of $\triangle A_1B_1C_1$, and therefore, they meet in one point, Q.E.D.

Remark. The point where the altitudes of a triangle meet is called its *orthocentre*.

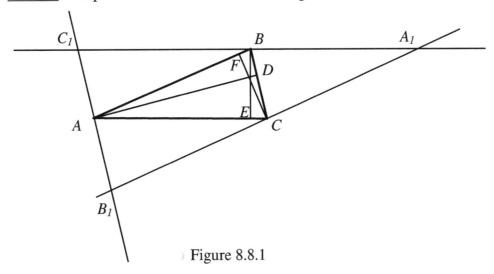

Figure 8.8.1

Theorem 8.8.2. ***The medians of a triangle meet in a point which is two thirds of the distance from any vertex to the middle point of the opposite side.***

Proof. In Fig. 8.8.2, medians *AE* and *BD* of $\triangle ABC$ intersect at some point *O*. (Why are we so sure that two medians do intersect?) Prove that $OD = \frac{1}{3}BD$, and $OE = \frac{1}{3}AE$.

Let *F* and *G* be the midpoints of *AO* and *BO* respectively. Connect consecutively *D*, *F*, *G*, *E* to create a quadrilateral *DFGE*.

FG connects the midpoints of the sides *AO* and *BO* in $\triangle ABC$, therefore *FG* // *AB*, and *FG* = ½ *AB*. Similarly, since *DE* connects the midpoints of *AC* and *BC* in $\triangle ABC$, it follows that *DE* // *AB*, and *DE* = ½ *AB*.

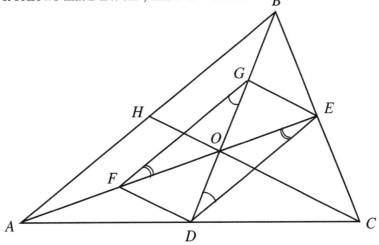

Figure 8.8.2

193

Since both *DE* and *FG* are parallel to *AB*, it follows that *DE //FG*; also each of these two segments is congruent to ½ *AB*, which means that *DE = FG*. Therefore *DFGE* is a parallelogram, whence *OF = OE*, and *OG = OD*.

As *F* and *G* are the midpoints of *AO* and *BO*, respectively, it follows that *AF = OF = OE*, and *BG = OG = OD*. Therefore, $OE = \frac{1}{3}AE$, and $OD = \frac{1}{3}BD$, Q.E.D.

If the third median, *CK*, cuts one of the considered medians, e.g., *AE*, at some point *O´* that does not coincide with *O*, then $O´E \neq \frac{1}{3}AE$. However we can apply the above consideration to medians *AE* and *CH* and obtain that $O´E = \frac{1}{3}AE$. The contradiction proves that *O´* coincides with *O*, which means that the medians are concurrent, Q.E.D.

It is known from physics (and it is easy to prove) that the *centre of mass* of a triangular homogeneous lamina (a flat plate made of some material of constant density) is located at the point of intersection of the medians of the triangle. That is why the point of intersection of the medians of a triangle is called the *centroid* of the triangle.

8.8 Questions and problems.

1. Which of the sides of a triangle is the closest to the circumcentre?

2. Which of the vertices of a triangle is the closest to the incentre?

3. Prove that for any triangle, the lines drawn through its vertices parallel to the opposite sides form a triangle in which the altitudes of the original triangle are the perpendicular bisectors of the sides.

4. Prove that the altitudes of a triangle meet in a point (the *orthocentre*).

5. Prove that the altitudes of a triangle are the bisectors in the triangle formed by the segments connecting the feet of the altitudes.* Use this result to show the existence of the orthocentre.

6. Which of the vertices of a triangle is the closest to the orthocentre

7. Which of the sides of a triangle is the closest to the orthocentre?

8. Which of the altitudes of a triangle is the least?

9. Prove that in a triangle, the points symmetric to the orthocentre about the sides of the triangle are located on the circle circumscribed about the triangle.

10. Construct a triangle, having given its side and two medians drawn to the other two sides.

11. Prove that the centre of mass (*centroid*) of a lamina in a shape of triangle is located at the point where the medians of the triangle meet. (Hereafter we call that point *the centoid of a triangle*).*

12. Given a segment and an exterior point, construct a triangle with the segment as a base and the given point as a centroid.

13. Construct a triangle, having given the three medians.

14. Through the vertex A of △ABC, a line is drawn in the exterior of the triangle (Figure 1). The projections of AC and AB on that line are segments b and c, respectively. Find the distances between the projections of the centroid and the vertices of the triangle onto that line.

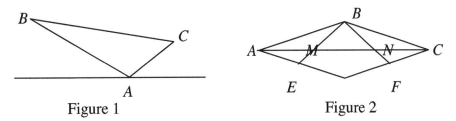

Figure 1   Figure 2

15. In rhombus ABCD in Figure 2, BE and BF bisect AD and DC, respectively. Find MN, having given that AC=l.

16. Prove that two segments, which connect a vertex of a parallelogram with the midpoints of the two sides that meet at the opposite vertex, divide the diagonal that joins the other two vertices into three congruent segments.

17. Which of the vertices of a triangle is the closest to the centroid?

18. Which of the sides of a triangle is the closest to the centroid?

19. Which of the medians is the least?

20. *Prove that the feet of the perpendiculars dropped from any point of the circle circumscribed about a triangle onto the three sides (or their extensions) are collinear (they lie in the so-called *Simpson's line*).

21. *Prove that the circumcentre, the ortocentre, and the centroid of a triangle are collinear (they lie in the so-called *Euler's line*).

9. SIMILARITY

> ...Rabbit had so many friends-and-relations, and of such different sorts and sizes, that he didn't know whether he ought to be looking for Small at the top of an oak-tree or in the petal of a buttercup.
>
> A.A.Milne, *The House at Pooh Corner*

9.1 The notion of measurement.

a) <u>The setting of the problem.</u>

In AXIOMS 3, 4 we have introduced the notion of *congruence of segments*. That has enabled us to compare segments. In particular, we can compare any segment with a given segment chosen to be the *unit of measurement*. Yet it has not been specified how one can determine the *length* of a segment.

In this section we state the problems of defining the notion of the *length* of a segment and expressing lengths in numbers. We shall use first letters of the alphabet (*a, b, ... h*, and sometimes *A, B, ..., H*) to denote segments and the mid-alphabet letters *(i, j, k, l, m, n, N,...)* for positive integers (natural numbers).

The notions of measurements of arcs, angles, and other *magnitudes* can be introduced in a similar manner.

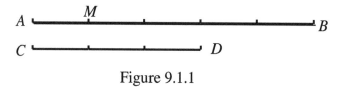

Figure 9.1.1

b) <u>Common measure.</u>

A segment is said to be a *common measure* of two segments, if it is contained an exact number of times in each of these segments.

For example, segment *AM* in Fig. 9.1.1, is contained exactly 5 times in *AB* and exactly 3 times in *CD*, thus *AM* is a common measure of *AB* and *CD*.

It is obvious that in Fig. 9.1.1, $\frac{1}{2}AM$ is also a common measure of *AB* and *CD* as well as $\frac{1}{3}AM$, $\frac{1}{4}AM$, and any other segment that is contained in AM an integer number of times. Therefore, *AB* and *CD* have infinitely many common measures. *AM* is the greatest of all these measures, because 3 and 5 do not have common factors.

In general, it is obvious that if some segment *c* is a common measure of segments *a* and *b*, then the segments $\frac{1}{2}c$, $\frac{1}{3}c$, ..., $\frac{1}{n}c$ are their common measures as well. Thus,

(i)   if two segments have a common measure, they have infinitely many common measures;

*(ii) one and only one of these common measures is their **greatest common measure**.*

(c)   Finding the greatest common measure.

In this subsection we shall discuss the method of finding the greatest common measure of two segments as proposed by Euclid. Nowadays it is called the *Euclidean algorithm* and is often used for funding the greatest common divisor of two natural numbers. The method is based on the following two theorems and one *continuity axiom*, called the *Archimedes' axiom*.

***Theorem 9.1.1** If the least of two segments is contained an integer number of times without a remainder in the greater of them, then the least segment is the greatest common measure of the two segments.*
Proof. For two segments, A and B, segment C is their common measure if $A = mC$ and $B = nC$. C is their greatest common measure when m and n are minimal possible integers (hereafter "integers" means "positive integers").

Suppose, for certainty, $A > B$. Then, in the case described in the hypothesis of the theorem, $A = mB$ and $B = nB$ where $n = 1$. Since n cannot be less than 1, the conclusion follows, Q.E.D.

***Theorem 9.1.2** If the least of two segments is contained an integer number of times, with a remainder, in the greatest segment, then their greatest common measure, if it exists, is also the greatest common measure of the least segment and the remainder.*
Proof. Two segments, A and B, are given, and suppose $A > B$. By the hypothesis,
$$A = nB + R. \qquad (1)$$
If C is the greatest common measure of A and B, then $A = kC$, $B = mC$, and the integers k and m are *relatively prime*, i.e. they do not have *nontrivial (that means different from 1)* common divisors, since had they had one, i.e. $k = k_1 j$ and $m = m_1 j$, where $j > 1$, then it would have entailed
$$A = k_1 jC \text{ and } B = m_1 jC,$$
which means that the segment jC, and not C, would be the greatest common measure.

Then it follows from (1) that $kC = nmC + R$, whence
$$R = (k - mn) C. \qquad (2)$$
It follows from (2) that C is a common measure of R and B. Is it the greatest common measure of R and B? – It will be, if the numbers m and $(k - mn)$ are relatively prime.

Suppose they are not, i.e. $k - mn = i p_1,$ (3)
$$m = i p_2 . \qquad (4)$$
Then $k = mn + i p_1 = i p_2 n + i p_1 = i(p_2 n + p_1)$.

Comparing the latter expression for k with formula (4), we see that i is a common factor of k and m that can only happen if $i = 1$. Therefore, m and $(k - mn)$ are relatively *prime*, and, consequently, C is the greatest common measure of B and R.

To complete the proof we have to show that if C is the GCM (greatest common measure) of B and R, then it is also the GCM of A and B. (It would be a good exercise for students).

Figures 9.1.2 (a) and (b) illustrate theorems 9.1.1 and 9.1.2, respectively. In Figure (a) B is the GCM of A and B; in Figure (b) the GCM of A, B, and R equals ½ R.

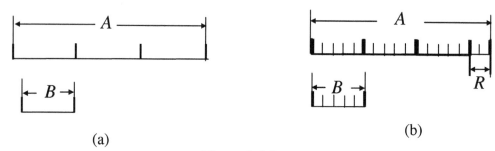

(a)　　　　　　　　　　　　　　(b)

Figure 9.1.2

I CONTINUITY AXIOM (Archimedes' Axiom). *For any two segments: either one of them is a multiple of the other, or there exists such a number n that the greater segment is less than n-multiple of the least segment.*

In other words, given any two segments, a and b, there are only two options:

(i)　　$a = nb$, i.e. a is a multiple of b (including the case $a = b$);

(ii)　　$(n - 1)b < a < nb$, i.e. a finite number of copies of the least segment b laid off sequentially on a ray covers the greatest segment a laid off on the same ray, and the remainder $(nb - a)$ is less than the least segment (b).

Figure 9.1.3 illustrates the second option:

$a = PM$ is the greatest of two segments; $b = PP_1 = P_1P_2 = \ldots = P_{n-1}P_n$ is the least of the two; $PP_1 + P_1P_2 + \ldots + P_{n-1}P_n = nb > PM = a$, and $MP_n = nb - a < b$.

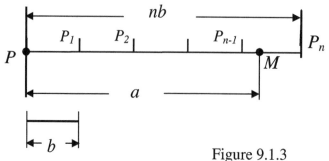

Figure 9.1.3

Now let us describe the procedure of finding the greatest common measure.

Suppose, we are looking for the GCM of two given segments, *AB* and *CD*, such that $AB > CD$ (Fig. 9.1.4).

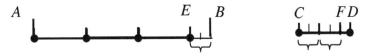

Figure 9.1.4

We lay off the least segment on the greatest one as many times as possible. Then, according to Archimedes' axiom, one of the following two options occurs: either

(i) a few *CD*s will cover the whole *AB* without a remainder, or

(ii) a remainder, $EB < CD$ will be obtained (like in Fig. 9.1.4); then, according to Theorem 9.1.2, the problem will be reduced to the finding of the GCM of *CD* and the first remainder, *EB*.

To find their GCM we apply the same procedure, which means that we lay off *EB* on *CD* as many times as possible without exceeding *CD*. Again, one of the following two cases takes place: either

(i) *EB* fits on *CD* an integer number of times without a remainder; then *EB* is the GCM sought for (why?), or

(ii) A remainder, $FD < EB$ is obtained, and the problem reduces to the finding of the GCM of *EB* and the second remainder, *FD*.

When the described procedure is applied consecutively, it leads to one of the following two outcomes:

1. no remainder left after another *iteration* (step of the procedure), or
2. the procedure never ends (in the assumption that we can lay off arbitrarily small segments, that is possible only theoretically, of course).

In the first case the last remainder will be the GCM sought for.

E.g., in Figure 9.1.4 *FD* is the GCM of *AB* and *CD*:

$AB = 3CD + EB$, $CD = 2EB + FD$, $EB = 2FD$, and using the backward substitution, we obtain $CD = 5FD$, and $AB = 15\,FD + 2FD = 17\,FD$.

(In a similar manner one can find the GCM of two arcs described by the same radius, of two angles, etc.)

In the second case the segments do not have a common measure. To see this let us suppose that they (*AB* and *CD*) have some common measure.

This measure, according to Th. 9.1.2, must be contained an integer number of times not only in *AB* and *CD* but also in the first remainder, *EB*, then, consecutively, in the second remainder, *FD*, in the third remainder, etc., etc.

As each next remainder is less than the previous one, the common measure will be contained in the next remainder fewer times than in the preceding one. Therefore, if the suggested common measure is contained *m* times in *EB*, it will be contained, at most, $(m-1)$ times in *FD*. In the next remainder it will be contained, at most, $(m-2)$ times, etc.

As the sequence $\{m, m-1, m-2, \ldots, 2, 1\}$ is finite, the process of the laying off segments will come to an end, i.e. after some iteration no remainder will be obtained. Hence, the suggestion of the existence of a common measure contradicts to the endlessness of the iteration procedure. Thus, if the described iteration procedure is endless, the segments do not have a common measure.

(d)   Commensurable and incommensurable segments

Two segments are called *commensurable* if they have a common measure and *incommensurable* if their common measure does not exist.

In practice one cannot get convinced that incommensurable segments do exist. Really, when performing the sequential laying off of segments, as suggested in the proposed above algorithm, one will obtain such a small remainder that will **apparently** fit an integer number of times in the previous remainder. Maybe a new remainder will exist but because of the inaccuracy of our tools (compass) and imperfection of our perception (eyesight) this tiny remainder will not be discovered.

Thus, incommensurability cannot be established experimentally. Yet, we can prove that *incommensurable segments do exist*.

Theorem 9.1.3. *A diagonal of a square is incommensurable with its side.*

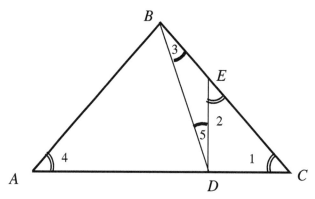

Figure 9.1.5

Inasmuch as a diagonal divides the square into two isosceles right triangles, the theorem may be reformulated as follows:

The hypotenuse of an isosceles right triangle is incommensurable with its leg.

Proof. In isosceles right $\triangle ABC$ (Fig. 9.1.5) lay off on the hypotenuse, AC, a segment AD congruent to the leg AB.

From D erect a perpendicular that intersects another leg, BC, at some point E. The obtained $\triangle CDE$ is isosceles as well as the original $\triangle ABC$, and segment BE laid off on the leg BC is congruent to segment DC laid off on the hypotenuse.

In order to see that, connect D with B and consider the angles of $\triangle CDE$ and $\triangle BED$. $\angle 1 = \angle 4 = 45°$ as acute angles of the right isosceles triangle ABC, and $\angle 2 = \angle 4$ since the sides of these angles are mutually perpendicular: $DE \perp AC$; $EC \perp AB$.

Therefore, $\angle 1 = \angle 2 = 45°$, and, hence, ΔCDE is isosceles: $DE = DC$, Q.E.D.

In ΔBDE, $\angle 3 = d - \angle ABD$, and $\angle 5 = d - \angle ADB$.
Since $AB = AD$, $\angle ABD = \angle ADB$, and, thus, $\angle 5 = \angle 3$. Therefore, $BE = DE$, and, taking into account $DE = DC$, we conclude that $BE = DC$, Q.E.D.

Now let us try to find a common measure of AB and AC. From triangle inequalities we know that $AC < AB + BC = 2AB$. On the other hand, $AC > AB$ as the hypotenuse of a right triangle. Therefore, $AB < AC < 2AB$. The latter means that AB can be laid off on AC only once. The obtained remainder is DC, and so, the problem is reduced to finding a common measure of DC and AB, or, which is the same, a common measure of DC and segment BC, which is congruent to AB.

Then we have to lay off DC on BC. As we have shown above, $DC = BE$, and hence $BC - BE = BC - DC$. Thus the problem of the finding a common measure of DC and BC is reduced to the finding a common measure of DC and EC. These segments are respectively a leg and the hypotenuse of ΔCDE, which is a right isosceles triangle.

Thus, the problem of finding a common measure of a leg and the hypotenuse of a right isosceles triangle is <u>reduced to the same problem</u> for a smaller right isosceles triangle. Then, it is obvious that the process of finding a common measure in this case is endless. Therefore, a common measure of AB and AC does not exist, <u>Q.E.D.</u>

(e)   <u>Measurement of segments. Rational and irrational numbers.</u>

To describe the measure (length) of a segment, i.e. to *measure* a segment, one has to compare it to some other segment, which is known, e.g. a segment equal one centimeter. This known segment, with which all other segments are compared, is called a *unit of length*. Hereafter, for the length of a segment a we shall use the symbol $m(a)$ (measure of a). For example, if $a = 4U$ (segment a contains the unit U exactly 4 times), we shall write $m(a) = 4$, which means: the length of a is 4.

When measuring a segment, one of the following two situations may occur: either the segment is commensurable with the unit, or incommensurable with it.

1.   *To measure a segment commensurable with a given unit means to find out, how many times the unit and / or maybe a fraction of the unit is contained in the segment without a remainder.*

In some cases the chosen unit itself (segment U) may be the greatest common measure of U and the measured segment, a. Then, $a = nU$, where n is a natural number, and the length (*measure*) of a is expressed by a natural number (positive integer) n: $m(a)=n$.

If the GCM of the unit, U, and the segment, a, which is to be measured, is a fraction of the unit, $\frac{1}{k}U$ (where k is a natural number, $k \in N$), then the length (*measure*) of A is expressed by an *irreducible* fraction, $\frac{n}{k}$ ($n \in N$).

For instance, in Fig.9.1.6 $a = \frac{9}{4}U$, and $m(a) = \frac{9}{4}$.

In any of these cases the *measure* of the segment is represented by a *rational* number.

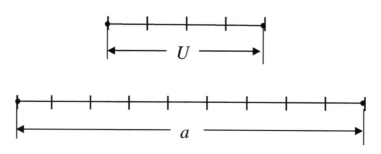

Figure 9.1.6

2. When a is incommensurable with U, the measure is determined as a result of consecutive estimates: instead of a one measures two segments, one of which, $\overline{a}$, is greater than a, and the other, $\underline{a}$, is less that A. These segments are commensurable with U and approximate *a within any given accuracy,* i.e. they differ from a by segments whose measures do not exceed any given (small) magnitude.

To find such segments, a certain procedure is applied. We shall illustrate it, at first, by means of an example.

Suppose, a is a segment incommensurable with the given unit U (Figure 9.1.7), and $U < a < 2U$. In the next step we are going to find such segments that are commensurable with U and differ from a by segments that are less than $\frac{1}{10}U$.

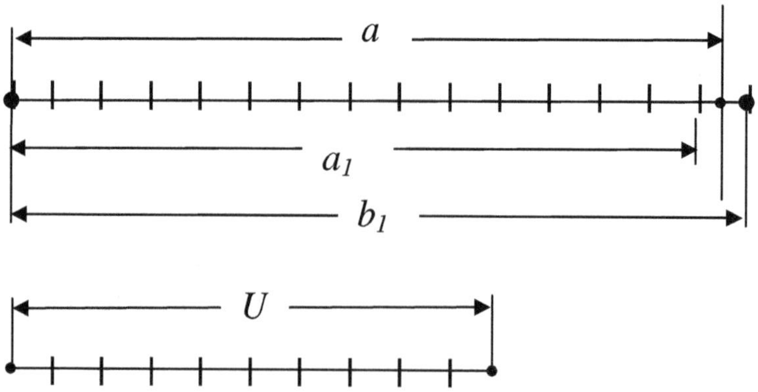

Figure 9.1.7

We divide U into 10 equal segments and sequentially lay off one of these segments on a, – as many times as possible. In our case, a contains 14 such segments (U itself and 4 segments equal $\frac{1}{10}U$) and a remainder that is less than $\frac{1}{10}U$. Thus, we obtain segment $a_1 = U + 4 \cdot \frac{1}{10}U = 1.4U < a$.

If we add to a_1 one more segment equal $\frac{1}{10}U$, we obtain another segment, $b_1 = U + 5 \cdot \frac{1}{10}U = 1.5U$. This segment is commensurable with the unit (as well as a_1). It is greater than a by some segment, which is less than $\frac{1}{10}U$.

The lengths of a_1 and b_1 are expressed respectively by the numbers $1 + \frac{4}{10} = 1.4$ and $1 + \frac{5}{10} = 1.5$. These numbers are a *lower* and an *upper estimates* of the length of a. Each of them expresses the length of a within an *accuracy of* $\frac{1}{10}$.

In order to obtain more accurate estimates of the length of a, we shall split the segment equal $\frac{1}{10}U$ into ten equal parts and use the obtained segments equal $\frac{1}{100}U$ to measure the difference between a and $1.4U$. Thus we shall obtain the next *lower* and *upper estimates* for the length of a.

Let us suggest, for certainty, that $a_1 + 5 \cdot \frac{1}{100}U < a < a_1 + 6 \cdot \frac{1}{100}U$. Then, the next two approximating segments are $a_2 = U + 4 \cdot \frac{1}{10}U + 5 \cdot \frac{1}{100}U = 1.45U$ and $b_2 = U + 4 \cdot \frac{1}{10}U + 6 \cdot \frac{1}{100}U = 1.46U$. The respective estimates of the length of segment a are 1.45 and 1.46: $1.45 < m(a) < 1.46$.

In a similar manner, we can continue such steps (called *iterations*) to estimate the length of a to an accuracy of 0.001, 0.0001, …, $\frac{1}{10^n}$, where n is any number. For instance, after two more iterations we shall obtain something like $a_4 = 1.4538U < a < 1.4539U = b_4$, and the respective lower and upper estimates of the length of a to an accuracy of 0.0001, are 1.4538 and 1.4539. Since, by choosing sufficiently large n, we can make $\frac{1}{10^n}U$ less than any given segment, we can estimate the length of a within any pre-assigned accuracy.

In this example we obtained each next unit of measurement by dividing the previous one into 10 equal parts. It is convenient, since this way we obtain our estimates of the length as decimals, however it is not necessary. In general, given a unit U, in order to find approximate measures of some segment a, we divide the unit into n equal parts and determine how many times $\frac{1}{n}$ - th of the unit is contained in a. If it is contained in segment a m times with a remainder less than $\frac{1}{n}U$, then numbers $\frac{m}{n}$ and $\frac{m+1}{n}$ are the approximate measures of a within an accuracy of $\frac{1}{n}U$. The first number is a lower and the second is an upper estimate.

These estimates will be improved if another number, n_1, greater than n, is chosen to replace n. n_1, in its own turn, may be replaced by a greater number, n_2, and this step will improve our estimates. This process can be continued until any given accuracy, for example, $\frac{1}{N}$, for the estimates of the measure of segment a is achieved.

Let us notice that the same method of finding approximations to the lengths of a is applicable in the case when a is commensurable with the unit. In this case we can, if we please, obtain the *exact* measure of a, and this measure will be represented by a *rational number*, whereas *in the case of the incommensurability, rational numbers will not provide the exact measure*.

Then, does the exact measure of a segment a, incommensurable with the unit, exist? - The answer is positive: the exact measure exists and it is expressed by a number called *irrational*. In order to obtain the number that represents the exact measure of a segment incommensurable with the unit, we shall apply the following procedure, generalizing the one described in the above example.

Let a be a segment incommensurable with the unit of length U. According to the Archimedes' axiom there exists such a natural number k_0 that $k_0 U < a < (k_0 +1)U$. The respective estimates for the measure of a are: $k_0 < m(a) < k_0 +1$.

The difference between the segments a and $k_0 U$ is less than U, then, in order to cover this difference by segments equal $\frac{1}{10}U$, we shall need fewer than 10 of them. Suppose k_1 is such a number that $k_1 \frac{1}{10}U < a - k_0 U < (k_1 +1)\frac{1}{10}U$. Then we shall have the following estimates for a and its length:

$$k_0 U + k_1 \frac{1}{10}U < a < k_0 U + (k_1 +1)\frac{1}{10}U ;$$

$$k_0 + \frac{k_1}{10} < m(a) < k_0 + \frac{k_1 +1}{10}, \text{ where } 0 \leq k_1 \leq 9.$$

The latter estimates can also be written in decimal notation:
$k_0 . k_1 < m(a) < k_0 . (k_1 +1)$

(For the above example it was $1.4 < m(a) < 1.5$, with $k_0 = 1; k_1 = 4$).

Let us continue this procedure in such a way that each subsequent step (iteration) improves the precision of the upper and lower estimates tenfold. After sufficient number of iterations we shall obtain the estimates that are accurate to the nth decimal place:

$$k_0 U + k_1 \frac{1}{10} U + ... + k_n \frac{1}{10^n} U < a < k_0 U + k_1 \frac{1}{10} U + ... + (k_n + 1) \frac{1}{10^n} U ;$$

$$k_0.k_1 k_2...k_n < m(a) < k_0.k_1 k_2...(k_n + 1).$$

The difference between the upper and lower estimates in the last inequality is equal to $\frac{1}{10^n}$. This difference tends to zero with indefinite growth of n. The procedure never ends (as we have proved in the subsection (c)) the following two sequences of the upper and lower estimates of the length will be obtained:

Lower estimates:

$$l_0 = k_0; \quad l_1 = k_0.k_1; \quad l_2 = k_0.k_1 k_2; \quad ... \quad l_n = k_0.k_1 k_2...k_n; ...$$

and

Upper estimates:

$$u_0 = k_0 + 1; \quad u_1 = k_0.(k_1 + 1); \quad u_2 = k_0.k_1(k_2 + 1); \quad ... \quad u_n = k_0.k_1 k_2...(k_n + 1); ...$$

Since, by choosing an appropriate n, the difference between the respective terms $u_n - l_n = \frac{1}{10^n}$ can be made less than any given number (we say in this case that the difference *tends to 0*), both sequences of estimates *converge* to the same number, which is defined to be the exact *numerical measure*, or *length* of the segment a.

This number is represented by an infinite (non-terminating and non-periodic) decimal, thus it cannot be transformed into an ordinary fraction (a rational number). Such numbers, represented by infinite decimals, are called *irrational*.

By means of sequences of lower and upper estimates, one can define the notions of equality and inequality as well as arithmetic operations for irrational numbers in such a way that these notions and operations possess all the features of their counterparts for rational numbers. All rational and irrational numbers together form the set of *real numbers*.

<u>Remark 1.</u> One might question the validity of the latter definition of the length of a segment. Really, in case when the length is being obtained as a result of an infinite procedure, it is not intuitive to see the length as the number that shows exactly how many segments U are needed to fill segment a.

Yet, by performing arithmetic operations as defined for irrational numbers, one can show that the notion of length as it is defined for the incommensurable case satisfies the following properties:

(i) the lengths of congruent segments are equal numbers;

(ii) the length of the sum of two segments equals to the sum of their lengths.

These properties, together with

(iii) a length of any segment is a positive number;

(iv) there is a segment of length 1,

constitute the most general definition of length accepted in advanced courses on foundations of geometry.

Remark 2. We have required a decimal representing a length in an incommensurable case to be not only non-terminating but also non-periodic. It can be shown by algebraic means that a periodic decimal can be transformed into an ordinary fraction and is, therefore, a rational number.

Remark 3. In the course of our procedure, we used the operation of dividing a segment into 10 equal parts. We know how to perform such a construction by using properties of segments intercepted on transversals by a few parallel lines. Therefore, when introducing the notion of length we have implicitly used the parallel postulate.

Let us notice, however, that we can carry out almost an identical procedure that involves the division of segments in halves instead of tens. The division of a segment into two congruent parts does not require the parallel postulate. Such a procedure would result in obtaining the numerical length in its *binary* representation instead of the *decimal* one.

Since such a procedure does not require the Euclidean parallel postulate, one can use it for defining the length in some other (non-Euclidean) geometries. It is important as in mathematics we always strive to develop as general an approach as possible.

Yet in our discussion suggested for inexperienced mathematicians, we have used a more customary decimal representation of numbers.

By providing an algorithm that allows us to evaluate the numerical measure of any segment, we have proved the following result:

Theorem 9.1.4. *To every segment there corresponds a positive real number which represents its length.*

Is the converse of this theorem true? If we introduce one more *continuity axiom*, called the *Cantor's principle of nested segments*, we will be able to prove the converse of this result:

Theorem 9.1.5. *Every real positive number is the length of some segment.*

It is very easy to construct a segment whose length is a given rational number. In this case, the solution is trivial (a good exercise for students). Let us concentrate on the case when an irrational number is given. In this case we shall need the *Cantor's principle of nested segments*.

A set of segments is called a *sequence (an infinite system) of nested segments* if each segment of this set is contained within the preceding segment and includes the next one.

In Figure 9.1.8, segment A_1B_1 contains A_2B_2, which, in its turn, contains $A_3B_3,…, A_nB_n$ contains $A_{n+1}B_{n+1}$ etc. With that, n is not bounded.
(In mathematical notations one would write briefly: $A_{n+1}B_{n+1} \subset A_nB_n$, $n = 1, 2, …$).

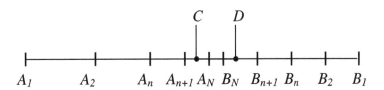

Figure 9.1.8

II CONTINUITY AXIOM (Cantor's Principle of Nested Segments). *For any sequence of nested segments there exists a point common to all of them.*

<u>Corollary</u>. *If in a sequence of nested segments there is a segment lesser than any given segment, then there is <u>one and only one point</u> common to the sequence of nested segments.*

<u>Proof.</u> Suppose, under the hypothesis of the corollary, there are two points, C and D in Fig. 9.1.8, common to all nested segments, and C does not coincide with D. Then C and D are the endpoints of some segment CD. According to the hypothesis, there is a segment in the sequence that is less than CD, and, thus, it cannot contain both C and D. (Such a segment either lies entirely within the interior of CD (like A_NB_N in Fig. 9.1.8) or includes only one of the two points, e.g. C, and is not sufficiently large to reach the other one, D). Thus, at least one of the two points is not common to all nested segments. We have come to the contradiction that proves the validity of the Corollary, <u>Q.E.D.</u>

In some texts the principle itself is stated in the form of our corollary: it asserts both the existence and uniqueness of the common point. In other versions of the corollary it is required that the lengths of the segments tend to zero. Such a formulation would be inappropriate in our case, as we are still struggling to introduce the notion of length. Yet we know how to compare segments, that is why we can talk about arbitrarily small segments.

Now let us prove the theorem (***Th.9.1.5***). Suppose some real positive number r is given in its decimal representation: $r = k_0.k_1k_2...k_n...$ and a certain unit of length U is chosen. Let us construct a segment a whose numerical length in the U-units is represented by r.

On some ray emanating from some point O, we shall lay off, starting from O, the following segments whose measures are the consecutive decimal approximations of the number r. We shall keep our old notation $l_1, l_2, ... l_n$ and $u_1, u_2, ... u_n$ for the upper and lower estimates, respectively.

We shall start with the construction of the segments whose measures are $l_1 = k_0.k_1$ and $u_1 = k_0.(k_1+1)$:

$$OA_1 = k_0 U + k_1 \frac{1}{10} U; \quad OB_1 = k_0 U + (k_1+1)\frac{1}{10} U.$$

As a result of this construction we have also obtained segment $A_1 B_1 = OB_1 - OA_1 = \frac{1}{10} U$. See Figure 9.1.9 for illustration (for the convenience, the scale of the figure is not realistic: the small segments are not much smaller than the big ones).

In the next step we shall construct the segments with the respective lengths $l_2 = k_0.k_1 k_2$ and $l_2 = k_0.k_1(k_2+1)$. These are

$$OA_2 = k_0 U + k_1 \frac{1}{10} U + k_2 \frac{1}{100} U; \quad OB_1 = k_0 U + k_1 \frac{1}{10} U + (k_2+1)\frac{1}{100} U.$$

Thus we have also obtained segment $A_2 B_2 = OB_2 - OA_2 = \frac{1}{100} U$. This segment is contained in $A_1 B_1$ (why?).

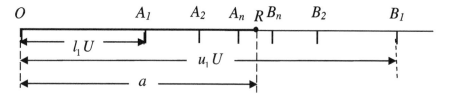

Figure 9.1.9

This process of constructing such segments will be continued indefinitely. As a result, we shall obtain a sequence of nested segments $A_1 B_1, A_2 B_2, \ldots A_n B_n, \ldots$. According to the Cantor's principle, these segments will have a common point. Also, $A_n B_n = OB_n - OA_n = \frac{1}{10^n} U$, and thus, by choosing a sufficiently large n $A_n B_n$ can be made less than any given segment. Therefore, by the corollary following from the principle, the constructed sequence of nested segments will have exactly one common point; let us denote it R. according to our construction, each lower and upper estimates of the length of OR will be the respective decimal estimates of the given number r, hence segment OR will have the length $k_0.k_1 k_2 \ldots k_n \ldots = r$. Thus, a segment of length r exists, <u>Q.E.D.</u>

<u>Remark 1</u>. It is clear from the construction that any other segment of the same length r will be congruent to OR (explain this in detail).

Remark 2. The recent two theorems are extremely important since they establish a one-to-one correspondence between the segments and real numbers and thus allow to introduce *coordinate systems* for lines, planes, and space.

(f) The numerical measure of a segment. The ratio of two segments.

The number obtained as a result of measuring a segment, is called the *numerical measure (length)* of the segment. If the segment is commensurable with the unit of measurement, its numerical measure is rational, otherwise (if the segment is incommensurable with the unit) it is an irrational number that can be represented by an infinite decimal fraction.

By the length of a segment we mean its numerical measure, given a certain unit of measurement. For example, given a unit U and a segment A, one can write $A = aU$, where a is the numerical measure of A.

It is worth noticing that we have introduced the notion of length based on our AXIOMS 1- 4 (axioms of straight lines and segments) and the axioms of continuity.

Now we can rigorously prove all our statements concerning lengths, which we substantiated at intuitive level in Chapter one. For example, we can prove that congruent segments have equal lengths and that the converse is true; the length of the sum of two segments equals the sum of their lengths.

We define the *ratio of two segments as the ratio of their numerical measures (lengths)*.

It is easy to see that *this ratio does not depend upon the choice of the unit* (otherwise the above definition would not make sense). Suppose U and u are two different units, $U = ku$, where k is some number. Then for two segments, A and B,

$A = aU = aku;$ also $A = \alpha u,$ and
$B = bU = bku;$ also $B = \beta u,$

where $a, b,$ and α, β are their measures in U or u - units, respectively.

Then, the ratio of their measures in U - units is $\dfrac{a}{b}$. At the same time,

$\dfrac{a}{b} = \dfrac{ka}{kb} = \dfrac{\alpha}{\beta},$ which is the ratio of their measures in u – units.

The two ratios of numerical measures in different units are equal, and that proves the validity of our definition.

9.1 Questions and Problems.

1. Three segments equal 4cm, 10cm, and 18 cm, respectively, are given.
   a) Is 1 cm their common measure?
   b) Is 0.5 cm their common measure?
   c) Is 0.3 cm their common measure?
   d) What is their greatest common measure (GCM)?

2. Let *U* be a known segment. Three segments equal *6U*, *9U*, and *15U*, respectively, are given.
   a) Is *U* their common measure?
   b) Is *0.25U* their common measure?
   c) Is *0.7U* their common measure?
   d) What is their greatest common measure (GCM)?

3. There are three segments, *A, B,* and *C*, such that *A=3B+C; B=5C*. Find the greatest common measure (GCM) of *A* and *B*.

4. There are three segments, *A, B,* and *C*, such that *A=4B+C; B=1.5C*. Find the greatest common measure (GCM) of *A* and *B* (express it in terms of *C*).

5. Illustrate the two possible outcomes postulated by Archimedes axiom by means of the pairs of segments shown in Figure 1 (a , b).

A	C
B	B
(a)	(b)

Figure 1

6. It is known about segments *A* and *B* that *3B<A<4B; 4B - A= C; C=¾B*. Find the GCM of *A* and *B* (express GCM in terms of *C*).

7. There are four segments, *A, B, R,* and *U*, such that *A=5B+R*, and *GCM(B,R)=U*. Find the greatest common measure (GCM) of *A* and *B*.

8. There are four segments *A, B, C,* and *D*, such that *A=2B+C; B=C+D; D=2/3C;* Find the greatest common measure of these segments and evaluate their lengths using that measure.

9. Use the Euclid's algorithm to find the greatest common measure of *A=27cm* and *B=111cm*.

10. Use the Euclidean algorithm to find the greatest common measure of *A=246U* and *B=16U*, where *U* is some segment.

11. *If you know some computer language, write in this language an algorithm that finds the greatest common measure of two segments , having given their expressions in some common measure (e.g., in *cm*).

12. Is it true or false that any two segments have a common measure? Can the falsity of that statement be established by means of experiments (e.g., measurements done by means of a ruler or more accurate devices)?

13. *A, B,* and *R* are segments. *A=3B+R*, and *2B=5R*. Is *A* commensurable with *R*?

14. *A, B,* and *R* are segments , and *m, n,* and *k* are natural numbers. *A= mB+R*, and *nB = kR*. Are *A* and *B* commensurable? Is *R* commensurable with each of them?

15. Adduce an example of two incommensurable segments. Try to prove their incommensurability by means of the Euclidean algorithm.

16. Prove that in an isosceles triangle with the vertical angle (angle opposed the base) equal 36°, the base is incommensurable with a lateral side.

17. Prove that in an isosceles triangle with the vertical angle (angle opposed the base) equal 108°, the base is incommensurable with a lateral side.

18. Is the length of a segment an undefinable notion in the given system of axioms? List the undefined notions and axioms we need to introduce the notion of length.

19. Prove that the ratio of the numerical lengths of two segments does not depend upon the choice of units.

20. In a given triangle, find all possible ratios of its sides to each other.

21. a, b, and c are the lengths of the sides of a triangle. Can it happen that
    a) $\dfrac{a}{c} + \dfrac{b}{c} < 1$?
    b) $\dfrac{a}{c} + \dfrac{b}{c} > 2$?
    Substantiate your answers.

22. Can a ratio of a leg to the hypotenuse in a right triangle be a) equal 1? b) more than 1? c) less than ½?

23. Segment AB is divided by point C in a ratio of 5:3 starting from A. Evaluate the length of AB if the length of CB is 12 cm.

24. Find the ratio of an altitude of an equilateral triangle to the circumradius.

9.2 Similarity of Triangles.

a) Preliminary remarks.

One can often encounter objects that have different sizes but similar shapes. For example, two different enlargements of the same photograph, two maps of the same country on different scales, a car and a toy model of the car, a cat and a cougar, etc. Such objects are called *similar*. Similarity is denoted by the ~ symbol (in handwriting the ↩ symbol is usually used to denote similarity).

We shall consider *similar plane geometric figures*. The ability to measure segments and consider their ratios allows us to define the notion of *geometric similarity* of figures and describe transformations that change sizes of figures without distorting their shapes. Such transformations we call *similarity transformations (or similarities)* of figures.

We start our discussion with the simplest figures, - triangles.

b) Corresponding sides; definition and existence of similar triangles.

In this section we consider triangles with respectively congruent angles. Such triangles are called mutually equiangular, and their equal angles are called *corresponding angles*. Let us agree to call the sides opposed to the equal angles the *corresponding sides*. Obviously, corresponding sides are also included between the pairs of respectively congruent angles.

<u>Definition</u>. *Two triangles are said to be similar if:*
    (i)   the angles of one triangle are respectively congruent to the angles of the other;
(ii)   the sides of one are proportional to the corresponding sides of the other.

The following lemma proves the existence of similar triangles.
<u>Lemma 9.2.1.</u> ***A line parallel to a side of a triangle, cuts off a triangle similar to the given one.***
Proof. In Figure 9.2.2, *DE // AC*. Prove that $\triangle DBE \sim \triangle ABC$, i.e. show that these triangles are mutually equiangular and their corresponding sides are proportional.
1.   The corresponding angles are congruent, since $\angle B$ is common and $\angle BDE = \angle BAC$, $\angle BED = \angle BAC$ as two pairs of corresponding angles formed by a pair of parallel lines, *AC* and *DE*, and their transversals, *AB* and *BC*, respectively.

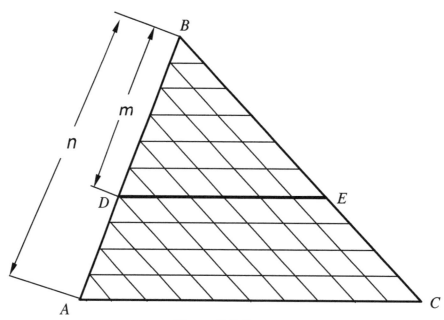

Figure 9.2.2

2.   Let us prove now that the corresponding sides are proportional, i.e.
$$\frac{BD}{BA} = \frac{BE}{BC} = \frac{DE}{AC}.$$
The following two cases are considered separately.
(i)   **AB and DB are commensurable.** Divide *AB* into parts equal to their common measure; then *BD* is also divided into an integer number of such parts. Suppose, *AB* and *DB* contain, respectively, *n* and *m* of these parts (Fig.9.2.2).

From each of the points of the partition draw a line parallel to *AC* and a line parallel to *BC*. The lines parallel to *AC* will divide *BC* and *BE* into equal parts: *BC* will consist of *n* parts, and *BE* will consist of *m* parts (why? – Refer to the theorem). Similarly *AC* and *DE* will consist of *n* and *m* mutually equal segments. Then,

$$\frac{BD}{BA}=\frac{m}{n};\quad \frac{BE}{BC}=\frac{m}{n};\quad \frac{DE}{AC}=\frac{m}{n},\quad \text{whence}\quad \frac{BD}{BA}=\frac{BE}{BC}=\frac{DE}{AC},\quad \text{Q.E.D.}$$

(ii) **AB and DB are incommensurable**. Find approximate estimates of $\frac{BD}{BA}$ and $\frac{BE}{BC}$, firstly within the accuracy of $\frac{1}{10}$, then - within the accuracy of $\frac{1}{100}$, etc. Each subsequent step of our estimation procedure will improve the accuracy tenfold.

At first divide *AB* into 10 equal parts and through the points of partition draw lines parallel to *AC* (Fig.9.2.3). These lines divide *BC* into 10 equal parts.

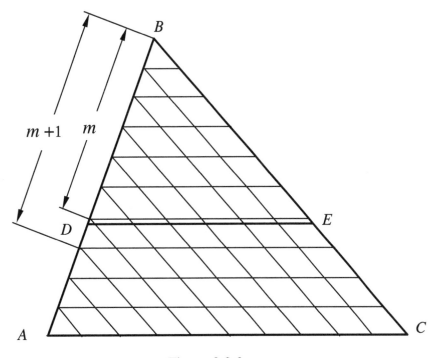

Figure 9.2.3

Suppose that *BD* contains the segment congruent to $\frac{1}{10}AB$ *m* times with some remainder, which is less than $\frac{1}{10}AB$. Then, by the property of parallel lines cutting the arms of an angle (∠*B* in our case), $\frac{1}{10}BC$ is contained in *BE* *m* times with a remainder less than $\frac{1}{10}BC$. Therefore, within the accuracy of $\frac{1}{10}$, $\frac{BD}{BA} = \frac{m}{10}$; $\frac{BE}{BC} = \frac{m}{10}$.

Then divide AB into 100 equal segments and suggest that $\frac{1}{100}AB$ fits into BD m_1 times. Using parallel lines drawn through the points of our new partition and intersecting the arms of $\angle B$, we can show that $\frac{1}{100}BC$ is contained in BE m_1 times. Thus, within the accuracy of $\frac{1}{100}$, $\frac{BD}{BA} = \frac{m_1}{100}$; $\frac{BE}{BC} = \frac{m_1}{100}$, and the two ratios are equal.

Repeating similar steps in such a way that each subsequent step improves the precision of the estimate (of BD) tenfold, we shall find that the corresponding estimates of the ratios $\frac{BD}{BA}$ and $\frac{BE}{BC}$ are equal when calculated with an arbitrary (but the same for both ratios) accuracy. Hence, the exact values of these ratios are represented by the same infinite decimal, i.e. $\frac{BD}{BA} = \frac{BE}{BC}$.

Similarly, by drawing lines parallel to BC through the points of partition, one can establish that $\frac{BD}{BA} = \frac{DE}{AC}$.

Therefore, $\frac{BD}{BA} = \frac{BE}{BC} = \frac{DE}{AC}$, Q.E.D.

Thus, the corresponding sides are proportional, and the angles (established earlier) are equal, therefore $\triangle DBE \sim \triangle ABC$, Q.E.D.

Remark 1. The established proportions of the sides of $\triangle ABC$ and $\triangle DBE$ may be rewritten as $\frac{BD}{BA} = \frac{BE}{BC}$; $\frac{BD}{BA} = \frac{DE}{AC}$; $\frac{BE}{BC} = \frac{DE}{AC}$.

By interchanging the mid-terms of these proportions one would obtain:
$\frac{BD}{BE} = \frac{BA}{BC}$; $\frac{BD}{BE} = \frac{BA}{AC}$; $\frac{BE}{DE} = \frac{BC}{AC}$.

The latter equalities mean that *if the sides of two triangles are proportional, the ratio of any two sides of one triangle equals to the ratio of the corresponding sides of the second triangle.*

9.2 Questions and Problems.

1. Triangles ABC and MNP are similar, and: $\angle A = \angle M$; $\angle B = \angle N$; $AB=3cm$; $AC=5cm$; $BC=6cm$; $\frac{AB}{MN} = \frac{2}{3}$. Evaluate the perimeter of $\triangle MNP$.

2. The sides of a triangle are in a ratio of $5 : 4 : 6$. Evaluate the greatest side of a similar triangle with a) the perimeter of 45cm; b) the sum of the sides that

include the least angle equal 121 cm; c) the difference between the two least sides equal 3cm.

3. The perimeters of two triangles are in a ratio of 3÷4, and in each of them the sides are in a ratio of 5:7:9. Find the sides of the greatest triangle (the one with the greater perimeter), if the sum of the two corresponding least sides equals 8cm.

4. In Figure 1, *DE // AB*. Write at least two ratios equal to a) $\dfrac{AB}{AC}$; b) $\dfrac{DE}{AB}$; c) $\dfrac{AC}{ED}$; d) $\dfrac{BD}{DC}$.

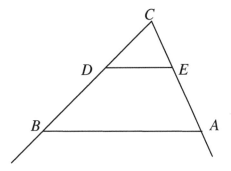

Figure 1

5. The extensions of the lateral sides, *AD* and *BC*, of trapezoid *ABCD* (with *AB>CD*) intersect at point *E*. a) Evaluate the greatest base, *AB*, of the trapezoid if the least base equals 6cm, *AE*=15cm, and *AD*=5cm. b) Evaluate *DE* if *AD*=6cm, and $AB = \dfrac{5}{4} CD$.

6. Evaluate the midline of trapezoid *ABCD (AB//CD)* if its lateral sides intersect at *E*, and *AE*=10cm, *AD*=8cm, and *AB*=20cm. (There are two solutions).

7. In Δ*ABC*, *D* and *E* are such points on the sides *AB* and *BC*, respectively, that *DE //AC*. Evaluate: a) *AC*, if *AD*= 5cm, *DB*= 4cm, and *DE*=6cm; b) *AB*, if *DE*= 4cm, *AD*= 3cm, and *AC*=12cm.

8. In a triangle, an angle of 120° is included between two sides equal, respectively, *a=3cm*, and *b=6cm*. Evaluate the bisector of that angle.

9. Given two rays that emanate from the same point, find the locus of points whose distance from one of the rays is twice the distance from the second ray.

10. In a given triangle, find a point whose distances from the three sides are in a ratio of 1:2:3.

9.3   Three tests for the similarity of triangles.

__Theorem 9.3.1.__ Two triangles are similar if
   (i) two angles of one triangle are respectively equal to two angles of the other triangle; or
   (ii) an angle of one is equal to an angle of the other and the including sides are proportional; or
   (iii) all their sides are respectively proportional.

Proof.

(i) Let $\triangle ABC$ and $\triangle A_1B_1C_1$ be triangles with $\angle A = \angle A_1$, $\angle B = \angle B_1$ (Fig. 9.3.1), and therefore $\angle C = \angle C_1$ (why?). Let us prove that $\triangle ABC \sim \triangle A_1B_1C_1$.

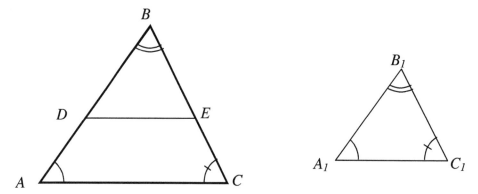

Figure 9.3.1

On BA lay off segment BD congruent to B_1A_1 and draw $DE \parallel AC$. In accordance with Lemma 9.2.1, $\triangle DBE$ is similar to $\triangle ABC$. On the other hand, $\triangle DBE = \triangle A_1B_1C_1$ by **ASA** condition (explain the latter in details). Therefore, $\triangle DBE \sim \triangle ABC$, Q.E.D.

(ii) In Fig. 9.3.2, $\angle B = \angle B_1$ and $\dfrac{AB}{A_1B_1} = \dfrac{BC}{B_1C_1}$. Prove that $\triangle ABC \sim \triangle A_1B_1C_1$.

Again, lay off on BA segment $BD = B_1A_1$ and draw $DE \parallel AC$. Let us prove that $\triangle DBE = \triangle A_1B_1C_1$.

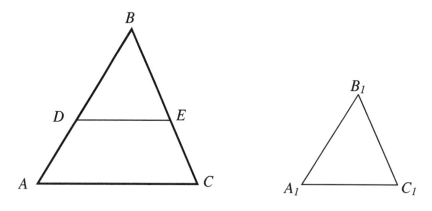

Figure 9.3.2

It follows from the lemma of the previous section that $\triangle ABC \sim \triangle DBE$, therefore

$$\frac{AB}{DB} = \frac{BC}{BE}. \tag{1}$$

By the hypothesis and taking into account $DB = A_1B_1$, one can obtain:

$$\frac{AB}{A_1B_1} = \frac{BC}{B_1C_1} \Rightarrow \frac{AB}{DB} = \frac{BC}{B_1C_1}. \tag{2}$$

It is easy to see from the comparison of (1) and (2) that $BE = B_1C_1$, hence $\triangle A_1B_1C_1 = \triangle DBE$ by *SAS* condition. Therefore, $\triangle A_1B_1C_1 \sim \triangle ABC$, Q.E.D.

(iii) In Fig. 9.3.2,
$$\frac{AB}{A_1B_1} = \frac{BC}{B_1C_1} = \frac{AC}{A_1C_1}. \tag{3}$$

Prove that $\triangle ABC \sim \triangle A_1B_1C_1$.

Let us use again the construction similar to the one used before: lay off $BD = A_1B_1$ and draw $DE \parallel AC$; then prove that $\triangle DBE = \triangle A_1B_1C_1$.

It follows from the similarity of $\triangle ABC$ and $\triangle DBE$ that
$$\frac{AB}{DB} = \frac{BC}{BE} = \frac{AC}{DE}. \tag{4}$$

As $DB = A_1B_1$, the first ratios of (3) and (4) are equal and, hence, so are the others. Particularly, $\frac{BC}{B_1C_1} = \frac{BC}{BE} \Rightarrow B_1C_1 = BE$, and $\frac{AC}{A_1C_1} = \frac{AC}{DE} \Rightarrow A_1C_1 = DE$.

Then $\triangle A_1B_1C_1 = \triangle DBE$ by *SSS* condition, and, therefore, this triangle is similar to $\triangle ABC$, Q.E.D.

9.4. Similarity of right triangles.

As all right angles are congruent, the first two of the following three tests of similarity are the direct corollaries of Theorem 9.3.1 (i), (ii).

Theorem 9.4.1. ***Two right triangles are similar if***
(i)   an acute angle of one is equal to an acute angle of the other, or
(ii)   the legs of one are proportional to the legs of the other, or
(iii)   a leg and the hypotenuse of one are proportional to a leg and the hypotenuse of the other.

Proof. (iii) In Fig. 9.4.1 $\angle B = \angle B_1 = d$, and
$$\frac{AB}{A_1B_1} = \frac{AC}{A_1C_1}. \tag{1}$$

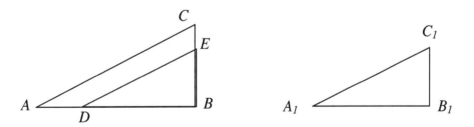

Figure 9.4.1

Lay off $BD=B_1A_1$ on BA and draw $DE//AC$. Let us prove that $\triangle DBE = \triangle A_1B_1C_1$.

It follows from the similarity of $\triangle ABC$ and $\triangle DBE$ that

$$\frac{AB}{DB} = \frac{AC}{DE} . \qquad (2)$$

Comparing this proportion to (1) and keeping in mind that $DB = A_1B_1$, we find that the left-hand sides of (1) and (2) are equal, hence $\dfrac{AC}{DE} = \dfrac{AC}{A_1C_1}$ whence $A_1C_1 = DE$.

Thus, $\triangle DBE$ and $\triangle A_1B_1C_1$ have a leg and the hypotenuse of one respectively equal to the corresponding leg and hypotenuse of the other, i.e. $\triangle DBE = \triangle A_1B_1C_1$; therefore the latter triangle is similar to $\triangle ABC$, Q.E.D.

Theorem 9.4.2. *The altitudes of two similar triangles are proportional to the bases.*

Proof. It easily follows from Theorem 9.4.1 (i) and should be carried out by students.

Question. Is it possible to formulate and prove statements analogous to Th. 9.4.2 concerning the bisectors and the medians in similar triangles?

Formulate such statements and prove them or show that they do not hold.

9.3-4 Questions and Problems.

1. Formulate and prove three tests of similarity for isosceles triangles.

2. In Figure 1, $AB=CD$ and $AC=BD$. Points $E, A,$ and C are collinear as well as $E, F,$ and D. List all similar triangles shown in the figure.

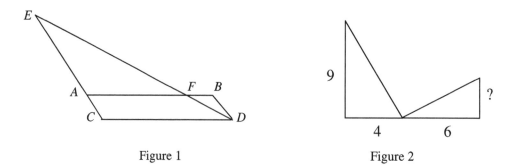

Figure 1   Figure 2

3. In Figure 2, two legs of two right triangles extend each other into a straight line, and the hypotenuses are perpendicular to each other. Find the unknown leg.

4. Prove the *transitivity* of similarity of triangles: if each of two triangles is similar to a third, then they are similar to each other.

5. In △ABC in Figure 3, ∠ABC=∠ACE. Evaluate AE, having given AB=18cm; AC=8cm.

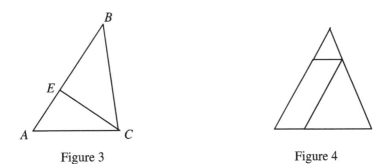

Figure 3   Figure 4

6. In Figure 4, a parallelogram is inscribed in a triangle in such a way that they have a common acute angle. The sides of the triangle that include that angle are equal, respectively, a=15cm, and b=12cm (b is horizontal in the figure). Evaluate the sides of the parallelogram if their ratio is 4÷1.

7. Prove that in similar triangles the bisectors of corresponding angles are in proportion with the corresponding sides.

8. Prove that in similar triangles the altitudes to corresponding sides are in proportion with the corresponding sides.

9. Find the height of a tree whose shadow is 11ft long at the moment when the shadow of a 6ft high by-passer is 4.5 ft long.

10. A method of determining the height of a high object is illustrated by Figure 5. Describe and substantiate the method.

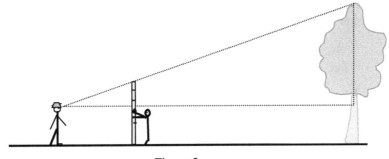

Figure 5

11. The diameter of the Sun is approximately $1.4 \cdot 10^6$ km, and the diameter of the Moon is approximately $3.5 \cdot 10^3$ km. The visual diameters of the both bodies are equal, i.e. they are seen from the Earth under the same angle. Estimate the average distance from the Earth to the Sun if the average distance from the Earth to the Moon is approximately $380 \cdot 10^3$ km.

12. In order to find the distance to an inaccessible point, N (Figure 6), set two stakes at convenient points, A and B, such that AM is perpendicular to MN, and AB is perpendicular to AM. Now, from stake B sight the point N and take note of the point C where the line of sight crosses AM. Show how to find MN by measuring distances between accessible points. Is it necessary to keep angles A and M right?

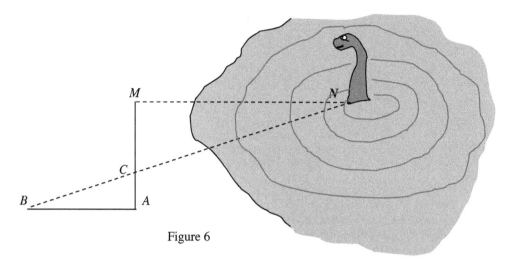

Figure 6

13. Prove that two isosceles triangles are similar if their altitudes to the bases are in proportion with the bases.

14. Prove that the triangle formed by the midlines of a given triangle is similar to the given triangle.

15. For each of the following pairs of triangles given by their sides determine whether the triangles are similar:
    a) 3cm; 4cm; 6cm and 9cm; 12cm; 18cm;
    b) 2cm; 5cm; 6cm and 8cm; 20cm; 22cm;
    c) 4m; 8m; 10m and 10m; 25m; 20m.

16. A circle is inscribed in an isosceles triangle with the base b and lateral side a. Find the distance between the points where the circle touches the lateral sides.

17. Prove that in similar triangles the corresponding bisectors (the bisectors of equal angles) are in proportion with the sides.

18. Prove that in similar triangles the corresponding medians (the medians drawn to the corresponding sides) are in proportion with the sides.

19. Prove that in similar triangles the corresponding altitudes (the altitudes dropped onto the corresponding bases) are in proportion with the sides.

20. In each of two given triangles a bisector is drawn. Each bisector splits the respective triangle into a pair of triangles such that one of the triangles of each pair is similar to a triangle of the second pair. Prove that the given triangles are similar.

21. In each of two given triangles a median is drawn. Each median splits the respective triangle into a pair of triangles such that one of the triangles of each pair is similar to a triangle of the second pair. Are the given triangles similar?

22. In the hypothesis of the previous problem replace the word "median" with the word "altitude". Is the conclusion of the obtained proposition true or false? – Prove or disprove it.

23. Given a triangle with sides a, b, and c, construct a triangle similar to the given one and such that: a) its sides are half the sides of the given triangle; b) its least side equals to a given segment; c) its perimeter is a given segment.

24. Construct a triangle, having given the ratio of two sides (2:3), the least of these sides, and their included angle.

25. Construct a triangle, having given the ratio of two sides (2:3), their included angle, and the third side.

26. Construct a right triangle, having given its hypotenuse and the ratio of the legs (4:5).

27. Construct a rhombus, having given its side and the ratio of the diagonals (2:3).

28. Construct a parallelogram, having given its greatest diagonal, an acute angle, and the ratio of the sides (3:4).

29. In a right triangle the hypotenuse equals $c=12cm$. Evaluate the leg whose projection onto the hypotenuse equals $a'=3cm$.

30. Two chords, *AB* and *MN* intersect at *P*. The endpoints of the chords are consecutively connected by straight segments. List all pairs of similar triangles in the obtained figure.

31. In Figure 7, diameters *AB* and *CD* are perpendicular to each other. Show that triangles *ABF* and *OAE* are similar.

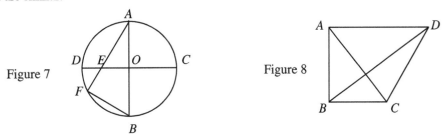

Figure 7

Figure 8

32. In Figure 8, the diagonals, *AC* and *BD*, of a right trapezoid, *ABCD* with ∠*A* right, are perpendicular to each other. Having given the greater base $a=8cm$ and the altitude $h=6cm$, find the least base *BC* and diagonal *BD* of the trapezoid.

33. In △*ABC*, *AE* and *BD* are the altitudes. Prove that points *C*, *D*, and *E* are the vertices of a triangle similar to *ABC*. Consider the following cases: a) the triangle is acute; b) the triangle is obtuse, and ∠*A* is acute.

34. In Figure 9, *PC* and *PB* are, respectively, a secant and a tangent emanating from point *P*. Prove that triangles *BCP* and *ABP* are similar.

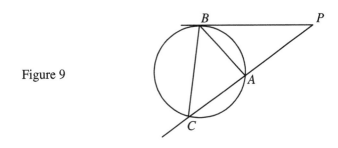

Figure 9

35. The altitudes of a parallelogram are in a ratio of $h:H = 2:3$, and its perimeter equals $P = 60cm$. Evaluate the sides of the parallelogram. (Try to obtain the formulae first, and then substitute the numbers).

36. The bases of a trapezoid are equal, respectively, *a* and *b*. A lateral side of the trapezoid is divided by a point in a ratio of *m:n*, and a line parallel to the bases is drawn through that point. Find a formula that represents the segment of this line that lies between the lateral sides.

37. The bases of a trapezoid are equal, respectively, *a =18cm* and *b=30cm*. A line parallel to the bases is drawn through the point of intersection of the diagonals. Evaluate the segment of this line that lies within the trapezoid.

38. Two circles touch externally, and a line drawn through their point of contact forms chords whose lengths are in a ratio *m:n* = 7: 4. Find the radii of the circles if the distance between their centres is 121mm.

39. Two circles touch internally, and a line drawn through their point of contact forms chords whose lengths are in a ratio *m:n* = 7: 4. Find the radii of the circles if the distance between their centres is 84mm.

40. A circle is inscribed in an isosceles triangle with the base *a =6cm* and a lateral side equal *b=10cm*. Find the distance between the points of contact on the lateral sides.

41. In Figure 10, $\triangle ABC$ is inscribed in a circle of radius *R*. *AK* is a diameter, and *AD* is an altitude of the triangle. Determine *CK*, having given that *AB=0.6R*, and *BD=1.5m*.

Figure 10

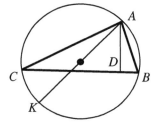

42. Each of two distinct circles touches two diagonals and a base of a trapezoid. Determine the radius of the greater circle if the radius of the least one is 2m, and the bases are equal 10m and 15m.

43. Given an acute-angled triangle, inscribe in it a square in such a way that each vertex of the square is situated on a side of the triangle. How many solutions does the problem have?

9.5 Similarity of polygons.

Two polygons are said to be *similar* if they have the same number of vertices, if their angles, taken in the same order, are respectively equal (i.e. these polygons are *mutually equiangular*), and if the respective pairs of sides adjacent to these equal angles are proportional.

(Obviously this definition is a generalization of the definition of similar triangles; it could be "invented" by students).

The equal angles are called the *corresponding angles*, the vertices at which angles are equal are called the *corresponding vertices* and the sides joining corresponding vertices are called the *corresponding sides*.

Two pentagons, *ABCD* and *A´B´C´D´* in Fig. 9.5.1, are similar if $\angle A = \angle A´, \angle B = \angle B´, \angle C = \angle C´, \angle D = \angle D´, \angle E = \angle E´$, and, in addition to these conditions, $\dfrac{AB}{A'B'} = \dfrac{BC}{B'C'} = \dfrac{CD}{C'D'} = \dfrac{DE}{D'E'} = \dfrac{EA}{E'A'}$.

The existence of similar polygons follows from the following problem.

<u>Problem 9.5.1.</u> *Given a polygon and a segment, construct a polygon similar to the given one, with the side corresponding to a given side of the given polygon equal to the given segment.*

<u>Solution.</u> Given a polygon *ABCDE*, construct a polygon $A_1B_1C_1D_1E_1$ similar to *ABCDE* with one side A_1B_1 congruent to a given segment *a*. In Fig. 9.5.1, $a < AB$; in case $a > AB$ the solution will be similar to the following below.

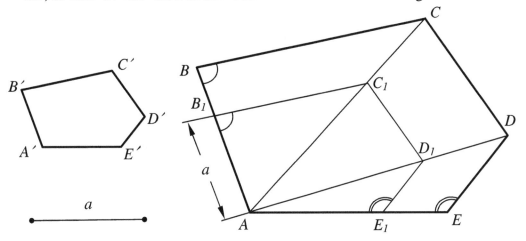

Figure 9.5.1

On *AB* lay off $AB_1 = a$. Draw all diagonals emanating from *A* (in our figure these will be *AC* and *AD*). Draw $B_1C_1 \parallel BC$, $C_1D_1 \parallel CD$, and $D_1E_1 \parallel DE$.

The constructed polygon $AB_1C_1D_1E_1$ is similar to *ABCDE*. Really, $\angle A$ is their common angle, $\angle B = \angle B_1$ and $\angle E = \angle E_1$ as corresponding angles formed by pairs of parallel lines; $\angle C = \angle C_1$ and $\angle D = \angle D_1$ as angles consisting of equal parts (why are these parts equal, e.g. why $\angle AC_1D_1 = \angle ACD$?). Thus, the corresponding angles of the polygons are congruent.

Also, we can show that the sides are proportional. The latter follows from the similarity of the constructed triangles:

$$\triangle AB_1C_1 \sim \triangle ABC \implies \frac{AB_1}{AB} = \frac{B_1C_1}{BC} = \frac{AC_1}{AC};$$

$$\triangle AD_1C_1 \sim \triangle ADC \implies \frac{AC_1}{AC} = \frac{C_1D_1}{CD} = \frac{AD_1}{AD};$$

$$\triangle AE_1C_1 \sim \triangle AEC \implies \frac{AD_1}{AD} = \frac{D_1E_1}{DE} = \frac{AE_1}{AE}.$$

Ratio $\dfrac{AC_1}{AC}$ is engaged in the first and second series of equalities, ratio $\dfrac{AD_1}{AD}$ connects the second and third ones; thus all 9 ratios are equal, therefore

$\dfrac{AB_1}{AB} = \dfrac{B_1C_1}{BC} = \dfrac{C_1D_1}{CD} = \dfrac{D_1E_1}{DE} = \dfrac{AE_1}{AE}$, and hence, taking into account the congruence of the corresponding angles, $AB_1C_1D_1E_1 \sim ABCDE$, Q.E.D.

Remark. As we have seen in triangles, the congruence of their angles implicates the proportionality of their sides and vice versa. It is not necessarily so in general polygons. Students should adduce counterexamples, e.g., (i) a square and a rectangle with non-congruent sides are not similar, - their sides are not proportional, (ii) a square is not similar to a rhombus with non-right angles, - they have proportional sides and non-congruent angles.

Theorem 9.5.1. Each of similar polygons can be decomposed into the same number of similar and similarly situated triangles.

Proof. *I method:* Draw all diagonals from two corresponding vertices of the similar polygons (as we did in problem 9.5.1 and in Fig. 9.5.1). The obtained triangles are respectively similar (prove!).

II method: Pick an arbitrary point, O, in the interior of one of the polygons, $ABCDE$ in Fig. 9.5.2, and join it with all vertices by straight segments. Then, the polygon is decomposed into a few triangles; the number of the triangles is equal to the number of sides of the polygon.

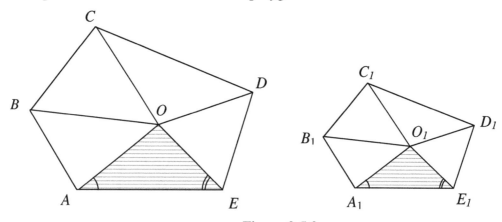

Figure 9.5.2

Pick one of these triangles, e.g. $\triangle AOE$, and in the second polygon on its side A_1E_1 corresponding to AE, build a triangle $\triangle A_1O_1E_1$, similar to $\triangle AOE$. One can do it, for instance, by drawing two rays: a ray emanating from A_1, forming an angle equal $\angle OAE$ with A_1E_1, and the second ray emanating from E_1 that forms an angle equal to $\angle OEA$ with E_1A_1. These rays intersect at some point O_1. Join O_1 with all the other vertices of the second polygon by straight segments. Let us show that all the triangles formed as a result of that construction are similar to the corresponding triangles of the first polygon.

According to our construction, $\triangle A_1O_1E_1 \sim \triangle AOE$. Hence,
$$\dfrac{AO_1}{AO} = \dfrac{AE_1}{AE}. \tag{1}$$

Since the polygons are similar,

$$\frac{A_1E_1}{AB} = \frac{A_1B_1}{AB}. \qquad (2)$$

Comparing (1) and (2), we conclude that

$$\frac{AO_1}{AO} = \frac{A_1B_1}{AB}. \qquad (3)$$

In addition, $\angle BAE = \angle B_1A_1E_1$ and $\angle OAE = \angle O_1A_1E_1$ whence $\angle BOA = \angle BAE - \angle OAE = \angle B_1A_1E_1 - \angle O_1A_1E_1 = \angle B_1A_1O_1$.

Thus, in $\triangle BOA$ and $\triangle B_1O_1A_1$, their proportional sides (see (3)) include congruent angles ($\angle BOA$ and $\angle B_1O_1A_1$), hence $\triangle B_1O_1A_1 \sim \triangle BOA$, Q.E.D.

The other triangles are considered in the same manner. Respectively similar triangles (e.g., $\triangle BOA$ and $\triangle B_1O_1A_1$, are constructed on the corresponding sides (like AB and A_1B_1); in that sense they are *similarly situated*.

The ratio of two corresponding sides of similar polygons is called the *similarity ratio*. (In some texts it is called the *magnification ratio*).

Theorem 9.5.2. *The perimeters of similar polygons are in the same proportion as their corresponding sides.*

Proof. Let a polygon with the sides $a_1, a_2, \ldots, a_n$ be similar to a polygon with the corresponding sides $b_1, b_2, \ldots, b_n$ with a similarity ratio of r:

$$\frac{a_1}{b_1} = \frac{a_2}{b_2} = \ldots = \frac{a_n}{b_n} = r.$$

Then $a_1 = b_1 r$, $a_2 = b_2 r, \ldots, a_n = b_n r$, and the perimeters are in a ratio of

$$\frac{a_1 + a_2 + \ldots + a_n}{b_1 + b_2 + \ldots + b_n} = \frac{b_1 r + b_2 r + \ldots + b_n r}{b_1 + b_2 + \ldots + b_n} = \frac{(b_1 + b_2 + \ldots + b_n)r}{b_1 + b_2 + \ldots + b_n} = r, \text{ Q.E.D.}$$

The construction of the polygon similar to a given polygon with a given similarity ratio is called a *similarity transformation* (or a *similarity*, or a *similitude*) of the given polygon.

The construction procedure discussed in problem 9.5.1 is a particular case of such a transformation. In general, a similarity transformation is carried out as described in the following problem.

Problem 9.5.2. *Apply a similarity transformation with a given similarity ratio, k, to a given polygon.*

Solution. Let us solve this problem for a quadrilateral, e.g. *ABCD* in Fig. 9.5.3. (The considered method is applicable for any other polygon).

Pick a point, O, in the plane, and draw rays from this point through the vertices of the polygon. From O on each of these rays lay off a segment that is a k-multiple of the segment joining O with the vertex of the polygon that lies on the same ray. Thus we obtain points A_1, B_1, C_1, and D_1 such that $OA_1 = k \cdot OA$, $OB_1 = k \cdot OB$, $OC_1 = k \cdot OC$, and $OD_1 = k \cdot OD$.

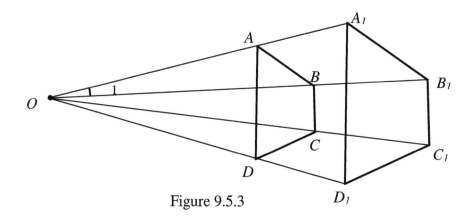

Figure 9.5.3

Join these points, A_1, B_1, C_1, and D_1 consecutively. In $\triangle AOB$ and $\triangle A_1OB_1$, $\angle 1$ is common and, by our construction, $\dfrac{OA_1}{OA} = k = \dfrac{OB_1}{OB}$.

Therefore, these triangles are similar, and hence, $\dfrac{A_1B_1}{AB} = k$. In addition, $\angle OAB = \angle OA_1B_1$, $\Rightarrow$ $A_1B_1 \parallel AB$.

Analogous consideration can be applied to $\triangle OBC$ and $\triangle OB_1C_1$, $\triangle OCD$ and $\triangle OC_1D_1$, $\triangle OAD$ and $\triangle OA_1D_1$, and as a result we obtain:
$\dfrac{B_1C_1}{BC} = \dfrac{C_1D_1}{CD} = \dfrac{A_1D_1}{AD} = k$, and
$\angle OBC = \angle OB_1C_1$, $\Rightarrow$ $B_1C_1 \parallel BC$;
$\angle OCD = \angle OC_1D_1$, $\Rightarrow$ $C_1D_1 \parallel CD$;
$\angle OAD = \angle OA_1D_1$, $\Rightarrow$ $A_1D_1 \parallel AD$.

Then, the (interior) angles of $A_1B_1C_1D_1$ are congruent to the corresponding angles of $ABCD$ (as angles with respectively parallel arms), and the sides of the two polygons are proportional with the coefficient k (by construction), therefore $A_1B_1C_1D_1$ is the result of the required similarity transformation.

Point O used in the above construction is called the *centre of similitude* (or the *centre of homothety*) of the considered polygons. To be more precise one would say that *if two similar polygons are placed with their corresponding sides respectively parallel, the point of intersection of the lines joining their corresponding vertices is called the centre of similitude of the two polygons*. Of course, to make this definition valid one has to prove that *if two similar polygons are placed with their corresponding sides parallel to each other, then the lines joining the vertices of one to the corresponding vertices of the other are concurrent.* (A good exercise for students).

Two polygons placed the way described above are said to be *similarly* (or *perspectively*) *situated*. It can be proved (a good exercise!) that any two similar polygons can be placed that way. A particular kind of a similarity transformation described in Problem 9.5.2 is called a *homothety*.

9.5 Questions and Problems.

1. Formulate and prove the necessary and sufficient conditions of the similarity for two
   a) rectangles; b) rhombi; c) parallelograms; d) isosceles trapezoids; e) right trapezoids.

2. Are quadrilaterals $ABCD$ and $A'B'C'D'$ similar if $\dfrac{AB}{A'B'}=\dfrac{BC}{B'C'}=\dfrac{CD}{C'D'}=\dfrac{DA}{D'A'}$?

3. Are quadrilaterals $ABCD$ and $A'B'C'D'$ similar if $\angle A=\angle A'$; $\angle B=\angle B'$; $\angle C=\angle C'$?

4. Adduce an example of equiangular but not similar pentagons.

5. Prove that two regular pentagons are similar.

6. Prove that two regular hexagons are similar.

7. Prove that two regular polygons with the same number of sides are similar.

8. Construct a quadrilateral similar to the one shown in a) Figure 1a; b) Figure 1b.

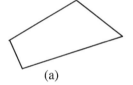
(a)
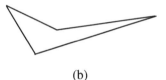
(b)

Figure 1

9. Given a trapezoid, construct a similar one with the greatest base equal half the greatest base of the given trapezoid.

10. Given a quadrilateral, construct a similar one with the perimeter thrice as big.

11. Determine the sides of a quadrilateral if they are in a ratio of $1:\dfrac{1}{4}:\dfrac{1}{2}:\dfrac{3}{4}$, and the perimeter of the quadrilateral equals 60dm.

12. A quadrilateral with sides equal 20mm, 16mm, 27mm, and 38mm is transformed into a similar one in which the sum of the least and the greatest sides equals 432mm. Find the magnification (similarity) ratio of the transformation.

13. A similarity transformation with the magnification ratio equal $\dfrac{3}{4}$ is applied to a rhombus with the perimeter equal 28m. Find the sides of the obtained rhombus.

14. Pentagon $ABCDE$ is transformed into a similar pentagon $A'B'C'D'E'$ by means of a homothety with the centre of similitude located at O (Figure 2). Find the perimeter of $A'B'C'D'E'$ if the perimeter of $ABCDE$ is 78mm, and $A'A=\dfrac{1}{3}OA'$.

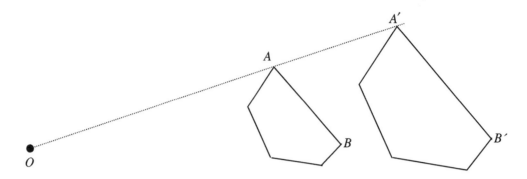

Figure 2

15. A centre of similitude (or: of *homothety*) is called internal if it lies between the corresponding points, and external otherwise (e.g., point O in Figure 2 is an example of an external centre of similitude). Two polygons are obtained as homothetic images of a given polygon with a given magnification ratio (k) and a given centre of similitude (O): for the first polygon O is an external, and for the second one – internal centre of similitude. Show that the two polygons (take quadrilaterals) are congruent. How one can transform one into the other?

16. Quadrilateral *ABCD* is transformed into a similar quadrilateral *A´B´C´D´* by means of a homothety with the centre of similitude located at some point O. A and A´, B and B´, etc. are the pairs of the corresponding vertices. AB=a=6cm, and AA´=3OA. Evaluate A´B´ (find all possible solutions).

17. Are two similar polygons necessarily similarly situated?

18. Consider two similar and not similarly situated polygons (e.g., triangles). Propose a sequence of transformations that will make them similarly situated. What kind of transformations are those?

19. Given two similar polygons that can be moved by means of isometries around the plane, how many centres of similitude they may have? What is common for all the homotheties that transform the first polygon into the second with different centres of similitude?

20. * Prove that if two similar and noncongruent polygons are similarly situated (placed with their corresponding sides parallel), then the lines joining the vertices of one to the corresponding vertices of the other are concurrent.

21. Prove that an isometry is a similarity transformation (i.e.: congruent polygons are similar).

22. * Prove that a composition of homotheties is a homothety.

23.* Prove that a composition of a homothety and an isometry is a similarity transformation.

24. Based on the results of the three preceding problems, give a definition of a general similarity transformation
    for polygons.

25. Prove that the corresponding sides of two similar polygons are in the same ratio as the corresponding diagonals (the diagonals that join pairs of corresponding vertices).

26. Propose a definition of what is to be called the corresponding points of similar polygons. (For instance, the vertices included between the corresponding sides of similar polygons are called the corresponding vertices).

27. Two segments lying in two similar polygons are said to be the corresponding segments if the endpoints of one are the corresponding points for the endpoints of the other. Prove that such segments are in the same ratio as the corresponding sides of the polygons.

9.6 Similarity of General Plane Figures.

a) <u>Construction of a figure similar to a given one.</u>

The idea of *similarity transformation* enables us to generalize the notion of similarity for non-rectilinear figures. We shall apply the similarity transformation, as it was described in the preceding section, to an arbitrary plane figure and, thus, we shall be able to obtain a figure similar to the given one.

Suppose some arbitrary figure, **A**, is given in the plane. Let us pick in the plane an arbitrary point, O, and draw rays emanating from O and passing through points M, N, P, ... of figure **A** (Figure 9.6.1). On these rays, OM, ON, OP, ... lay off segments OM_1, ON_1, OP_1, ... such that $\frac{OM_1}{OM} = \frac{ON_1}{ON} = \frac{OP_1}{OP} = \ldots = k$, where k is some (real and positive) number.

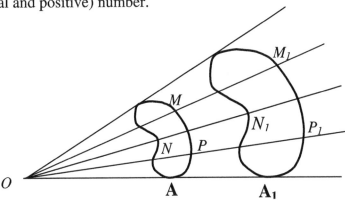

Figure 9.6.1

The more points M, N, P, ... of figure **A** we use, the more points M_1, N_1, P_1, ... of figure $\mathbf{A_1}$ will be obtained.

To construct the whole figure $\mathbf{A_1}$, one has to draw the corresponding rays through **all** the points of **A** and construct on these rays the corresponding points of $\mathbf{A_1}$. Then $\mathbf{A_1}$ is said to be *similar* to the original figure **A**.

Thus, a figure $\mathbf{A_1}$ *similar to a given one* **A**, is introduced (defined) as an image of the original figure under a *similarity transformation* performed as described above. That particular kind of similarity transformation, described in the beginning of the section and illustrated in Fig. 9.6.1, is called a *homothety* (also *dilation or dilatation*). We have already used it for polygons.

In general, a similarity transformation is a *composition* (i.e. a sequential combination) *of homotheties and isometries*. To support the latter statement the following propositions are to be proved:

Prop. 9.6.1. *An isometric image of a figure is a figure similar to the original one,*

or briefly: *congruent figures are similar (with the similarity ratio k = 1).*

Prop. 9.6.2. *A composition of an isometry and a homothety is a similarity transformation.*

Prop. 9.6.3. *A composition of two homotheties is a similarity transformation.*

As we have already mentioned, *two figures are called similar if one of them is an image of the other as a result of a similarity transformation.*

The proofs of Prop. 9.6.1 and 9.6.2 are trivial as any figure is similar to itself with $k = 1$. The proof of Prop. 9.6.3 is more difficult and may be suggested as a project for stronger students.

One can check that similarity is an equivalence relation.

In some particular (but still interesting) cases it is not necessary to draw rays through all points of figure **A** in order to construct a similar figure A_1: one can construct just a few points and then use particular properties of the figure to restore all other points. For instance, in order to construct a polygon similar to a given one, it would be enough to construct its vertices (and then to connect them by straight segments).

For curvilinear figures the situation is more complicated with the exception of circles, since circles possess a remarkable symmetry.

b)   Similarity of circles.

Theorem 9.6.4. *(i) A figure similar to a circle is a circle and*
(ii) any two circles are similar,
   i.e. *all circles and only them are similar to each other.*

Proof. (i) Given a circle **O** of radius R with the centre at some point O, one can easily construct a concentric circle, O_1 of radius $r = kR$ where k is a given ratio of similarity.

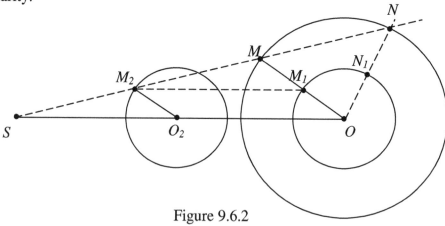

Figure 9.6.2

Now in Figure 9.6.2, $OM = R$; $OM_1 = r = kR$ (in our figure $k < 1$).

For any other pair of points, N and N_1, lying on the same ray emanating from the centre O and the two circles respectively, $\dfrac{ON_1}{ON} = \dfrac{r}{R} = k$.

Therefore, the interior circle (if $k > 1$ it would be exterior) is a similarity (homothety) image of the given circle, with the similarity ratio k and the similitude centre at O.

Let us show that if any other point, S, is chosen as a centre of similitude, the similarity image of **O** will be a circle congruent to $\mathbf{O_1}$.

Join S with O and M by straight segments and translate OM_1 parallel to itself by moving point O along OS until point M reaches SM. Now OM_1 has taken position $O_2 M_2$.

According to our construction, $O_2 M_2 \parallel OM$; hence $\dfrac{SM_2}{SM} = \dfrac{O_2 M_2}{OM} = k$, i.e. M_2 is the similarity image of M with point S as the centre of similitude.

For any other point, e.g. point N, of the original circle, the above procedure (that has been applied to point M) creates its similarity image, N_2, lying at the distance $r = kR$ from some point O_3 on SO. *It does not follow directly from the construction that O_3 coincides with O_2!* However, it follows from the similarity of $\triangle SN_2 O_3$ and $\triangle SNO$ that $\dfrac{SO_3}{SO} = k; \Rightarrow SO_3 = kSO;$ also (according to our construction) $SO_2 = kSO$, and therefore, $SO_3 = SO_2$, i.e. SO_3 and SO_2 do coincide. Thus, the similarity images of the points of the circle **O** will fill the circle of radius $r = kR$ with the centre at O_2.

Therefore, with any choice of the centre of similitude, the image of a circle is a circle of radius kR, where k is the similarity ratio, Q.E.D.

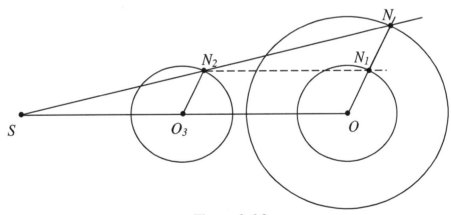

Figure 9.6.3

(ii) Now let us prove that any two circles are similar. Given two circles, of radii R and r respectively, apply an isometry to one of them so as to make the circles concentric. Draw all possible rays emanating from the common centre; two points of the two circles that lie on the same ray are similarity (homothetic) images of one another. For instance, a point (N_1) lying on the circle with radius r is a homothetic image of the corresponding point (N) lying on the other circle if the common centre of the circles is the centre of similitude and the similarity ratio is $k = \dfrac{r}{R}$.

Thus, any circle can be transformed into any other circle by means of a composition of an isometry and a homothety, i.e. by a similarity transformation, Q.E.D.

9.6 Similarity of general plane figures.

1. Draw a simple non-rectilinear figure and construct approximately a similar figure with the magnification ratio $k=2$. Upon which parameters of your procedure does the accuracy of the approximation depend?

2. Extend the definition of a similarity transformation of polygons given in #24 of the preceding section for general (non-rectilinear) plane figures. Then propose a definition of similar plane figures.

3. Define the corresponding points and corresponding segments of similar figures.

Characteristic size of a figure.

A segment that joins two points lying on the boundary of a figure may be used as a *characteristic linear size of the figure*. For example, a diagonal of a polygon or a chord in a circle may serve as a characteristic linear size. Having given two similar figures, one can find the ratio of their linear sizes (dimensions) as a ratio of any two corresponding segments (their lengths in the same units).

Yet, as one can see from Figure 1, not all diagonals and not all chords are equally good representatives to characterize the linear sizes of the corresponding figures. In each of the considered cases the dotted segment is much shorter than the continuous one (they are incommensurate).

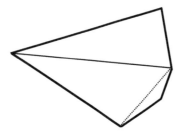
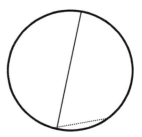

Figure 1

That is why the notion of the *diameter of a figure* is introduced. The *diameter of a (plane) figure* is the greatest segment joining two points of the figure (it is obvious that these points would be located on the boundary of the figure). The diameter is usually chosen as a characteristic linear dimension of a figure. Then the ratio of the diameters characterizes the comparative sizes (dimensions) of two **similar** figures.

4. Use the maps of continents to estimate approximately the diameters of Chile, Algeria, and Mongolia. Does the country with the greatest diameter also have the longest boundary? – the greatest territory? Why the diameter could not be used as a characteristic to compare the three countries?

5. Adduce examples of figures that can and cannot be compared based on their diameters from everyday life, sciences, etc.

6. Is there a plane figure for which the diameter would be a complete characteristic of its size?

7. Prove that the locus of points that divide all segments joining a given point with the points of a given circle in a given ratio is a circle. (In other words: a curve perspectively similar to a circle is a circle).

8. * Two circles in a plane can always be considered as perspectively similar figures with two centres of similitude: external and internal. (It follows from this statement that any two circles are similar to each other).

Remark: It follows from the results of # 7 & 8 that all circles and only them are similar to each other (Theorem 9.6.4).

9. Prove that if each of two circles lies in the exterior of the other, their external centre of similitude coincides with the point of intersection of their common exterior tangents.

10. Prove that if each of two circles lies in the exterior of the other, their internal centre of similitude coincides with the point of intersection of their common interior tangents.

11. Describe the respective location of two circles for which the external and internal centres of similitude coincide. Substantiate your answer.

12. * Pantograph is a devise designed for drawing similar figures. The simplest pantograph is depicted in Figure 2. The rods are hinged at each joint, denoted by a letter, so that they can turn easily, and the lengths are arranged so that CT=CV, AP=AV, BC=PA, and CA=BP. The endpoint V of the rod CV is fixed, i.e. it cannot move from V, though the rod can rotate about the point. A tracing point is placed at T, and a pencil at P. Then, if T traces the outlines of a picture, P draws it on a reduced scale. If the picture is to be enlarged, the tracing point is placed at P, and the pencil – at T. Explain, why would the drawn picture be similar to the traced one. How one would change the magnification ratio?

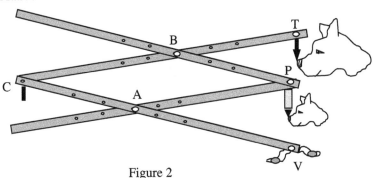

Figure 2

9.7. Similarity method in construction problems.

The essence of the method is in the construction of a figure similar to the required one, followed by a similarity transformation delivering the sought for figure.

The method is particularly convenient when only one of the given data is length and all the others are angles or ratios. For example: *construct a triangle given its angle, side and the ratio of the other two sides*, or: *construct a triangle given its two angles and some segment (e.g., a bisector or a median, or an altitude)*, or: *construct a square given the sum of (or the difference between) its diagonal and its side*, etc.

__Problem 9.7.1.__ *Construct the triangle given its angle C, the ratio of the sides including this angle, and the altitude, h, dropped from vertex C onto the opposite side.*

__Solution.__ Suppose $\frac{AC}{BC} = \frac{m}{n}$, where m and n are two given numbers. Construct $\angle C$ (Fig. 3.7.1.) and from its vertex lay off on its sides segments CA_1, CB_1, proportional to m and n respectively. If m and n are segments, then choose $CA_1 = m$, $CB_1 = n$. If m and n are numbers, then choose an arbitrary segment l and construct $CA_1 = m\,l$, and $CB_1 = n\,l$.

In any of these cases, $\frac{CA_1}{CB_1} = \frac{m}{n}$, and, obviously, ΔCA_1B_1 is similar to the sought for triangle.

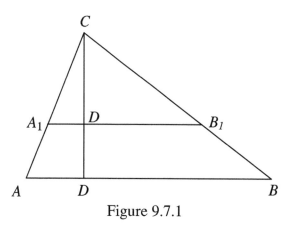

Figure 9.7.1

To construct the required triangle, draw the altitude, CD_1, of ΔCA_1B_1 and denote it h_1. Then choose a centre of similitude and construct a triangle similar to ΔA_1B_1C with the similarity ratio equal $\frac{h_1}{h}$, where h is the given altitude of the sought for triangle.

It is convenient to choose C as the centre of similitude. In that case, in order to construct ΔABC, just lay off $CD = h$ on the ray CD_1 and draw through D a line parallel to A_1B_1. This line cuts arms of $\angle C$ at some points A and B. ΔABC is the required one (prove that!).

In some problems of that kind the location of the required figure is not fixed; in the others it is required to construct a figure situated in a certain way with respect to given points and lines. In the latter case it often happens that if one ignores one of the given conditions (keeping the others), infinitely many figures similar to the required one are obtained. Then the ignored condition is to be imposed, and that will single out the solution(s). In the described situation the similarity method may turn out to be useful.

Let us consider a few examples.

<u>Problem 9.7.2.</u> *Inscribe a circle in a given angle so as to make the circle pass through a given point.*

<u>Solution.</u> Suppose ∠*ABC* in Fig. 9.7.2 is the given angle, and *M* is the given point. Temporarily ignore the requirement saying that the circle passes through *M*. Then, infinitely many circles with their centres on the bisector of ∠*ABC* will satisfy the conditions.

Choose one of these circles, with its centre at some point, *o*. Ray *BM*, that emanates from *B* and passes through *M* cuts that circle at two points, *m* and m_1. Draw radius *mo* and then through *M* draw a line parallel to *mo*. It will cut the bisector at some point *O*. Let us show that *O* is the centre of a circle that satisfies the requirements of the problem.

Drop perpendiculars *on* and *ON* onto *AB*. △*MBO* ~ △*mBo*, and △*NBO* ~ △*nBo*, whence: $\frac{MO}{mo} = \frac{BO}{bo}$; $\frac{NO}{no} = \frac{BO}{bo}$, and therefore, $\frac{MO}{mo} = \frac{NO}{no}$, ⇒ *MO* = *NO*, since *mo* = *no* as radii of the same circle.

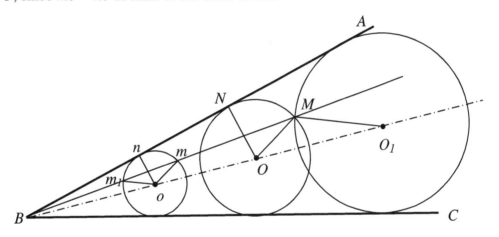

Figure 9.7.2

Therefore, *ON* is also a radius of the circle with the centre at *O*, and *N* is the point where this circle touches *AB*. By the axial symmetry property of a bisector, this circle also touches *BC*. Thus, *O* is the centre of a circle that satisfies the given conditons.

If m_1 (not *m*) is chosen as the corresponding point for *M*, we shall find another centre of the sought for circle, point O_1. Thus, the problem has two solutions.

<u>Problem 9.7.3.</u> *Given a triangle, △ABC, inscribe in this triangle a rhombus with a given acute angle and one of its (rhombus) sides lying on the base, AB, whereas two vertices of the rhombus lie on the sides AC and BC.*

<u>Solution.</u> For the beginning disregard the condition requiring one of the vertices of the rhombus to lie on *BC*. Then we can construct infinitely many rhombuses satisfying the remaining conditions. Let us construct one of them.

Pick on *AC* some point, *M* (Fig. 9.7.3). Construct an angle equal to the given one with its vertex at *M* and one side parallel *AB*. The other side of this angle intersects *AB* at some point *N*.

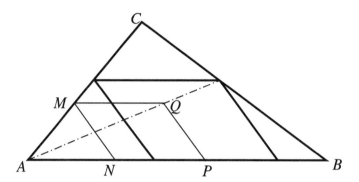

Figure 9.7.3

From *N*, lay off on *AB* a segment *NP* = *MN* and from points *M* and *P* draw lines parallel to *AB* and *MN*, respectively. These lines intersect at some point *Q*. *MNPQ* is a rhombus (why?) and it is similar to the sought for one. (It is easy to show that two rhombuses are similar if they have two corresponding angles congruent).

Choose *A* as a centre of similitude and construct a rhombus similar to *MNPQ* with one vertex (corresponding to *Q*) lying on *BC*. This vertex is found as a point of intersection of ray *AQ* with *BC*. To complete the construction draw from this vertex lines parallel to *MQ* and *QP*. One more side of the required rhombus is parallel to *QP* and emanates from the vertex located on *AC*. The last (fourth) side lies on *AB*.

The following problems can be solved by the similarity method:
1. Construct a triangle given its two angles and the radius of the inscribed circle.
2. Construct a triangle given the ratio of its altitude to the base, the angle of the vertex and a median drawn to the lateral side.
3. Given $\angle AOB$ and point *C* in the interior. On the side *OB* find point *M* equidistant from *OA* and *C*.

9.7 Questions and Problems.

1. Inscribe in a circle a triangle similar to a given one.

2. Construct a triangle, having given the ratio of an altitude to the base, the vertical angle, and a median to a lateral side.

3. Given an angle and a point in its interior, find on one side of the angle a point equidistant from another side of the angle and the given point.

4. Construct a triangle, having given two angles and the inradius.

9.8 Some theorems on proportional segments.

a) <u>Segments formed by parallel lines and their transversals.</u>

Theorem 9.8.1. *If a series of parallel lines intersects two sides of an angle, the corresponding segments cut on the two sides by the parallel lines are proportional.*

Proof. In Fig. 9.8.1, $DD_1 \parallel EE_1 \parallel FF_1$. Show that $\dfrac{BD}{B_1D_1} = \dfrac{DE}{D_1E_1} = \dfrac{EF}{E_1F_1}$.

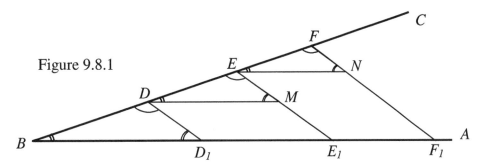

Figure 9.8.1

Draw auxiliary lines DM and EN parallel to BA. $\triangle DBD_1$, $\triangle EDM$, and $\triangle FEN$ are similar as mutually equiangular, hence $\dfrac{BD}{BD_1} = \dfrac{DE}{DM} = \dfrac{FE}{EN}$.

Since $D_1E_1 = DM$ and $E_1F_1 = EN$ (why?), the latter proportion is equivalent to the conclusion of the theorem.

Theorem 9.8.2. *If two parallel lines are met by a pencil (bundle) of lines, the corresponding segments cut off on the parallel lines are proportional.*

Proof. In Figure 9.8.2, $MN \parallel M_1N_1$. Therefore:

$\triangle OAB \sim \triangle OA_1B_1 \Rightarrow \dfrac{AB}{A_1B_1} = \dfrac{OB}{OB_1}$; $\triangle OBC \sim \triangle OB_1C_1 \Rightarrow \dfrac{BC}{B_1C_1} = \dfrac{OB}{OB_1}$. Hence $\dfrac{AB}{A_1B_1} = \dfrac{BC}{B_1C_1} = ...$, Q.E.D.

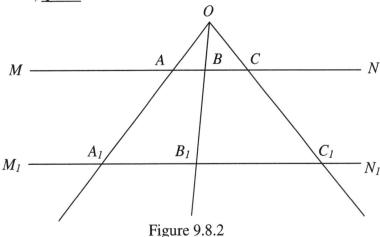

Figure 9.8.2

The solutions of the following problems follow directly from the latter two theorems.

Problem 9.8.1. *Given a segment, divide it into three parts in a given ratio m:n:p where m, n, and p are given numbers or segments.*

Problem 9.8.2. *Find the **fourth proportional** to three given segments, i.e. given three segments, a, b, c, find such a segment x that $\frac{a}{b} = \frac{c}{x}$.*

    b)   <u>Properties of bisectors.</u>

Theorem 9.8.3. *The bisector of an angle of a triangle divides the opposite side into two segments which are proportional to the sides that include the angle.*

Proof. In Figure 9.8.3, BD is the bisector of $\angle B$, i.e. $\angle ABD = \angle DBC$. The conclusion of the theorem asserts that $\frac{AD}{DC} = \frac{AB}{BC}$.

Through point C, draw a line parallel to BD. That line meets the extension of AB at some point E. Then

$$\frac{AD}{DC} = \frac{AB}{BE} \qquad (1) \text{ (why?)}.$$

On the other hand, $\angle ABD = \angle BEC$ (as corresponding angles with $BD \parallel EC$), and $\angle DBC = \angle BCE$ (as alternate), which leads to $\angle BCE = \angle BEC \Rightarrow BC = BE$.

Then, proportion (1) with BE replaced by BC transforms into $\frac{AD}{DC} = \frac{AB}{BC}$, Q.E.D.

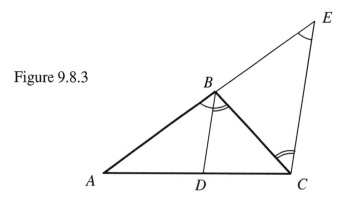

Figure 9.8.3

Theorem 9.8.4. *(the property of an exterior angle of a triangle).* ***The bisector of an exterior angle of a triangle meets the extension of the opposite side at such a point that the distances from that point to the endpoints of the opposite side are proportional to the two respective sides of the triangle ending at these points if these sides are not congruent.***

Proof. In Figure 9.8.4, BD is the bisector of $\angle CBF$; it meets the extension of AC at D. The theorem asserts that $\frac{DA}{DC} = \frac{BA}{BC}$.

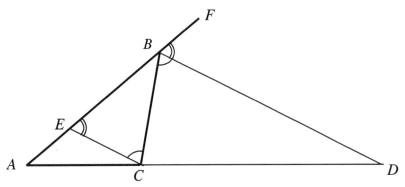

Figure 9.8.4

Draw $CE \parallel BD$; then

$$\frac{DA}{DC} = \frac{BA}{BE}.\qquad(1)$$

$\angle BEC = \angle FBD$ as corresponding; $\angle ECB = \angle CBD$ as alternate, and $\angle FBD = \angle CBD$ by the hypothesis. Therefore, $\angle BEC = \angle ECB$, $\Rightarrow \triangle EBC$ is isosceles: $BE = BC$.

In proportion (1) replace BE by BC and obtain the conclusion of the theorem.

Remark. The hypothesis of the theorem suggests that the bisector of the considered exterior angle intersects with the extension of the opposite side. It is not so in the case of an isosceles $\triangle ABC$ with $AB = BC$ (why?), i.e. $\dfrac{AB}{BC} = 1$.

Yet, even in that case one can say that point D is located *infinitely far* from points A and C and, in a certain sense, $DA = DC$ (the difference between the two segments is negligible comparing to any of them), i.e. $\dfrac{DA}{DC} = 1 = \dfrac{AB}{BC}$.

9.8 Questions and Problems.

1. Point D on the side AB of $\triangle ABC$ is 2cm distant from A. AB=6cm; BC=9cm. Evaluate the segments into which BC is divided by a line passing through D parallel to AC.

2. In Figure 1, $AB \parallel CD \parallel EF \parallel KL$. AC=25cm, CE=35cm, EK=20cm, and BL=128cm. Find BD, FL, and DL.

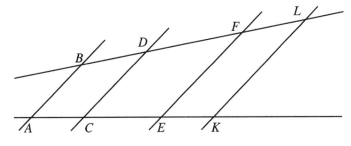

Figure 1

3. The sides of ∠A are cut by two parallel lines: BC and DE such that BC is closer to the vertex of the angle than DE, and B and D are located on one side of the angle, and C and E – on the other. Determine:
   b) AB, if AC:CE = $\frac{3}{4}$:0.6, and BD=12m;
   b) BD, if AD+AB=32m, and AC:CE =1.5.

4. The bases of a trapezoid are equal *a*=9cm and *b*=16 cm. Find the length of the segment that divides the trapezoid into two similar trapezoids and the ratios into which it divides each of the lateral sides.

5. Construct two circles that touch each other externally with their centers at given points and the ratio of their radii equal 2:3.

6. Construct two circles that touch each other internally with their centers at given points and the ratio of their radii equal 2:3.

7. Given a segment, divide it into three parts in a ratio *m* : *n*: *p*, where *m*, *n*, and *p* are given segments or numbers.

8. Find the fourth proportional to three given segments, i.e., given segments *a*, *b*, and *c*, construct such a segment x, that $\dfrac{a}{b} = \dfrac{c}{x}$.   (Propose two different methods #).

9. Prove that a straight line passing through the midpoints of the bases of a trapezoid is concurrent with the extensions of the lateral sides.

10. Figure 2 illustrates a method of evaluating the distance between two inaccessible points. Explain and substantiate the procedure. Describe it as an algorithm suggesting the minimum number of measurements.

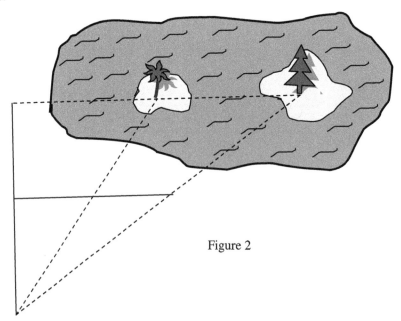

Figure 2

11. Determine the base of an isosceles triangle if a bisector drawn to a lateral side divides it into segments equal 6dm and 9dm.

12. Evaluate the lateral side of an isosceles triangle if its base equals 9cm, and the inradius equals one fifth of the altitude drawn to the base.

13. The sides of a triangle are equal 7cm, 9cm, and 12cm. Evaluate the segments into which the longest side is divided by the bisector of the opposite angle.

14. It is easy to show that in an isosceles triangle the bisector of an exterior vertical angle is parallel to the base. Show that. How would you interpret that result in terms of Theorem 9.8.4?

9.9 Metric relations in triangles and some other figures.

a) <u>Mean proportionals in right triangles.</u>
Segment x is said to be the *mean proportional* between segments a and b if $\dfrac{a}{x} = \dfrac{x}{b}$ or (that is equivalent) $x^2 = ab$.

Lemma 9.9.1. *In a right triangle, the altitude drawn to the hypotenuse divides the triangle into two similar triangles, and each of them is similar to the original triangle.*

<u>Proof.</u> In $\triangle ABC$ in Fig. 9.9.1 $\angle C = d$ and $CD \perp AB$. Then $\angle 1 = \angle 4$, $\angle 2 = \angle 3$ as angles with respectively perpendicular sides, whence the conclusion of the lemma follows (the three triangles, ABC, ACD, and CBD are mutually equiangular, and, thus, similar).

Figure 9.9.1

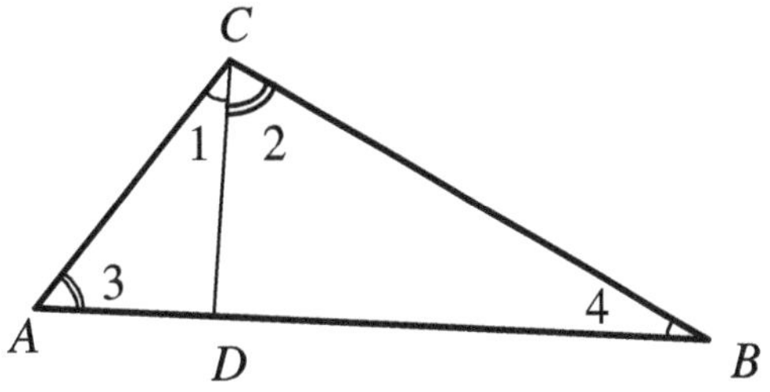

Theorem 9.9.2. *In a right triangle,*
(i) the altitude drawn to the hypotenuse is the mean proportional between the segments into which it divides the hypotenuse, and
(ii) each leg is the mean proportional between its projection onto the hypotenuse and the hypotenuse.

<u>Proof.</u> The theorem asserts that in Figure 9.9.1, where $\angle ACB = d$, and $CD \perp AB$

(i) $\dfrac{AD}{CD} = \dfrac{CD}{DB}$ and

(ii) $\dfrac{AD}{AC} = \dfrac{AC}{AB}$; $\dfrac{BD}{BC} = \dfrac{BC}{BA}$.

The above three proportions follow immediately from the similarity of the three triangles that was proved in Lemma 9.9.1.

Corollary 1 **In a circle,**
(i) *a perpendicular dropped from a point on a circle onto a diameter is the mean proportional between the segments into which it divides the diameter, and*
(ii) *a chord connecting a point on a circle with an endpoint of a diameter is the mean proportional between the diameter and the projection of the chord onto the diameter.*

Proof. It follows directly from the theorem, taking into account the fact that an inscribed angle standing on a diameter is right.

Problem 9.9.1. *Construct the mean proportional between two given segments.*

Solution. **I method** (based on Corollary 1 (i)). On a line, lay off sequentially two given segments, a, b, and draw a circle with $a + b$ as a diameter (Fig. 9.9.2). (Let students recall how to do that). From the common endpoint of a and b erect the perpendicular to the diameter. The segment of the perpendicular between its foot and the point of its intersection with the circle is the sought for segment (by Corollary 1 (i)).

Figure 9.9.2

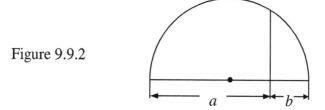

II method (based on Corollary 1 (ii)). The construction should be completed by students.

b). <u>The Pythagorean Theorem.</u>

In this subsection we shall consider a remarkable relation between the lengths of the three sides of a right triangle. This relation proved (most likely) by Pythagoras, a Greek mathematician, who lived about V B.C., enables us to find the length of a side of a right triangle if the other two sides are given. We shall also derive some corollaries of the theorem.

Theorem 9.9.3. (Pythagorean Theorem) *In a right triangle, the square of the length of the hypotenuse is equal to the sum of the squares of the lengths of the legs.*

More often the following short formulation is used. *The square of the hypotenuse equals to the sum of the squares of the legs of a right triangle.*

(We shall often use for brevity such expressions as *the square of the segment* instead of *the square of the length of the segment*).

Proof. Suppose in right $\triangle ABC$ (Fig. 9.9.3) the lengths of its hypotenuse, AB, and sides, AC and BC, are expressed by numbers a, b, and c: $AB = c$, $BC = a$; $AC = b$. We also suggest that, in the same units of length the altitude, CD, drawn to the hypotenuse has length h, and the measures of the projections of the legs onto the hypotenuse are a' and b', respectively.

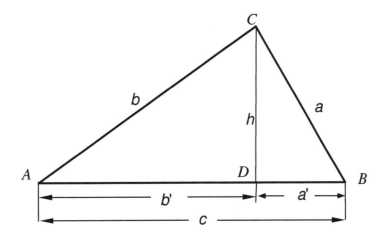

Figure 9.9.3

It follows from Th.9.9.2 that $a^2 = a'c$ and $b^2 = b'c$, whence
$a^2 + b^2 = a'c + b'c \Rightarrow a^2 + b^2 = (a' + b')c$.
Since $a' + b' = c$, the latter means: $a^2 + b^2 = c^2$,   Q.E.D.

Example. If in a right triangle the legs (measured by a common measure) have lengths 3 and 4, the length of the hypotenuse is $c = \sqrt{3^2 + 4^2} = 5$. A right triangle with sides equal (measuring) 3, 4, and 5 is called an *Egyptian triangle*. It was used by ancient Egyptians in construction and land measurement. (An example will be considered later).

Right triangles with their sides lengths expressed by integers are called *Pythagorean triplets*. It can be proved that the hypotenuses, c, and the respective legs a and b of such triangles can be expressed as follows:
$c = m^2 + n^2$; $a = 2mn$; $b = m^2 - n^2$, where m and n are positive integers, and $m > n$.

Theorem 9.9.4. *In a right triangle, the ratio of squares of its legs is equal to the ratio of the corresponding projections of the legs upon the hypotenuse.*

Proof. It follows from Th.9.9.2 that (see Fig. 9.9.3) $\dfrac{a^2}{b^2} = \dfrac{a'c}{b'c} = \dfrac{a'}{b'}$, Q.E.D.

Remark 1. We have established that in a right triangle its legs, a, b, hypotenuse, c, the altitude, h, drawn to the hypotenuse, and the projections, a', b', of the legs onto the hypotenuse are connected by the following relations:

1. $a'c = a^2$,
2. $b'c = b^2$,
3. $a^2 + b^2 = c^2$,
4. $a' + b' = c$,
5. $h^2 = a'b'$.

Equality (3), as we have seen, follows from (1), (2), and (4), therefore, only four out of five relations are independent. These four relations connect the measures of six segments, hence, given two of these segments, we can find all of them.

Example 1. Suppose the projections of the legs of some right triangle on the hypotenuse are given: $a' = 3$ cm; $b' = 4$ cm; then $c = a' + b' = 5$ cm;
$a = \sqrt{a'c} = \sqrt{15}$ cm ≈ 3.9 cm;
$b = \sqrt{b'c} = \sqrt{20}$ cm ≈ 4.5 cm;
$h = \sqrt{a'b'} = \sqrt{12}$ cm ≈ 3.5 cm.

Remark 2. Throughout our discussion we suggest that the lengths of the segments under consideration are determined with respect to the same unit (unless it is specified otherwise) and we loosely say (for brevity) "the square of the side" instead of "the square of the number expressing the length of the side in the given units", "the product of segments" instead of "the product of the numbers representing the measures of segments", etc.

The following two theorems generalize the Pythagorean Theorem for the cases of triangles that are not right.

Theorem 9.9.5. *In any triangle the square of the side opposite an acute angle is equal to the sum of the squares of the other two sides diminished by twice the product of one of these sides and the projection of the other upon it.*

Proof. In Fig. 9.9.4 (a, b) $\angle C$ is acute. As usually, let a, b, c denote the sides opposed to $\angle A$, $\angle B$, $\angle C$, respectively. h denotes the altitude drawn from B, and a', c' denote the projections of a and c respectively upon b (or upon the extension of b when $\angle A$ is obtuse as in Fig. 9.9.4 (b)).

We are going to prove that $c^2 = b^2 + a^2 - 2ba'$.

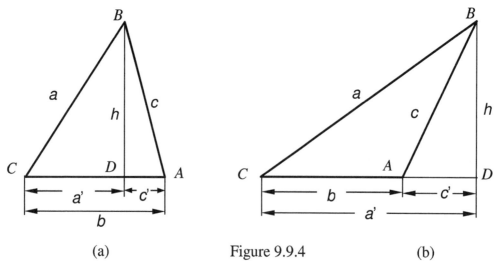

Figure 9.9.4

From right $\triangle BDC$ one can find: $c^2 = h^2 + (c')^2$. (1)
Let us determine each of the squares, h^2 and $(c')^2$.
In right $\triangle BCD$ $h^2 = a^2 - (a')^2$. (2)
On the other hand either $\quad c' = b - a'$ (Fig. 9.9.4 (a)) or
$\quad\quad\quad\quad\quad\quad\quad\quad\quad c' = a' - b$ (Fig. 9.9.4 (b)),
and in any case
$(c')^2 = (b - a')^2 = b^2 - 2ba' + (a')^2$
$(c')^2 = (a' - b)^2 = (a')^2 - 2ba' + b^2$. (3)
After substituting (2) and (3) in (1), we obtain:
$c^2 = a^2 - (a')^2 + b^2 - 2ba' + (a')^2 = a^2 + b^2 - 2ba'$, Q.E.D.

Theorem 9.9.6. *In any obtuse triangle the square of the side opposite the obtuse angle is equal to the sum of the squares on the other two sides increased by twice the product of one of those sides and the projection of the other upon it.*
Proof. (May be left to students).
In Fig. 9.9.5, $\angle C$ is obtuse. Using our standard notations, we find from $\triangle ABD$ and $\triangle CBD$: $c^2 = h^2 + (c')^2 = a^2 - (a')^2 + (a' + b)^2 =$
$= a^2 - (a')^2 + (a')^2 + 2ba' + b^2 = a^2 + b^2 + 2ba'$,
Q.E.D.

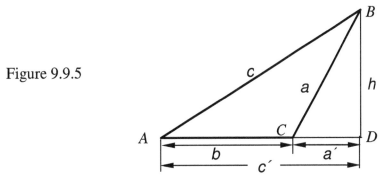

Figure 9.9.5

Corollary 1 *The square of a side of a triangle is less, equal, or greater than the sum of the squares of the two other sides if the angle opposite to that side is, respectively, acute, right, or obtuse.*
It is easy to prove (e.g., by contradiction) that the converse of the latter result is true as well:
Corollary 2 *An angle of a triangle is acute, right, or obtuse, depending on whether the square of the opposite side is less, equal, or greater than the sum of the other two sides.*
Problem. Ancient Egyptians used a rope divided into equal parts by knots to construct right angles. Reinvent their method. (Hint: use Corollary 2).
Theorem 9.9.7. *The sum of the squares of the diagonals of a parallelogram is equal to the sum of the squares of the four sides.*
Proof. (may be left to students). From *B* and *C* (Fig. 9.9.6) drop perpendiculars, *BE* and *CF* onto *AC* (and its extension).

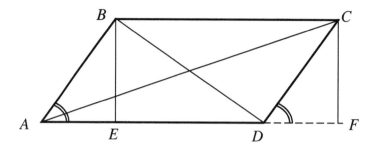

Figure 9.9.6

From $\triangle ABD$ and $\triangle ACD$ we find:
$$BD^2 = AB^2 + AD^2 - 2\,AD\cdot AE; \quad (1)$$
$$AC^2 = AD^2 + CD^2 + 2\,AD\cdot DF \quad (2)$$
$\triangle ABE = \triangle DCF$ (why?), therefore, $AE = DF$. Taking that into account, add (1) and (2). Terms $-2\,AD\cdot AE$ and $2\,AD\cdot DF$ eliminate each other and we obtain:
$$BD^2 + AC^2 = AB^2 + AD^2 + AD^2 + CD^2 =$$
$$= AB^2 + BC^2 + AD^2 + CD^2, \text{ Q.E.D.}$$

Problem. *Given the sides of a triangle, find its altitudes.*
Solution. (The solution may be left to students). Let *a*, *b*, and *c* be the sides of $\triangle ABC$, and let h_a denote the altitude drawn from the vertex *A* to the corresponding side $BC = a$.

Then use Th.9.9.5 and the Pythagorean theorem to obtain:

$$h_a = \sqrt{c^2 - \left(\frac{a^2 + c^2 - b^2}{2a}\right)^2}.$$

The other altitudes are found similarly.

9.9 Questions and Problems.

Hereafter we often use the following notations for the elements of right triangles:

c – the hypotenuse;

a, b – the legs;

a′, b′ – the projections of the respective legs upon the hypotenuse;

h – the altitude dropped onto the hypotenuse.

In problems suggesting numerical answers:
(i) In each problem start with introducing notations for known and unknown magnitudes and deriving algebraic expressions for the unknowns; then substitute the numbers;
(ii) Approximate the answers by decimals with one significant digit after the decimal point.
(iii) When extracting square roots, the following approximate formula may be helpful:

$$\sqrt{1+x} \approx 1 + \frac{x}{2},$$ where *x* is substantially less than 1. Substantiate that formula.

An example of the use of the above formula follows:

$\sqrt{10} = \sqrt{9+1} = \sqrt{9\left(1+\frac{1}{9}\right)} = 3\sqrt{1+\frac{1}{9}} \approx 3\left(1+\frac{1}{18}\right) \approx 3.16$. The accuracy is of order $\left(\frac{1}{9}\right)^2 \approx 0.012$.

1. Construct the mean proportional between two given segments. (Propose at least two methods).

2. Find the length of a chord which is perpendicular to a diameter and divides it into two parts equal, respectively, 12cm and 3 cm.

3. *Prove geometrically that the product of two positive numbers with a fixed sum is maximal when these numbers are equal.

4. A chord forms a 60° angle with a diameter in a circle. Given the length of the diameter, c, find the length of the chord that joins the second endpoint of the given chord with the second endpoint of the diameter.

5. Write down four independent relations between the (six) elements of a right triangle.

6. Given two elements of a right triangle in some units of length, find all the remaining (four) elements of the right triangle : (i) *a=16, b=12*; (ii) *a=7, b=24*; (iii) *a=5, c=13*; (iv) *b=8, c=10*; (v) *b=17, h=15*; (vi) *c=122, a′=50*; (vii) *a′=1, b′=9*; (viii) *c=20, h=8*; (ix) *a=3, a′=1.8*; (x) *h=18, b′=12*.

7. The altitude divides the hypotenuse into the segments equal 5.0cm and 7.2 cm. Evaluate the altitude and the legs of the triangle.

8. One of the legs of a right triangle is 4m long, and the sum of the other leg and the hypotenuse equals 6m. Evaluate the hypotenuse and the unknown leg.

9. The ratio of the legs in a right triangle is 6:5, and the hypotenuse is 11cm long. Determine the projections of the legs upon the hypotenuse.

10. Evaluate the side of a rhombus whose diagonals are 10 cm and 12 cm, respectively.

11. The bisector of an acute angle of a right triangle divides the leg to which it is drawn into two segments equal, respectively, 10 and 8 cm. Evaluate the hypotenuse.

12. Find the sides of a rectangle inscribed in a circle of radius equal 15m, if the sides of the rectangle are in a ratio of 3:4.

13. Find the distance between the opposite sides of a rhombus with the side equal 10 cm and one of the angles equal 120°.

14. An altitude of a right triangle divides the hypotenuse into two segments one of which is 3m greater than the other. Find the legs and their projections upon the hypotenuse, if the altitude equals 2m.

15. In a right trapezoid, the greatest lateral side equals 5dm, the greatest diagonal equals 6dm, and the difference between the bases is 30cm. Find the bases of the trapezoid.

16. Determine the segments into which the bisector of the right angle divides the hypotenuse in a right triangle with the legs respectively equal *a* and *b*.

17. The hypotenuse of a right triangle equals 4.25 dm, and one of the legs is 34cm. Find the segment of the hypotenuse included between the foot of the altitude dropped onto the hypotenuse and the point where the bisector of the right angle cuts the hypotenuse.

18. Find the sides of an isosceles triangle if its lateral side is in a ratio of 25:48 to the base, and the median bisecting the base is 14 cm long.

19. A rectangle with the sides equal 4dm and 2.4 dm, respectively, is folded along a diagonal. Find the segments into which a longest side cuts the opposite longest side under such a transformation.

20. Find the side and the apothem (the distance from the circumcentre to a side) of a regular triangle inscribed in a circle of radius $R = 1$.

21. Find the side and the apothem (the distance from the circumcentre to a side) of a regular hexagon inscribed in a circle of radius $R = 1$.

22. An isosceles triangle is inscribed in a circle of radius equal 5m. The base of the triangle is 4m distant from the centre of the circle. Find the lateral side of the triangle. (Note: there are two cases).

23. Two circles, of radii equal 15cm and 20 cm, respectively, intersect. Find the distance between the centres if the length of their common chord is 24cm. (There are two cases.)

24. Figure 1 illustrates a method of determining the diameter of a rod by means of a slide gauge. Find the diameter in terms of the known magnitudes, a and h.

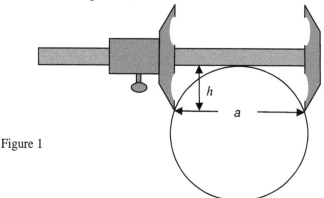

Figure 1

25. A circle of radius 2m is inscribed in an isosceles triangle with the base equal 6m. Determine the lateral side of the triangle. (In order to avoid cumbersome algebraic expressions, one may use the numerical values from the very beginning).

26. In a circle of radius 5cm, two chords are parallel to each other. Their lengths are 6cm and 8cm, respectively. Find the distance between the chords. (Note: there are two possible cases).

27. Find the radius of the circle circumscribed about an isosceles trapezoid with the bases equal, respectively, 6cm and 8cm and the altitude equal 7cm.

28. Two sides of a triangle are equal, respectively, 3m and 5m and include an angle of 120°. Find the third side.

29. Two sides of a triangle are equal, respectively, 36m and 25m and include an angle of 60°. Find the third side.

30. Two sides of a triangle are equal, respectively, 7m and $\sqrt{18}$ m and include an angle of 45°. Find the third side.

31. Evaluate whether a triangle is acute, right, or obtuse if its sides are in a ratio of
    a) 4:13:14; b) 12:14:19; c) 24:7:25; d) 13:30:33; e) 11:41:45.

32. Ancient Egyptians used a long rope divided into equal parts by knots to construct right angles. Reinvent and substantiate their method.

33. Given the sides of a triangle, find its altitudes (obtain the formula for one altitude; the other two can be found similarly).

34. Find the altitude of a trapezoid if its bases are equal 7cm and 2cm, and the lateral sides are equal 5cm and 6cm, respectively.

35. BM and BE are, respectively, a median and an altitude drawn to the side AC in $\triangle ABC$. Evaluate AC if $AB=14$cm; $BC=6$cm; $ME=4$cm.

36. The greatest of two angles including the base of a triangle equals 60°. Evaluate the side opposite to that angle if the foot of the altitude divides the base into segments equal 25cm and 75 cm, respectively.

37. Find the sides of a parallelogram if their difference is 2cm, and the diagonals are 6cm and 7cm, respectively.

9.10 Proportional segments in circles.

We have already discovered some interesting ratios in the preceding section. Here we are going to discuss two more results.

Theorem 9.10.1. *If in a circle a chord and a diameter intersect each other, the point of intersection divides them into segments such that the product of the segments of the chord equals to the product of the segments of the diameter.*

Proof. In Fig. 9.10.1, chord AB and diameter CD intersect at M. Let us draw two auxiliary chords, AC and BD. $\triangle AMC$ and $\triangle DMB$ are mutually equiangular (why?), hence they are similar.

Therefore, $\dfrac{AM}{DM} = \dfrac{MC}{MB}$, whence $AM \cdot MB = DM \cdot MC$, Q.E.D.

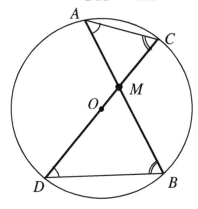 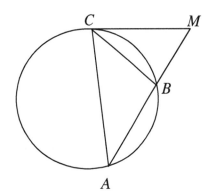

Figure 9.10.1   Figure 9.10.2

Corollary 1. ***Given a point in the interior of a circle, the product of segments into which the point divides chords passing through the point, is invariant (i.e., it is the same for all chords passing through that point).***

Proof. As there is one and only one straight line passing through two given points (AXIOM 1), there is one and only one diameter, *CD*, passing through a given point, *M* (Fig. 9.10.1). Therefore, according to Th.9.10.1, for any chord passing through *M* the product of the segments will be equal to *DM·MC*, i.e. it will be the same for all chords, Q.E.D.

Theorem 9.10.2. ***Given a circle and a point in the exterior of the circle, for any secant emanating from that point the product of the segments connecting the given point with the points where the secant meets the circle is equal to the square of the segment of the tangent drawn to the circle from the same point.***

Proof. We are going to prove that in Fig. 9.10.2, where *MC* is the tangent and *MA* is a secant drawn from *M*, $MA \cdot MB = MC^2$.

Draw auxiliary chords *AC* and *BC*. $\triangle MAC \sim \triangle MBC$ since $\angle M$ is their common angle, and $\angle MCB = \angle CAB$ as each of them is measured by ½∪*BC*. From the similarity, we obtain for the corresponding sides: $\dfrac{MA}{MC} = \dfrac{MC}{MB}$, whence $MA \cdot MB = MC^2$, Q.E.D.

Remark 1. It is convenient to call segments *MC*, *MA*, and *MB* the tangent, the secant, and the exterior part of the secant, respectively.

Corollary 1. *For any secant drawn to a circle from an exterior point, the product of the secant and its exterior part is invariant. (This invariant value equals to the square of the tangent drawn to the circle from the same point.)*

9.10 Questions and Problems

1. A chord equal 30 cm is perpendicular to a diameter and divides it into a ratio of 5:3. Find the diameter of the circle.

2. A chord is divided by some other chord into parts equal 8cm and 2cm and bisects the other chord. Find the length of the latter.

3. A perpendicular dropped from a point of the circle onto a diameter divide it into segments whose difference is 18cm. The perpendicular is 12cm long. Evaluate the diameter.

4. Two secants are drawn to a circle from the same point. Their lengths are 12m and 9m, respectively. The exterior part of the shortest secant equals 3m. Find the exterior part of the longest secant.

5. The greatest secant drawn to a circle from some point equals 12m. Find the length of the tangent drawn to the circle from the same point, if the radius of the circle is 3.5m.

6. A point is 13cm distant from the centre of a circle of radius 5cm. Find the length of the secant that emanates from that point and is bisected by the circle.

7. A secant is $1\frac{7}{9}$ as long as its exterior part. Find the ratio of the tangent drawn from the same point to the exterior part of the secant.

8. James Bond is planning his escape from a secret plant. He is going to escape through a round sewerage pipe whose radius is 1.2m. He knows that the level of water in the pipe is exactly at ground level. In order to understand whether he will be able to walk (the superhero cannot swim) in the pipe, James measures the length of a board leaned against the pipe between the point of contact with the pipe and the endpoint located on the ground (2.9m) and the distance between that endpoint of the board and the pipe on the ground (2.0m) (see Figure 1). The height of the superagent from the heels to the chin is 1.7m. Help James to make the necessary estimates. Is he going to walk or drown?

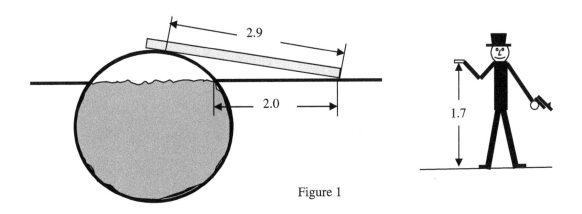

Figure 1

9. The length of a chord is 12cm. Through an endpoint of the chord a tangent is drawn, and the distance from the other endpoint of the chord to that tangent is 8cm. Find the radius of the circle.

10. The radii of two concentric circles are 9cm and 7cm, respectively. A chord of the largest circle is divided by the smallest circle into three equal parts. Evaluate the chord.

9.11. Trigonometric functions of acute angles.

a) Definitions.

Let α be an acute angle. Pick a point, M, on its side and drop a perpendicular onto the other side. Thus, a right triangle, ΔAMN is constructed (Fig. 9.11.1).

Consider the following ratios of the sides of the constructed triangle:

1) $\dfrac{MN}{AM}$, that is the ratio of the leg opposite to angle α to the hypotenuse;

2) $\dfrac{AN}{AM}$, that is the ratio of the leg adjacent to angle α to the hypotenuse;

3) $\dfrac{MN}{AN}$, that is the ratio of the leg opposite to α to the leg adjacent to that angle; and the reciprocals of these three ratios:

4) $\dfrac{AM}{MN}$, 5) $\dfrac{AM}{AN}$, 6) $\dfrac{AN}{MN}$.

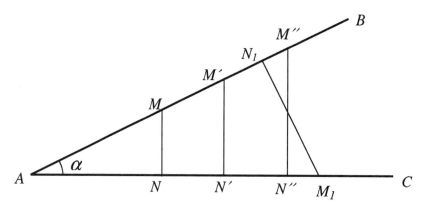

Figure 9.11.1

We can prove that *the value of each of these (six) ratios does not depend upon the location of point M on the side of the angle.*

Really, if any other point, e.g., M', M'', etc, be chosen and the corresponding perpendiculars, $M'N'$, $M''N''$, etc. drawn, all obtained triangles are similar to each other and $\triangle AMN$, hence, $\dfrac{MN}{AN} = \dfrac{M'N'}{AN'} = \dfrac{M''N''}{AN''} = \ldots$, $\dfrac{AN}{AM} = \dfrac{AN'}{AM'} = \dfrac{AN''}{AM''} = \ldots$, etc.

Also, these ratios do not depend upon the choice of the side of the angle on which point M is located. If, for instance, M_1 is picked on AC (not on AB) and M_1N_1 is perpendicular to AB, $\triangle AM_1N_1$ is similar to $\triangle AMN$, and, therefore,

$$\dfrac{M_1N_1}{AM_1} = \dfrac{MN}{AM} = \dfrac{M'N'}{AM'} = \ldots, \text{ etc.}$$

Thus, all of the above ratios, (1), (2), (3), and their reciprocals, do not change if point M changes its position on the sides of the angle; they will change, however, if the value of α changes.

For a given α, each of the above ratios ((1 – 3) and their reciprocals (4-6)) will have a certain value.

In that sense each ratio is *a function of the angle* only and thus each ratio characterizes the value of the angle. These ratios are called *trigonometric functions of angles*. The following four functions are used more often than others (Fig. 9.11.2):

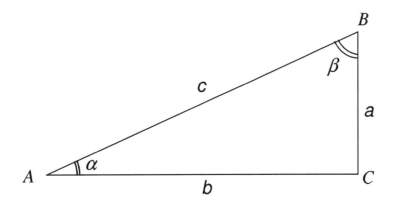

Figure 9.11.2

1) $\sin \alpha = \dfrac{a}{c}$ is the sine of the angle $= \dfrac{\text{opposite leg}}{\text{hypotenuse}}$;

2) $\cos \alpha = \dfrac{b}{c}$ is the cosine of the angle $= \dfrac{\text{adjacent leg}}{\text{hypotenuse}}$;

3) $\tan \alpha = \dfrac{a}{b}$ is the tangent of the angle $= \dfrac{\text{opposite leg}}{\text{adjacent leg}}$;

4) $\cot \alpha = \dfrac{b}{a}$ is the cotangent of the angle $= \dfrac{\text{adjacent leg}}{\text{opposite leg}}$.

As any leg of a right triangle is less than the hypotenuse, the *sine* and *cosine* of an angle are always less than one, whereas the *tangent* and *cotangent* may be less than one, equal to one, or greater than one.

It follows from the definitions (1) – (4) that if angle β is complimentary to α, i.e. $\alpha + \beta = d$, then

$\sin \alpha = \cos \beta$, $\sin \beta = \cos \alpha$; $\tan \alpha = \cot \beta$; $\cot \alpha = \tan \beta$.

<u>Problem 9.11.1</u> Find the basic trigonometric functions of 30°, 60°, 45°.

<u>Solution.</u> If in Fig. 9.11.2 $\alpha = 30°$, then $\beta = 60°$. It is known that $a = \dfrac{1}{2}c$, therefore,

$$\sin 30° = \dfrac{1}{2}.$$

To find the cosine and tangent of 30° we have to relate the third side, b, to a and c. Using the Pythagorean theorem, we obtain

$$b^2 = c^2 - a^2 = c^2 - \left(\dfrac{1}{2}c\right)^2 = \dfrac{3}{4}c^2.$$

Then $b = \dfrac{\sqrt{3}}{2}c$, and

$$\cos 30° = \frac{b}{c} = \frac{\sqrt{3}}{2}; \quad \tan 30° = \frac{a}{b} = \frac{1}{2}c \div \frac{\sqrt{3}}{2}c = \frac{1}{\sqrt{3}};$$

$$\cot 30° = \frac{b}{a} = \sqrt{3}.$$

It is easy to obtain:

$$\sin 60° = \cos 30° = \frac{\sqrt{3}}{2}; \quad \cos 60° = \sin 30° = \frac{1}{2};$$

$$\tan 60° = \cot 30° = \sqrt{3}; \quad \cot 60° = \tan 30° = \frac{1}{\sqrt{3}}.$$

The right triangle with an acute angle equal 45° is isosceles (why?), i.e. $a = b$.
Then $c = \sqrt{a^2 + b^2} = \sqrt{2a^2} = a\sqrt{2}$, whence

$$\sin 45° = \cos 45° = \frac{a}{a\sqrt{2}} = \frac{1}{\sqrt{2}}; \quad \tan 45° = \cot 45° = \frac{a}{a} = 1.$$

b) **Basic trigonometric identities.**

It is easy to see from the definitions of the basic trigonometric functions and Figure 9.11.2 that:

$$\tan\alpha = \frac{a}{b} = \frac{a}{c} \div \frac{b}{c} = \frac{\sin\alpha}{\cos\alpha};$$

$$\cot\alpha = \frac{a}{b} = \frac{1}{\tan\alpha};$$

$$\sin^2\alpha + \cos^2\alpha = \left(\frac{a}{\sqrt{a^2+b^2}}\right)^2 + \left(\frac{b}{\sqrt{a^2+b^2}}\right)^2 = 1.$$

The above three equalities relating trigonometric functions of the same angle and valid for any angle, are called the *basic trigonometric identities*. Let us list them again:

$$\tan\alpha = \frac{\sin\alpha}{\cos\alpha}; \qquad (1)$$

$$\cot\alpha = \frac{1}{\tan\alpha}; \qquad (2)$$

$$\sin^2\alpha + \cos^2\alpha = 1 \qquad (3)$$

Identities (1) and (2) are often considered as definitions of the tangent and cotangent. All the other (secondary) identities between trigonometric functions of the same variable can be derived from (1) – (3). For instance, let us find the relation connecting the cosine and tangent of the same angle.

After dividing (3) by $\cos^2\alpha$, we obtain:

$$\frac{\sin^2 \alpha}{\cos^2 \alpha} + \frac{\cos^2 \alpha}{\cos^2 \alpha} = \frac{1}{\cos^2 \alpha}; \quad \text{hence,}$$

$$\tan^2 \alpha + 1 = \frac{1}{\cos^2 \alpha} \qquad (4).$$

c) Construction of angles given their trigonometric functions.

Let us notice first that if any trigonometric function of an angle is known, all the others can be derived from the identities (1) – (3).

<u>Problem 9.11.2.</u> *Construct angle α having given $\sin \alpha = \frac{3}{4}$.*

<u>Solution.</u> Let us choose a segment, *l*, and construct a right triangle with its hypotenuse $c = 4l$ and one leg $a = 3l$. Then, the angle opposite to *a* is the required angle.

In order to construct such a triangle, describe a semicircle with a segment AB congruent to $4l$ as a diameter (Fig. 9.11.3). From one of the endpoints, B, of the diameter draw a circle of radius $3l$. Connect the point C of intersection of the two circles with the other endpoint, A, of the diameter. ∠BAC is the required angle, since $\triangle ACB$ is right with the hypotenuse $AB = 4l$ and the leg opposed ∠BAC equal $3l$, which leads to $\sin \angle BAC = \frac{3l}{4l} = \frac{3}{4}$.

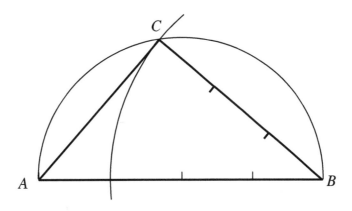

Figure 9.11.3

<u>Problem 9.11.3</u> *Construct an angle x such that $\tan x = \frac{2}{3}$.*

(To be solved by students).

d) Behaviour of trigonometric functions of acute angles.

To observe the behaviour of sine and cosine we suggest that the hypotenuse is fixed and equals to the unit of length, and the legs are changing.

Describe a quarter-circle by radius $OA = 1$ and consider the central angle, $\angle AOB = \alpha$, inscribed in that arc (Fig. 9.11.4).

Drop a perpendicular, BC, from B onto OA. Then
$$\sin \alpha = \frac{BC}{OB} = \frac{BC}{1} = \text{numerical value of } BC;$$
$$\cos \alpha = \frac{OC}{OB} = \frac{OC}{1} = \text{numerical value of } OC.$$

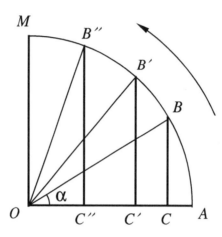

Figure 9.11.4

Imagine that OB rotates counterclockwise starting from the position coinciding with OA and ending at the position OM. Then α increases from 0° to 90°. Simultaneously, BC increases from 0 to 1, and OC decreases from 1 to 0. Then, as it follows from the above formulae, *when angle grows from 0° to 90°, its sine increases from 0 to 1, and its cosine decreases from 1 to 0.*

To observe the behaviour of tangent, it is convenient to suggest that the adjacent leg of considered triangles is fixed and equals the unit segment. By changing $\alpha = \angle BOA$ (see Fig. 9.11.5) from 0° to 90°, we generate a sequence of right triangles: $\triangle AOB$, $\triangle AOB'$, ..., such that each subsequent angle formed by OA and the hypotenuse (OB, OB', OB'', etc.) is greater than the preceding one.

According to the definition, $\tan \alpha = \dfrac{AB}{OA} = \dfrac{AB}{1} =$ the numerical value of AB. Then, $\tan \angle AOB < \tan \angle AOB' < \tan \angle AOB''...$, i.e. **when an angle grows from 0° to 90°, its tangent increases indefinitely, starting from 0.**

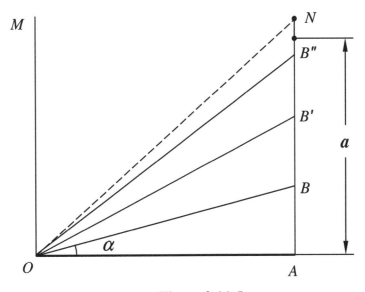

Figure 9.11.5

The words "increases indefinitely" mean that under the described circumstances, the tangent of the considered angles can reach a magnitude that will be greater than any given number.

Really, for any given number, *a*, one can choose such a point, *N*, on the perpendicular to OA erected from A, that $AN > a$, i.e. $\tan\angle AON > a$ (Figure 9.1.15). Thus, the tangent of an acute angle can be done greater than any given number by choosing the angle to be sufficiently great (sufficiently close to 90°).

Mathematicians say that the value of the tangent of an acute angle *tends to infinity* when the angle tends to 90°. In short notation it is expressed as follows:

$$\alpha \to 90^0 - \quad \Rightarrow \quad \tan\alpha \to \infty.$$

Symbol "∞" is being read *infinity*; it does not represent a certain number but a magnitude (variable) that *grows indefinitely*. The minus standing right after 90^0 signifies that the angle that tends to the right angle, is always acute (its measure is less than 90^0).

As $\cot\alpha = \dfrac{1}{\tan\alpha}$, it is clear that when $\tan\alpha \to 0, \Rightarrow \cot\alpha \to \infty$,

and when $\tan\alpha \to \infty, \Rightarrow \cot\alpha \to 0$.

Thus, **when angle grows from 0° to 90°, its cotangent decreases from ∞ to 0.**

e) <u>**Solving triangles using trigonometry.**</u>

Let us start with right triangles. As we know, one side and one angle of a right triangle completely define the triangle.

If the hypotenuse, c, and an angle, α, are given, one can find the legs of the triangle:

$a = c \sin \alpha$ (if a is opposite to α), $b = c \cos \alpha$.

The other acute angle, β can be found as $\beta = 90° - \alpha$. Thus, all the elements of the triangle are determined.

If only one trigonometric function of α, e.g., $\sin \alpha$, is given, one can find the legs as

$a = c \sin \alpha$ *(if a is opposite to α)*, $b = c \cos \alpha = c\sqrt{1 - \sin^2 \alpha}$.

Acute angles α and β can be found by means of trigonometric tables or a calculator by their sine and cosine, respectively, since $\sin \alpha$ is known, and $\cos \beta = \sin \alpha$.

If a leg, a, and an angle, α, are given, one can find the hypotenuse, c, and the other leg, b, as $c = \dfrac{a}{\sin \alpha}$; $b = a \cot \alpha$, and the other angle, $\beta = 90° - \alpha$, also can be found by any of its trigonometric functions.

In the case of a non-right triangle, at least three elements of a triangle, - two sides and an angle or two angles and a side should be given to solve the triangle. (The *SSA* case is ambiguous, - there are two solutions, i.e. there are two distinct triangles, for each *SSA* set that allows solutions; that is why we said "at least"). Then the following two theorems, called the laws of cosines and sines, respectively, are helpful.

When formulating these theorems we shall use, for our convenience, the following standard notations: the sides of a triangle are denoted *a, b, c,* and the opposite angles, respectively, $\angle A = \alpha$, $\angle B = \beta$, $\angle C = \gamma$. The circumradius (the radius of the circumscribed circle) will be denoted *R*.

The following theorem of cosines is just another (trigonometric) formulation of theorems 9.9.5 and 9.9.6 (see Fig. 9.9.4 and Fig. 9.9.5).

Theorem 9.11.1 *(Law of cosines) In a triangle with the sides a, b, and c, the square of the side opposite $\angle C = \gamma$ is expressed as*
$c^2 = a^2 + b^2 - 2ba\cos\gamma$ *if $\angle C = \gamma$ is acute, and*
$c^2 = a^2 + b^2 + 2ba\cos(180° - \gamma)$ *if $\angle C = \gamma$ is obtuse.*
Proof. The conclusion follows directly from Th.9.9.5 and 9.9.6 where a' is replaced by $a\cos\gamma$ or $a\cos(2d - \gamma)$, respectively (see Fig. 9.9.4, 9.9.5).

Remark 1. The above two expressions of the law of cosines can be united in a single *universal* (i.e. suitable for all cases) formula if we introduce the notion of the projection of a segment onto an axis with direction and generalize the notion of cosine for obtuse angles.

It is convenient to use the expression ***a*cosθ** for the ***projection of segment a onto an axis that forms angle θ with the segment***.

Let us consider the projections of segments onto an axis that is assigned a certain *direction*, which we shall call *positive*. We shall say that the *projection of a segment onto the axis is positive* if the segment forms an acute angle with the positive direction of the axis, and we shall say that the *projection of a segment onto the axis is negative* if the angle between the positive direction of the axis and the segment is obtuse.

These definitions emerge naturally, as one can see from the diagram below.

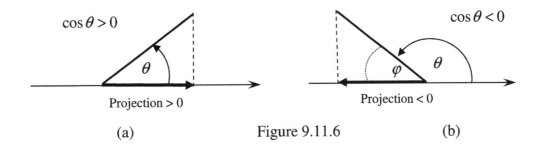

(a)   Figure 9.11.6   (b)

When a segment forms an acute angle θ with the positive direction of the axis (Figure 9.11.6a), the projection can be viewed as directed towards the positive direction. When the angle is obtuse, as in Figure 9.11.6b, the direction of the projection is opposite to the direction of the axis, and in this case we can say that the projection is negative.

In either case we define the projection to be equal to $a\cos\theta$, where a is the length of the segment.

Since in case (b), when the angle is obtuse, the projection is negative, we assume that the cosine of an obtuse angle is negative. As one can determine from the triangle in Figure 9.11.6b, the length of the projection in this case is equal to $a\cos\varphi$, where φ is the supplement of θ. Hence the projection in case (b) is equal to $a\cos\theta = -a\cos\varphi$, it follows that $\cos\theta = -\cos\varphi$.

Therefore, in order to have a universal formula $a\cos\theta$ for the projection of a segment onto an axis, we shall assume by definition that ***the cosine of an obtuse angle is equal to the negative of the cosine of its supplement:***
$$\cos\theta = -\cos(180^0 - \theta).$$

Now the Law of Cosines can be reformulated in the following universal form:

Theorem 9.11.1' *(Law of cosines) In a triangle with the sides **a**, **b**, and **c**, the square of the side opposite* $\angle C = \gamma$ *is expressed as*
$$c^2 = a^2 + b^2 - 2ba\cos\gamma.$$

The Law of Sines (also called *the Theorem of Sines*) is another powerful tool for solving triangles. In order to be able to formulate the *theorem of sines* for an arbitrary triangle, we have to introduce the *sine function for obtuse angles*. This notion generalizes the notion of the sine of an angle in a right triangle.

Let us extend our quarter-circle that we used for studying sine and cosine, in Fig. 9.11.4, into a semicircle and consider the sine of some angle $\gamma = \angle AOE > d$ (Fig.9.11.7).

Through point E, draw a line parallel to the diameter. The line intersects OM at D and the circle at some point B. Drop perpendiculars EF and BC upon the diameter.

$EF = DO = BC$, and $\triangle EFO = \triangle ODE = \triangle ODB = \triangle BCO$ (why?).

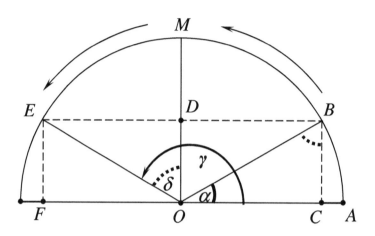

Figure 9.11.7

Therefore, $\angle FOE = \angle AOB = \alpha$, and $\sin \angle FOE = \sin \alpha =$ numerical value of EF. We shall accept *by definition* that $\sin \angle AOE = EF = \sin \alpha$. This definition is natural since both angles, $\angle FOE$ and $\angle AOE$ (α and its supplement), pertain to the same vertical segment, and the latter defines the sine for acute angles. Thus, by definition, we assume:

$\sin (180° - \alpha) = \sin \alpha$,

or,

$\sin \gamma = \sin \alpha$ if γ and α are the supplements of each other.

We shall also use the following corollary of the given definition. In Fig. 9.11.7, $\angle EOD = \delta = \angle OBC = 90° - \alpha$. Then $\cos \delta = \cos(90° - \alpha) = \sin \alpha$, or taking into account that $\delta = \gamma - 90°$, $\cos(\gamma - 90°) = \sin \alpha = \sin \gamma$ (the latter equality follows from the above definition of the sine of an obtuse angle).

Thus, for an obtuse angle γ, $\cos(\gamma - 90°) = \sin \gamma$.

Theorem 9.11.2. *(Law of sines). **In any triangle with the sides a, b,** and **c, opposed the angles α, β, and γ respectively, and circumradius R,** $\dfrac{a}{\sin \alpha} = \dfrac{b}{\sin \beta} = \dfrac{c}{\sin \gamma} = 2R$.*

<u>Proof.</u> We consider separately the cases of acute (Fig. 9.11.8(a)) and obtuse (Fig. 9.11.8(b)) triangles. In any of these cases we start with circumscribing a circle about $\triangle ABC$ and drawing a diameter, BD, through one of the vertices, B, of the triangle.

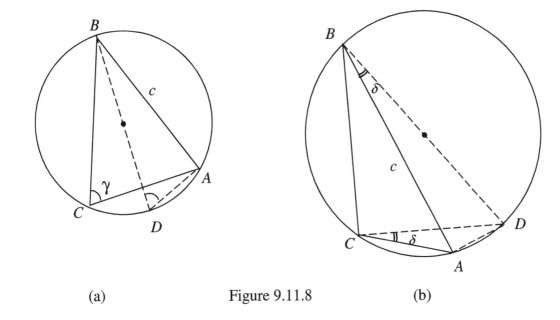

(a)　　　　　　Figure 9.11.8　　　　　(b)

When all three angles are acute (Fig. 9.11.8 (a)), in right ΔDAB, $c = BD \cdot \sin \angle BDA = 2R \sin \gamma$ since $\angle BDA$ and γ are inscribed angles intercepting the same arc ($\cup BA$). Therefore,

$$\frac{c}{\sin \gamma} = 2R.$$

The consideration of the other two sides, $BC = a$ and $AC = b$, is similar.

If one of the angles, $\angle C$ in Fig. 9.11.8 (b), is obtuse, then one can find from right ΔDBA ($\angle DAB = 90°$ as it stands on a diameter) : $c = AB = BD \cdot \cos \delta = 2R \cos \delta$, where $\delta = \angle DBA$ (why?). Then, $c = 2R \cos (\gamma - \angle BCD) = 2R \cos (\gamma - 90°) = 2R \sin \gamma$.

Therefore, in case of obtuse $\angle C = \gamma$, we still have $\dfrac{c}{\sin \gamma} = 2R$, Q.E.D.

9.11　Questions and Problems.

The following standard notations are usually used for elements of triangles.
Sides: a, b, c (c is usually the hypotenuse in a right triangle);
Their opposite angles: $\angle A \equiv \alpha$; $\angle B \equiv \beta$; $\angle C \equiv \gamma$;
The respective medians: m_a, m_b, m_c;
　　altitudes: h_a, h_b, h_c;
　　bisectors: l_a, l_b, l_c;
The radii of the
　　circumscribed circle: R;
　　inscribed circle: r;
　　escribed circles: ρ_a, ρ_b, ρ_c;
Perimeter of the triangle: $P = 2p$, where p stands for the semiperimeter.

To solve a triangle usually means to find all unknown sides and angles based on the given set of data.

1. The legs of a right triangle are equal 5cm and 12cm, respectively. Use the definitions to find the four (basic) trigonometric functions of the least angle of the triangle. Verify the basic trigonometric identities for the received values.

2. The altitude of an isosceles triangle equals 8cm, and the base is 12cm long. Evaluate the four trigonometric functions of an angle at the base of the triangle.

3. Use a protractor and a ruler to find the four trigonometric functions of the following angles approximately with two significant digits: a) 24°; b) 40°; c) 75°.

4. Construct an angle if:
   a) its tangent equals   (i) $4/3$; (ii) 0.6;
   b) its sine equals   (i) $3/4$; (ii) 0.25;
   c) its cosine equals   (i) 0.4;   (ii) 0.75.
   d) its cotangent equals (i) 1.25;   (ii) $\sqrt{5}$.

5. Find the basic trigonometric functions for each of the following angles: a) 30°; b) 60°; c) 45°.

6. *Find the tangent and the other basic trigonometric functions of an angle containing 15°.

7. *Find the basic trigonometric functions of an angle containing 75°.

8. Construct right triangle ABC, if $\angle C = d$, and:
   a) AB=6cm; sin A=0.6 ;
   b) AC=10cm; cos A=0.8;
   c) AB=15cm; tan A=3/4;

9. In a right triangle ABC $\angle C$=d; sin A=4/5; and AC=15cm. Find the projections of the legs onto the hypotenuse.

10. Find the basic trigonometric functions of an acute angle formed by the diagonals of a rectangle whose sides are, respectively, 8m and 12m long.

11. Prove that for any (acute) angle θ, $\tan^2\theta + 1 = \dfrac{1}{\cos^2\theta}$; $\cot^2\theta + 1 = \dfrac{1}{\sin^2\theta}$
    a) geometrically, by means of the Pythagorean theorem;
    b) algebraically, departing from the basic trigonometric identity.

12. Verify the formulae established in #11 for 15° angle.

13. Prove that for any angle θ,
    a) $\sin 2\theta = 2\sin\theta\cos\theta$;   b) $\cos 2\theta = 1 - \sin^2\theta = \cos^2\theta - \sin^2\theta = 2\cos^2\theta - 1$.

14. *Prove that for any (acute) angles α, β
    a) $\cos(\alpha+\beta) = \cos\alpha\cos\beta - \sin\alpha\sin\beta$;
    b) $\sin(\alpha+\beta) = \sin\alpha\cos\beta + \sin\beta\cos\alpha$.
    (restrict yourself to the case when $\alpha+\beta$ is acute as well).

Remark: the above formulae (proved in #13 & #14) **are valid for any** (not necessarily acute) **angles** if the definitions are extended properly.

In the following problems use calculators or trigonometric tables only **where necessary**. In each problem that suggests a numerical answer, make it clear whether the answer obtained is **exact or approximate**. Wherever possible, obtain the exact expression first, and then – an approximate one. To approximate lengths, keep one significant digit after the decimal point unless a finer accuracy is required.

15. Find acute angle α in each of the following cases: a) $\sin \alpha = 0.24$; b) $\cos \alpha = 0.689$; c) $\tan \alpha = 7.2$; d) $\cot \alpha = 1.56$.

16. Find the complement to acute angle α in each of the following cases: a) $\sin \alpha = 0.564$; b) $\cos \alpha = 0.89$; c) $\tan \alpha = 0.32$; d) $\cot \alpha = 12$.

17. Find the acute angles of a triangle if one of its legs is 0.51 of the hypotenuse.

18. A segment forms a 40° angle with a straight line, and the endpoints of the segment are, respectively, 16cm and 12cm away from the line. Find the length of the segment and its projection upon the line.

19. An isosceles triangle with the vertical angle of 70°24′ is inscribed in a circle of radius 10cm. Evaluate the height of the segment cut off by the base from the circle.

20. Find an angle formed by two tangents drawn from the same point to the same circle of radius 90mm, if the length of each of the tangents is 12cm.

21. Given two elements of a right triangle, find all unknown sides and angles:
    a) $c=12$; $\angle A=64°$; b) $a=5$; $\angle A=10°30′$; c) $a=8$; $b=6$; d) $c=13$; $a=5$;
    e) $h=10$; $\angle A=22°30′15″$; f) $a′=6$; $\angle A=22°30′15″$.

22. A chord that subtends an arc of 120° in a circle, is divided into four equal parts. The points of the partition are joined to the centre of a circle by straight segments. Find the angles of the obtained triangles.

23. Evaluate the diagonals of a rhombus if its side is $a = 6$dm, and one of its angles is $\delta = 110°$.

24. Evaluate the unknown side in a triangle with two sides equal, respectively, 3cm and 8cm, and their included angle 75°.

25. Evaluate the unknown sides and angles of a triangle with two sides equal, respectively, 4cm and 10cm, and their included angle 120°.

26. Evaluate the bisector drawn to the longest side in the triangle of problem #25.

27. *Evaluate the longest median in the triangle of problem #25.

28. Find the sine of each of the following angles: a) 120°; b) 135°; c) 122°11′ d) =108°20′15″.

29. Evaluate the unknown side and the circumradius of the triangle in which two sides, equal 10m and 6m, respectively, include a 150° angle.

30. Evaluate the unknown sides, angle, and the circumradius of the triangle in which one side equals 10m and it is included between the angles equal 135° and 15°.

31. *Find the unknown sides and angles of a triangle where side *a* =6cm is included between two angles such that one of them is half the other, and the sine of the least of the two angles equals $3/5$.

9.12. Applications of algebra in geometry.

a) <u>The construction of the golden ratio.</u> We shall start the section by applying algebraic methods to the problem of finding the so-called *golden ratio* of a segment.

<u>Problem 9.12.1.</u> *Given a segment, divide it into extreme and mean ratio, i.e. divide the segment into two parts such that the ratio of the greater part to the lesser part is equal to the ratio of the whole segment to the greater part.*

<u>Solution.</u> Given a segment, *AB* in Fig. 9.12.1, we have to find point *P* on the segment, such that $\frac{AP}{PB} = \frac{AB}{AP}$.

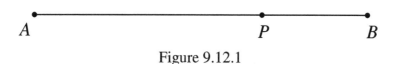

Figure 9.12.1

Let us find *AP*, the part that is the mean proportional between the lesser part and the whole segment. Suggest, for the beginning, that we are talking about the determining the numerical value of that segment. We denote that value *x*, and the length of the whole segment is denoted *a*. Then the length of PB equals *a* − *x*, and the required proportion can be rewritten as $\frac{x}{a-x} = \frac{a}{x}$.

The latter equation can be reduced to $x^2 + ax - a^2 = 0$, with the solutions

$$x = -\frac{a}{2} \pm \sqrt{\frac{a^2}{4} + a^2} \qquad (1)$$

Rejecting the negative solution as geometrically meaningless, we find:

$$x = -\frac{a}{2} + \sqrt{\frac{a^2}{4} + a^2} = -\frac{a}{2} + \sqrt{\frac{5a^2}{4}} = \frac{a(\sqrt{5}-1)}{2} = a \cdot 0.61803... \quad (2)$$

Thus, the problem is solvable and its solution is unique. The length of the segment representing the solution is given by formula (2).

Now let us construct that segment. It is easier to construct the segment using the positive root of the expression (1): $x = \sqrt{\left(\dfrac{a}{2}\right)^2 + a^2} - \dfrac{a}{2}$. (3)

Expression $\sqrt{\left(\dfrac{a}{2}\right)^2 + a^2}$ represents the hypotenuse of a triangle with its legs equal a and $\dfrac{a}{2}$. To obtain x, we construct that hypotenuse and then subtract from it a segment equal $\dfrac{a}{2}$.

Let in Fig. 9.12.2, $AB = a$. From B erect a perpendicular to AB and lay off on it a

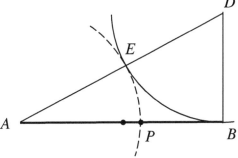

Figure 9.12.2

segment $BD = \dfrac{a}{2}$. Draw AD. By Pythagorean theorem, $AD = \sqrt{\left(\dfrac{a}{2}\right)^2 + a^2}$. From point D describe an arc of radius $\dfrac{a}{2}$. It will cut off on AD a segment (DE) equal $\dfrac{a}{2}$.

Then $AE = AD - DE$ is the segment described by formula (3). From A as a centre, describe an arc of radius $r = AE$. It will cut off on AB segment $AP = AE = x$.

Therefore, point P divides AB in the extreme and mean ratio.

<u>Remark 1.</u> *Golden ratio* is used for the construction of a regular decagon inscribed in a given circle (see Chapter 10).

b). <u>General principles of the use of algebraic methods for geometric problems.</u>

We solved the above problem using the algebraic approach. In particular, we wrote the equation relating the unknown segment with the known one. Then we solved the equation, rejected a non-interpretable solution, and, finally, constructed the segment represented by the relevant solution.

Thus, in general, one can say that the algebraic method of solving geometric problems consists of the following steps:

(1) composition of the equation(s);

(2) algebraic solution;

(3) analysis of the solution(s);

(4) construction.

c). <u>The constructions of some elementary formulae.</u>

Let us show a few constructions of segments represented by simple algebraic formulae. Hereafter *a, b, c, d,* ... denote given segments and *x* stands for unknown segments.

(1) Formulae $x = \dfrac{a}{2}$, $x = \dfrac{a}{3}$, ..., $x = \dfrac{3}{4}a$, ..., etc. are constructed by means of dividing segment *a* into a few equal segments and, if necessary, adding a few of the obtained parts.

(2) Formula $x = \dfrac{ab}{c}$ expresses the so-called *fourth proportional* to segments *a, b,* and *c,*. (One can rewrite the formula as $\dfrac{c}{a} = \dfrac{b}{x}$). The solution follows directly from Th. 9.8.1.

(3) Formula $x = \dfrac{a^2}{c}$ represents the fourth proportional to *a, a,* and *c*. $\left(\dfrac{c}{a} = \dfrac{a}{x}\right)$. The solution is the same as the one of the preceding problem.

(4) Formula $x = \sqrt{ab}$ represents the *mean proportional* of *a* and *b*. The construction is described in section 9.9 (Problem 9.9.1).

(5) Formula $x = \sqrt{a^2 + b^2}$ represents the hypotenuse of a right triangle with legs *a* and *b.*

(6) Formula $x = \sqrt{c^2 - b^2}$ represents the second leg of a right triangle with hypotenuse c and one of the legs equal b.

(7) $x = a\sqrt{\dfrac{2}{3}}$. We can transform it into $x = \sqrt{a^2 \cdot \dfrac{2}{3}} = \sqrt{a \cdot \dfrac{2}{3}a}$.

Now x can be constructed as the mean proportional between a and $\dfrac{2}{3}a$.

(8) $x = \sqrt{a^2 + b^2 - c^2 + d^2}$. This problem can be solved by students.

Hint: for problems like that the Pythagorean theorem should be applied a few times.

We conclude the section with the following statement (that is not proved in this course): *With the use of a compass and a straightedge one can construct only those algebraic expressions that are obtained from known magnitudes by means of a few rational operations and the extraction of square roots.*

9.12 Questions and Problems.

1. Given segments a and b, construct $\sqrt{a^2 + b^2}$.

2. Given segments a and b ($a > b$), construct $\sqrt{a^2 - b^2}$.

3. Given segments a, b, and c, construct segment x such that $\dfrac{x}{a} = \dfrac{b}{c}$ (the fourth proportional to three known segments).

4. Given segments a and b, construct segment x such that $\dfrac{x}{a} = \dfrac{b}{x}$ (the mean proportional of given segments: $x = \sqrt{ab}$).

5. Given segment a, construct $a\sqrt{\dfrac{2}{3}}$.

6. Given segment a, construct $a\sqrt{5}$. (There are a few different ways to do this).

7. Given segment a, construct $a\dfrac{\sqrt{5} - 1}{2}$.

8. Given a segment, divide it into extreme and mean ratio (also called the *golden ratio*) i.e. find a point which partitions the segment into two parts such that the greatest part is to the least part as the whole segment to the greatest part.

9. *Construct an isosceles triangle with its vertical angle equal 36°, having given its lateral side a. (Use only a straightedge and a compass!).

10. Construct $\sqrt{a^2+b^2-c^2+d^2}$.

11. *Construct $\sqrt{a^2+bc}$.

9. Review

Prove the following propositions:

1. A straight line drawn through the midpoints of the bases of a trapezoid, passes through the point of intersection of the diagonals and through the point of intersection of the lateral sides.

2. If a median and a bisector of a triangle emanate from a vertex common to unequal sides, the bisector is less than the median.

3. If segments cut off from the vertex on the sides of an angle are in proportion, then the lines that join the corresponding endpoints are parallel.

4. If two circles touch externally, then the segment of their common tangent with the endpoints at the points of contact is the mean proportional between the diameters.

5. If a square is inscribed in a right triangle in such a way that one of its sides lies on the hypotenuse, then the side of the square is the mean proportional between the two segments joining the two vertices of the square which are located on the hypotenuse with the respectively closest endpoints of the hypotenuse.

6. If two segments, AB and CD, or their extensions intersect at point E such that $BE \cdot EA = DE \cdot EC$, then points A, B, C, and D are located on a circle.

7. A circle, Γ, and two points, A and B, are given. A few circles which either touch or intersect Γ are drawn through A and B. Each chord common to Γ and one of these circles is extended as well as each tangent common to Γ and one of these circles. Prove that all above extensions have a common point that lies on the extension of AB.

8. Two circles in a plane are given and two radii, one in each circle, are revolving in such a way that they are parallel to each other. Prove that the line joining the endpoints of the radii always cuts the centre line at the same point, and this point is the centre of similitude of the circles.

9. A median of a triangle bisects all segments cut by the sides including the median from the lines parallel to the third side (which is bisected by the median).

10. Prove that a median of a triangle can be expressed through its sides as $m_a = \frac{1}{2}\sqrt{2(b^2+c^2)-a^2}$.

11. Three lines are concurrent. If a point moves along one of them, the ratio of the distances from this point to the other two lines is a constant.

12. If two circles are concentric, then the sum of the squares of the distances from a point on one of them to the endpoints of any diameter of the other is a constant.

13. A triangle is divided into four triangles by the segments joining the feet of the altitudes. Prove that   a) the triangles that have common vertices with the original triangle are similar to the

latter; b) the altitudes of the original triangle are the bisectors of the triangle formed by the segments joining the feet of these altitudes.

14. *AB*, a diameter of a circle, is extended beyond *B*, and *C* is a point on this extension. *M* is an arbitrary point on the line that passes through *C* perpendicularly to *AB*. *AM* cuts the circle at some point *P*. Prove that *AM·AP* is a constant (i.e., this value does not depend upon the location of *M* on the line).

15. Prove that altitudes in a triangle are inversely proportional to the corresponding sides.

Find the following loci:

16. The midpoints of all chords passing through a given point on a circle.

17. The points that divide all chords passing through a given point on a circle in a given ratio, *m:n*.

18. The points lying in the interior of a given acute angle whose distances from the sides of the angle are in a given ratio, *m:n*.

19. The points for which the sum of the squared distances from given two points is a constant.

20. The points for which the difference between the squares of the distances to given two points is a constant.

21. The points that divide in a given ratio, *m:n*, all straight segments connecting the points of a given circle with a given point (that may lie in the interior or in the exterior of the circle).

Construction problems.

22. Given a point in the exterior or in the interior of a given angle, draw such a line that the segments included between the point and the sides of the angle are in a given ratio, *m:n*.

23. In the interior of a given triangle, find a point whose distances from the sides of the triangle are in a given ratio, *m:n:p*.

24. Construct a triangle, having given an angle, one of the sides forming this angle, and the ratio of that side to the side opposed the angle. (How many solutions the problem may have?).

25. Construct a triangle, having given its base, its vertical angle (the one opposed the base), and the ratio of the base to one of the lateral sides.

26. Construct a triangle, having given its altitude, the vertical angle and the ratio of the projections of the lateral sides upon the base.

27. Construct a triangle, having given its base, the vertical angle, and the point where the bisector of that angle cuts the base.

28. Construct a triangle, having given two angles and the sum of the base and the corresponding altitude.

29. Construct a triangle, having given two angles and the difference between the base and the corresponding altitude.

30. Construct an isosceles triangle, having given its vertical angle and the sum of the base and a lateral side.

31. Given two points, *A* and *B*, find on their common straight line such a point *C* that *CA:CB = m:n*, where *m* and *n* are given segments or numbers. How many solutions does the problem have?

32. Inscribe a triangle with the given base and the ratio of the lateral sides in a given circle.

33. Inscribe a triangle with a given base and a median to a lateral side in a given circle.

34. Inscribe a square in a given segment of a circle in such a way that two vertices of the square are on the chord of the segment, and the other two – on the subtended arc.

35. Inscribe a square in a given triangle in such a way that one side of the square lies on the base of the triangle, and the other two – on the lateral sides.

36. Inscribe a rectangle with a given ratio of sides (*m:n*) in a given triangle in such a way that one side of the rectangle lies on the base of the triangle, and the other two – on the lateral sides.

37. On a circle, two points are given. Find on this circle a point whose distances from the given points are in a given ratio.

38. Given a square, inscribe it in a triangle (see #35) similar to a given triangle.

39. *Construct a triangle, having given two sides and the included bisector.

40. Given segments *m* and *n*, construct a segment which is in a ratio $\frac{m^2}{n^2}$ to a given segment, *a*.

41. In the exterior of a given circle, find such a point that the (segment of the) tangent to the circle from this point is half the secant drawn from this point through the circle.

42. Given a point in the exterior of a circle, draw a secant which is cut by the circle in a given ratio.

43. Given segments *a, b, c, d,* and *e*, construct segment $x = \frac{abc}{de}$.

44. Given segments *a, b,* and *c*, construct segment $x = \sqrt{4bc - a^2}$.

10. REGULAR POLYGONS AND CIRCUMFERENCE.

> He had explained this to Pooh and Christopher Robin once before, and had been waiting ever since for a chance to do it again, because it is a thing which you can explain twice before anybody knows what you are talking about.
>
> A.A.Milne, *The House at Pooh Corner*

10.1 Regular polygons

A broken line is called *regular* it if satisfies the following three conditions:
(i) its sides are congruent;
(ii) its angles formed by pairs of neighbouring sides, are congruent;
(iii) for any three sequential sides of the line, the first and the third sides are situated on the same side of the line on which the second side lies.

For instance, in Fig.10.1.1 lines *ABCDE* and *FGHKL* are regular, whereas *MNPQR* is not regular since it does not satisfy the third condition.

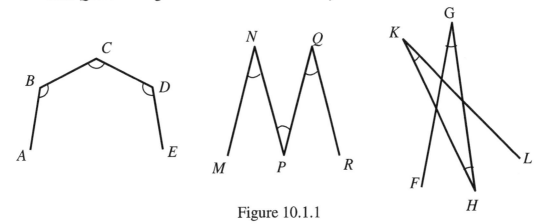

Figure 10.1.1

A polygon bounded by a regular broken line is equilateral and equiangular, i.e. it is a *regular polygon*.

In Fig. 10.1.2 two regular pentagons, convex (a) and non-convex (b), are shown. (Recall the definitions of a convex broken line, given in chapter 4, and of a convex (closed) figure given in chapter 3). We are going to consider only *convex* regular polygons, thus wherever we mention a *regular polygon*, a *convex regular polygon* is suggested unless specified otherwise.

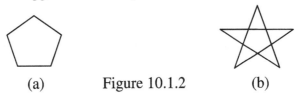

(a)   Figure 10.1.2   (b)

As we shall see from the following theorems, construction of regular polygons is closely related to the problem of division of circles into a number of equal parts.

Theorem 10.1.1. *If a circle is divided into (more than two) equal parts, then*
(i) the figure obtained by joining the neighbouring points of the partition by consecutive chords, is a regular polygon (inscribed in the circle);
(ii) *the figure obtained by drawing tangents touching the circle at the points of the partition and extending each of these tangents till its intersection with the neighbouring two, is a regular polygon (circumscribed about the circle).*

Proof. In Fig. 10.1.3 $A, B, C, D, \ldots$ are the points of the aforesaid partition, i.e. $\cup AB = \cup BC = \cup CD = \ldots$, and $SM, MN, NP, PQ, \ldots$ are the segments of tangents touching the circle at $A, B, C, D, \ldots$, respectively.
Then,
(i) The inscribed polygon $ABCD\ldots$ is regular since $AB = CD = \ldots$ as chords subtending congruent arcs, and all angles of the polygon are congruent as inscribed angles standing on congruent arcs.

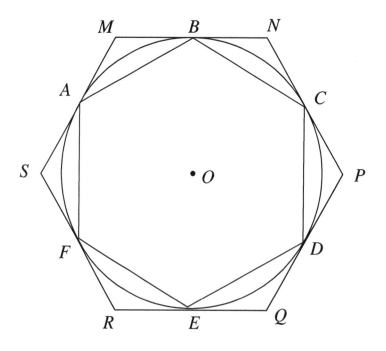

Figure 10.1.3

(ii) To prove the regularity of the circumscribed polygon, $SMNP \ldots$, let us consider
$\triangle AMB$ and $\triangle BNC$.

$AB = BC$ as chords subtending congruent arcs, and $\angle MAB = \angle MBA = \angle NBC = \angle NCB$ since each of these angles is measured by ½ $\cup AB$ = ½ $\cup BC$. Therefore, each of these triangles is isosceles and they are congruent, as well as $\triangle CPD$, $\triangle DQE$, ...

Hence, $\angle M = \angle N = \angle P = ...$, and $MN = 2BN = 2NC = NP = ... = PQ = ...$, and, thus, $MNPQ$ ... is regular, Q.E.D.

Remark 1. Let A, B, C, D, ... in Fig. 10.1.4 divide the circle with its centre at O into equal arcs, and points M, N, P, ... be the points of intersection of the circle with the bisectors of AB, CD, ... drawn from O (these bisectors are perpendicular to the respective chords – why?).

Draw the tangents touching the circle at M, N, P, ... and extend them till they cut each other. Let A_1, B_1, C_1, D_1,... denote the points of intersections. Then (it is easy to prove), $A_1B_1C_1D_1$... is a regular polygon whose sides are respectively parallel to the sides of $ABCD$..., and each triple of points such as OAA_1, OBB_1, ... is collinear: O, A, A_1 are located on the bisector of $\angle SOM$; O, B, B_1 are located on the bisector of $\angle MON$, etc.

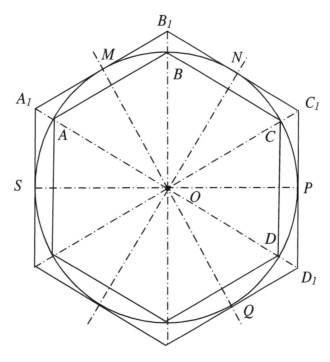

Figure 10.1.4

Theorem 10.1.2. *If a polygon is regular, then*
    (i)   a circle can be circumscribed about it;
    (ii)   a circle can be inscribed in it.

Proof. Draw a circle through three consecutive points, A, B, C, of regular polygon $ABCD$..., in Figure 10.1.5, and prove that the circle passes through the next consecutive vertex, D.

Drop a perpendicular, *OK*, onto chord *BC* and connect *O* with *A* and *D*. Superimpose quadrilateral *ABKO* onto *DCKO* by reflecting it in *OK*. Then *KB* falls along *KC* (since both ∠*OKB* and ∠*OKC* are right), point *B* falls onto *C* (since *BC* is bisected by *K*), *BA* falls along *CD* (since ∠*B* = ∠*C*), and, finally, *A* falls onto *D* (since *BA* = *CD*).

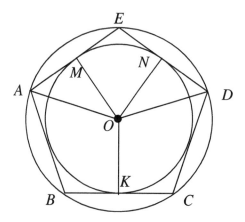

Figure 10.1.5

The latter means that *OA* coincides with *OD*, whence *OA* = *OD* follows, i.e. points *A* and *D* are equidistant from the centre, *O*, of the circle. Thus, *D* lies on the circle passing through *A*, *B*, and *C*. The proof can be repeated for the next (after *D*) consecutive vertex, etc.

(i) It follows from the above proof that the sides of a regular polygon can be considered as equal chords of the circumscribed circle. Those chords are equidistant from the centre of the circle, hence all perpendiculars, *OM*, *ON*, *OK*, ..., dropped from the centre onto the sides of the polygon are equal. Therefore, the circle described from *O* with radius *OM* is inscribed in the polygon.

<u>Corollary 1.</u> *The incentre and circumcentre of a regular polygon coincide.*
<u>Corollary 2.</u> *To find the incentre/circumcentre of a regular polygon, one can*
(i)   *find the point of intersection of the perpendiculars bisecting any two sides of the polygon, or*
(ii) *find the point of intersection of the bisectors of two (interior) angles of the polygon, or*
(iii) *find the point of intersection of a bisector of an (interior) angle of the polygon with the perpendicular bisecting a side of the polygon.*
(Corollary 2 should be proposed to students as a construction problem).
<u>Corollary 3.</u> *A bisector of an interior angle of a regular polygon is a symmetry axis of the polygon.*
<u>Corollary 4.</u> *A perpendicular that bisects any side of a regular polygon is a symmetry axis of the polygon.*
<u>Corollary 5.</u> *For a regular polygon with even number of sides its incentre (circumcentre) is the centre of symmetry.*
(Corollaries 3-5 should be proposed to students as problems).

Let us introduce a few definitions.

The in/circumcentre of a regular polygon is called the *centre of the polygon*.

The radius of the circle circumscribed about a polygon is called the *radius of the polygon*.

The radius of the circle inscribed in a polygon is called the *apothem of the polygon*.

The angle formed by two radii drawn from the centre to the endpoints of one side of a regular polygon, is called the *central angle* of the polygon. Evidently, all central angles of a regular polygon are congruent, and each of them measures $\frac{4d}{n} = \frac{360°}{n}$, where n is the number of sides of the considered polygon.

Each interior angle of a regular polygon with n sides measures $\frac{2d(n-2)}{n} = 180° \cdot \frac{n-2}{n}$ (prove that).

For instance, an interior angle of a regular octagon measures $\frac{2d(8-2)}{8} = \frac{12d}{8} = \frac{3}{2}d = 135°$.

Theorem 10.1.3. *Any two regular polygons with the same number of sides are similar, and their sides, their apothems, and their radii are in the same ratio.*

Proof. In Fig. 10.1.6 regular polygons $ABCD\ldots$ and $A_1B_1C_1D_1\ldots$ have the same number of sides (n). $\angle A = \angle B = \angle C = \ldots = \angle A_1 = \angle B_1 = \angle C_1 = \ldots$ since each of these angles measures $180° \cdot \frac{n-2}{n}$. In addition, $AB = CD = \ldots$, and $A_1B_1 = C_1D_1 = \ldots$; hence

$$\frac{AB}{A_1B_1} = \frac{CD}{C_1D_1} = \ldots$$

Therefore $ABCD\ldots \sim A_1B_1C_1D_1\ldots$, Q.E.D.

If $OM \perp AB$ and $O_1M_1 \perp A_1B_1$, i.e. OM and O_1M_1 are the corresponding apothems, these apothems are the corresponding altitudes in two similar triangles, $\triangle AOB$ and $\triangle A_1O_1B_1$, therefore, $\frac{AO}{A_1O_1} = \frac{AB}{A_1B_1} = \frac{OM}{O_1M_1} = \ldots$, Q.E.D.

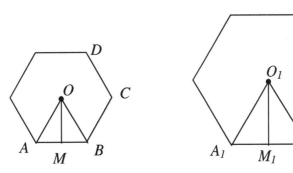

Figure 10.1.6

Corollary 1. *The perimeters of two regular polygons are in the same ratio as their radii or apothems.* (To be proved by students).

For a regular polygon with n sides we use (if it is not specified otherwise) notations a_n for its side, p_n (or P_n) for its perimeter and h_n for the apothem.

<u>Problem 10.1.1.</u> *Given a circle of radius R, find the side of an inscribed*
(i)   square,
(ii)   regular hexagon,
(iii)   regular triangle.
<u>Solution.</u>
(i)   ABCD is a square inscribed in the circle (Fig. 10.1.7a). From right $\triangle AOB$:
$AB^2 = AO^2 + OB^2 = 2R^2$, therefore, $a_4 = R\sqrt{2}$.

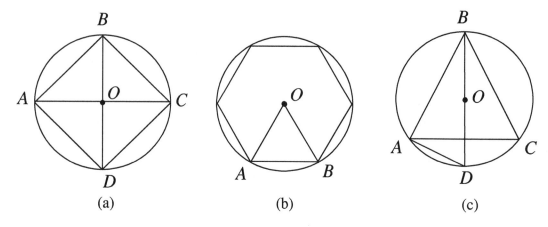

Figure 10.1.7

(ii)   Each central angle of a regular hexagon measures $\dfrac{360^0}{6} = 60^0$, hence any triangle formed by its side and two radii (such as $\triangle AOB$ in Fig.10.1.7b) is isosceles. Therefore, $a_6 = R$.

(iii) $\triangle ABC$ in Fig. 10.1.7c is regular. Draw a diameter, BD, bisecting AC (and, also, perpendicular to AC). Then, from right $\triangle ABD$ we find:
$AB = BD \cos 30° = 2R \cdot \dfrac{\sqrt{3}}{2} = \sqrt{3}R$, i.e.
$a_3 = R\sqrt{3}$.

<u>Problem 10.1.2.</u> *Double the number of sides of an inscribed regular polygon, i.e.*
(i)   Given a regular polygon with n sides inscribed in a circle, inscribe in the circle a regular polygon with 2n sides;
(ii)   Express a_{2n} in terms of a_n and R.
<u>Solution.</u>
(i)   Let AB in Fig. 10.1.8 be a side of a regular n-gon (polygon with n sides) inscribed in a circle with the centre at O. Draw radius $OC \perp AB$ and connect A

with C. $\cup AB$ is bisected at point C, therefore, AC is a side of a regular inscribed polygon with $2n$ sides.

Figure 10.1.8

(ii) In $\triangle ACO$, $\angle O$ is always acute, since $\cup AB$ is less than semicircle, therefore $\cup AC$ is less than quarter circle. Then, by Th.9.9.5, $AC^2 = OA^2 + OC^2 - 2\, OC \cdot OD$, i.e. $a_{2n}^2 = R^2 + R^2 - 2R \cdot OD = 2R^2 - 2R \cdot OD$.

OD can be found from right $\triangle AOD$:

$$OD = \sqrt{AO^2 - AD^2} = \sqrt{R^2 - \left(\frac{a_n}{2}\right)^2} = \sqrt{R^2 - \frac{a_n^2}{4}}.$$ Therefore,

$$a_{2n}^2 = 2R^2 - 2R\sqrt{R^2 - \frac{a_n^2}{4}}.$$

The latter is the required *doubling formula*.

Example 1. Calculate the length of a side of a regular 12-gon inscribed in a circle of radius $R = 1$.

Solution. As we know, $a_6 = R$, then $a_{12}^2 = 2 - 2\sqrt{1 - \frac{1}{4}} = 2 - \sqrt{3}$, whence $a_{12} = \sqrt{2 - \sqrt{3}} = 0.517...$

*Problem 10.1.3. *Inscribe a regular decagon in a given circle and express its side in terms of the radius.*
Solution. If AB in Fig.10.1.9(a) is a side of a regular decagon, then in $\triangle AOB$: $\angle O = 36°$, and $\angle A = \angle B = 72°$ (why?). Let AC bisect $\angle A$; then $\angle OAC = \angle CAB = 36°$, whence $\angle OAC = \angle O$, and $\angle ACB = 180° - \angle CAB = \angle B = 72°$.

Therefore, $\triangle OAC$ and $\triangle CAB$ are isosceles with $AC = OC$ and $AC = AB$; then it follows from these equalities that $AB = OC$.

By the property of the bisector of an angle, $\dfrac{OA}{AB} = \dfrac{OC}{CB}$.

By replacing in the latter equality OA and AB by their respectively congruent OB and OC, we obtain: $\dfrac{OB}{OC} = \dfrac{OC}{CB}$, i.e. OC is the greater part of segment OB divided by *C* into the extreme and mean ratio (see Problem 9.12.1 concerning the *golden section*).

Hence, if *x* denotes the length of a side *AB* of the required decagon, and as we have just established *OC* = *AB*, the length of *OC* is also *x*, and

$$\frac{R}{x} = \frac{x}{R-x}, \text{ whence } a_{10} = x = R\frac{\sqrt{5}-1}{2} = R \cdot 0.61803...$$

In order to construct the decagon, we divide a radius, e.g. *OA* in Fig. 10.1.9(b), into the extreme and mean ratio and then by radius $AC = a_{10}$ we partition the circle into 10 congruent parts: ∪*AB*, ∪*BD*, ∪*DE*, ... Then *ABDE*... is the required regular decagon.

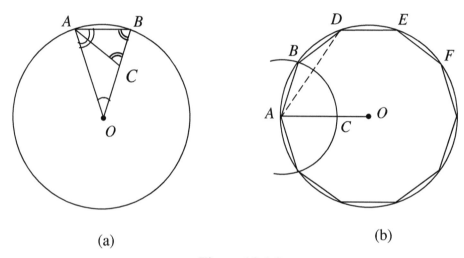

(a)   (b)

Figure 10.1.9

Remark 1. To inscribe a regular pentagon into a given circle, one can repeat the same construction and, at the very end, connect consecutively each second point of the obtained division, e.g. draw *AD*, then *DF*, ...

Remark 2. It is easy to divide a circle into 15 equal parts. Let us notice that $\frac{1}{15} = \frac{1}{6} - \frac{1}{10}$. Then, we can construct an arc equal to $\frac{1}{6}$ of the whole circle and subtract from it an arc equal to $\frac{1}{10}$ of the circle.

Summarizing the above results, we can state that we know how to divide a circle into certain numbers of equal parts (using a straightedge and a compass). These numbers are represented by the following table:

2,	2·2,	2·2·2,...	2^n
3,	3·2,	3·2·2,...	$3·2^n$
5,	5·2,	5·2·2,...	$5·2^n$
15,	15·2,	15·2·2,...	$15·2^n$

It has been proved by Karl Gauss (a German mathematician, who lived in the 19th century) that with a straightedge and a compass alone one can divide a circle into a *prime* number of equal parts if the prime number can be represented as $2^{2^n}+1$. (For instance, $17 = 2^{2^2}+1$, $257 = 2^{2^3}+1$).

It has also been proved that one can divide a circle into a *composite* number of equal parts if the composite number does not contain any prime factors but the ones of the form $2^{2^n}+1$ and 2. (For instance, one can inscribe in a circle a regular 170-gon, as $170 = 2 \cdot 5 \cdot 17 = 2(2^2+1)(2^{2^2}+1)$.

In practice, to divide a circle into a number of equal parts not represented in our table, one can use a protractor and solve the problem *approximately* by constructing the respective central angles. For instance, to divide a circle into 7 equal parts (inscribe a regular 7-gon), one has to divide the perigon with the vertex at the centre into 7 congruent central angles, each of them containing $\frac{360^0}{7} \approx 51.4^0$.

10.1 Questions and Problems.

1. Adduce, if possible, an example of an equiangular but not regular
   a) quadrilateral;   b) pentagon;   c) hexagon.

2. Adduce, if it is possible, an example of an equilateral but not regular polygon with *n* sides for
   a) *n*=4;   b) *n*=5;   c) *n*=6.

3. Determine the (interior) angles of a regular   a) triangle;   b) quadrilateral;
   c) pentagon;   d) hexagon;   e) octagon;   f) decagon;   g) dodecagon (12-gon).

4. Determine the central angle that subtends a side of a regular   a) triangle;   b) quadrilateral;   c) pentagon;
   d) hexagon;   e) octagon;   f) decagon;   g) dodecagon.

5. What kind of regular polygon has a central angle of a) 30° ?   b) 72° ?   c) 36° ?   d) 24° ?

6. Is an inscribed polygon necessarily regular?

7. Is a circumscribed polygon necessarily regular?

8. Is an inscribed equilateral polygon necessarily regular?

9. Is an inscribed equiangular polygon necessarily regular?

10. Is a circumscribed equilateral polygon necessarily regular?

11. Is a circumscribed equiangular polygon necessarily regular?

12. Three congruent circles are inscribed in a circle of radius *R* in such a way that each of the three touches the other two and the circle in which they are inscribed. Find the radii of the inscribed circles.

13. Find the angle between two non-adjacent and non-parallel sides of a regular hexagon.

14. Prove that the midpoints of the sides of a regular polygon are the vertices of a regular polygon with the same number of sides.

15. Prove that the points that divide in a given ratio when traversed in the same direction the sides of a regular polygon are the vertices of a regular polygon with the same number of sides.

16. In order to check if a cross-section of a cable is a right octagon, a worker measures angles between the faces as shown in Figure 1. What is the expected value of these angles?

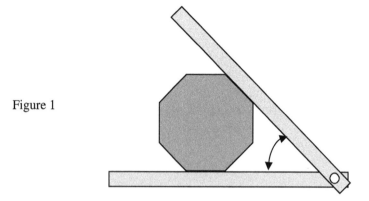

Figure 1

17. The picture of a regular polygon had been incompletely erased: one side and a part of a neighbouring side have survived (see Figure 2). Draw the inscribed and circumscribed circles of the former polygon (without drawing the polygon).

Figure 2

18. Prove that the perimeters of two regular polygons with the same number of sides are in the same ratio as their respective radii and respective apothems.

19. Find the side of a regular a) triangle; b) square; c) hexagon inscribed in a circle of radius R.

20. Find the side of a regular a) triangle; b) square; c) hexagon circumscribed about a circle of radius R.

21. Find the side of a regular a) octagon; b) 16-gon inscribed in a circle of radius R.

22. Find the side of a regular a) 12-gon; b) 24-gon inscribed in a circle of radius R.

23. Prove that a regular n-gon has n axes of symmetry, and all of them pass through its centre.

24. Prove that the centre of a regular polygon with the even number of sides is its centre of symmetry.

25. Let AB, BC, and CD be three consecutive sides of a regular polygon with the centre at some point O. Point F is the point of intersection of the extensions of AB and CD. Prove that quadrilateral $AOFC$ is inscribable.

26. *Prove that in a regular pentagon, two diagonals which do not have common endpoints cut each other in the extreme and mean ratio.

27. Prove that a side of an inscribed regular decagon is the greatest segment of the two obtained by dividing the radius into extreme and mean ratio.

28. Find the side of a regular decagon inscribed in a circle of radius R. Inscribe a regular decagon into a given circle.

29. Inscribe a regular pentagon into a given circle.

30. *Inscribe a regular 15-gon into a given circle.

31. Find the radius and the apothem of a regular hexagon if their difference equals l.

32. Construct a regular octagon, having given its side.

33. Construct a regular decagon, having given its side.

34. Construct the following angles: a) 120°;   b) 72°;   c) 75°;   d) 18°;   e) 22°30′.

35. A regular polygon is circumscribed about a given circle. Use this polygon to inscribe in the same circle a regular polygon with the number of sides twice the number of sides of the given polygon.

36. Find the radius of the circle, having given the difference, a, between the sides of regular inscribed and circumscribed n-gons.

37. Prove that the perimeter of a circumscribed regular n-gon is greater than the perimeter of an inscribed in the same circle regular n-gon.

38. Express the perimeter, P_n, of a circumscribed about a circle regular n-gon through the radius of the circle.

39. Express the perimeter, p_n, of an inscribed in a circle regular n-gon through the radius of the circle.

40. Prove that for any acute angle θ, $\sin\theta < \tan\theta$.

10.2 The circumference of a circle.

> " I shall do it," said Pooh after waiting a little longer, "by means of a trap. And it must be a Cunning Trap..."
> A.A.Milne, *Winnie-The-Pooh*.

The circumference of a circle can be perceived as the boundary of the circle, similarly to the perimeter of a polygon, which is defined as the sum of all the sides of the polygon. There is however a substantial difference between the notions of the circumference of a circle and the perimeter of a polygon. The latter is the sum of straight segments, and therefore it can be identified as a straight segment itself.

A straight segment can be compared to another straight segment chosen as a unit of length. As a result of such a comparison, we can determine the length of any straight segment. Thus, the length of the perimeter of a polygon is the length of a straight segment congruent to the perimeter. Measuring the length of the circumference of a circle is quite a different task.

Two arcs of the same circle (or of two circles of equal radii) can be compared, since two arcs of the same circle can be superimposed, one upon the other. Yet, an arc of a circle (or some other curve) cannot be superimposed upon a part of a straight line, and by that reason we are not able to compare an arc with a straight segment. Particularly, it is not defined (yet) how one can compare an arc with a unit of length, which is a straight segment.

Therefore, in order to introduce a notion of the length of an arc as compared to a given (straight) unit of length, we must first identify an arc with a straight segment.

That is why first of all we are going to *define the circumference* of a circle. The definition will identify the circumference of a circle with a straight segment. Then we will be able to measure this segment; that is, to measure the circumference.

Since the procedure of defining the circumference is rather lengthy, we shall first outline the plan we are going to follow and the major ideas behind our actions.

It is intuitively clear that the perimeters of inscribed and circumscribed convex polygons provide good approximations for the circumference. See, for instance, Figure 10.2.1, showing the regular octagons, inscribed in a circle of a given radius and circumscribed about it.

 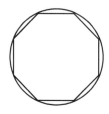 

Figure 10.2.1

If we take a polygon with a greater number of sides, it will look even more similar to the inscribed or circumscribed circle. It seems (and we shall prove it is really so) that as we increase the number of sides of inscribed and circumscribed convex polygons we obtain better and better approximations for the circumference by means of their perimeters. The circle always remains in between the inscribed and circumscribed polygons, so it feels like the circumference must be somewhere in between their perimeters.

Thus our plan is the following. We shall prove that

the perimeter of a regular convex polygon inscribed in a circle increases if the number of sides of the polygon increases;

the perimeter of a circumscribed polygon decreases when the number of sides increases;

the perimeter of an inscribed convex regular polygon is less than the perimeter of a circumscribed regular polygon;

by choosing a polygon with a sufficiently great number of sides, we can make the difference between the circumscribed and inscribed perimeters as small a segment as we please;

then it will follow from Cantor's principle that there is exactly one segment that is greater that any inscribed perimeter and less than any circumscribed perimeter.

This segment will be defined as the circumference.

For convenience, let us make some agreements concerning the terminology and notations. As we have already mentioned in the beginning of the chapter, we are going to consider only *convex* polygons; so the word *polygon* stands for *convex polygon* unless specified otherwise.

We shall often consider sets that we call *sequences*. A *sequence* is a set of objects that are ordered by means of natural numbers: each object is assigned a certain positive integer, which is its ordinal number. Usually these objects, called the *terms of the sequence*, are quantities of some kind: segments, lengths, etc., and the ordinal numbers are shown as subscripts. For example, in Chapter 9 we considered the sequences $\{l_n\}$ and $\{u_n\}$ of the lower and upper estimates of the length of a segment. The terms l_1 and u_1, for instance, are the lower and upper estimates accurate to the first decimal. Another example from Chapter 9: a sequence of nested segments mentioned in Cantor's principle.

We suggest that a circle of some radius R with its centre at some point O is given. Two *sequences* of segments will be considered: the sequence $\{p_n\}$ of the segments congruent to the perimeters of regular convex polygons inscribed in the circle and the sequence $\{P_n\}$ of the segments congruent to the perimeters of regular convex polygons circumscribed about the circle. For brevity we shall often say *the perimeter of an n-gon* instead of *a segment congruent to the perimeter of a polygon with n sides*. The perimeter of an inscribed regular polygon with n sides will be denoted p_n and its side will be denoted a_n; the respective notations for a circumscribed polygon of n sides will be P_n and b_n.

For example, it follows from the results of Problem 10.1.1 that

for an inscribed square $a_4 = R\sqrt{2}; \Rightarrow p_4 = 4a_4 = 4R\sqrt{2}$;

for an inscribed regular hexagon $a_6 = R; \Rightarrow p_6 = 6R$.

One can easily find the following expressions for the sides and perimeters of a circumscribed square and of a regular hexagon:

$b_4 = 2R; \quad P_4 = 8R; \quad b_6 = \frac{2}{\sqrt{3}}R; \quad P_6 = 4\sqrt{3}R$. (Find these as an exercise).

We shall call a sequence *monotonically increasing* if each subsequent term of the sequence is greater than the preceding one. A sequence will be called *monotonically decreasing*, if every term of the sequence, starting with the second one, is less than its preceding term.

A sequence of segments will be called *bounded above* if there exists a segment that is greater than any segment of the sequence. Similarly, a sequence of segments is said to be *bounded below* if there exists a segment that is less than any term of the sequence.

Examples: (1) The sequence of sums of all broken lines joining two distinct points is bounded below, since each of them is greater than the straight segment connecting these points. (2) The sequence of perimeters of all the convex polygons contained in the interior of a given polygon, is bounded above since each of the perimeters of the sequence is less than the perimeter of the given polygon (prove that!).

Now we are ready to proceed with our plan that will result in defining the circumference of a circle.

Lemma 10.2.1. *The sequence of the perimeters of inscribed convex regular polygons obtained by doubling the number of sides of a polygon is monotonically increasing.*

Proof. Let $ABCD...$ be a regular n-gon inscribed in a circle centered at O (Figure 10.2.2). We can construct a regular $2n$-gon by bisecting each of the arcs $\cup AB, \cup BC, \cup CD,...$ by means of radii $OM, ON, OP,...$ and consecutively joining the vertices of the n-gon with the points bisecting the arcs. This way we shall create a $2n$-gon $AMBNCPD...$ inscribed in the circle.

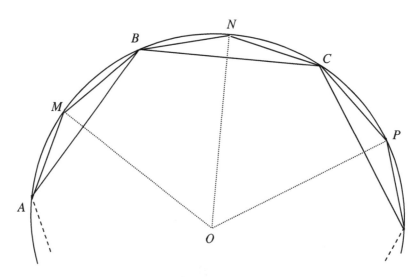

Figure 10.2.2

It follows from the triangle inequality that

$AM + MB > AB$;

$BN + NC > BC$;

$CP + PD > CD$;

...

By adding these inequalities, we obtain
$AM + MB + BN + NC + CP + PD + ... > AB + BC + CD + ...,$ which means $p_{2n} > p_n$, Q.E.D.

Remark. It is easy to see that a similar proof can be applied to non-regular convex polygons. The triangle inequalities $AM + MB > AB$, and the others used in the proof are true regardless of whether AM is equal to MB or not. Also, we have never used in the proof the fact that $ABCD...$ is a regular. The whole proof is based on the fact that we have created a new polygon, with a greater number of vertices. Thus a more general result is true:

Theorem 10.2.2. The sequence of the perimeters of inscribed convex polygons is increasing if each subsequent polygon is obtained from the preceding one by adding new vertices.

Still, for our goal of defining the circumference it will suffice to consider only regular polygons, whose perimeters, by the way, are easier to estimate.

Lemma 10.2.3. The sequence of the perimeters of circumscribed convex regular polygons obtained by doubling the number of sides of a polygon is monotonically decreasing.

Proof. Let $ABCD...$ be a circumscribed convex regular n-gon, which touches the circle at the points $T_1, T_2, T_3, ...$ (Figure 10.2.3).

We can double the number of sides of the polygon by adding new points of tangency: S_1 lying between T_1 and T_2, S_2 between T_2 and T_3, etc., and drawing straight lines that touch the circle at these new points of tangency.

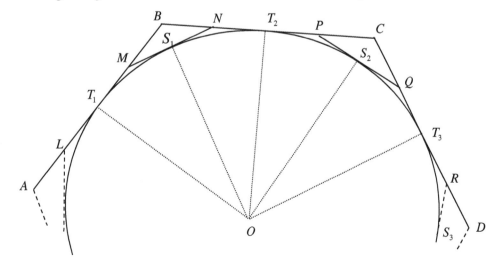

Figure 10.2.3

These tangent lines cut the sides of the original *n*-gon at some points *L*, *M*, *N*, *P*, *Q*, … . These points will be the vertices of a new circumscribed polygon, whose sides are tangent to the circle at the points $T_1, S_1, T_2, S_2, T_3, S_3,...$. Thus, we shall create a circumscribed polygon *LMNPQR*…with twice as many points of tangency as that of the original polygon *ABCD*… . Since each vertex of a circumscribed polygon is followed by its point of contact with the circle, the number of the vertices of a circumscribed polygon is equal to the number of the points of tangency. Therefore, the number of vertices of the new polygon *LMNPQR*… will be twice the number of sides of the original polygon *ABCD*… .

One can say that we have created a 2*n*-gon by "cutting the corners" of the original *n*-gon: vertex *B* with the broken line *MBN* have been replaced with two vertices *M* and *N* and the segment that joins them; vertex *C* with *PCQ* is replaced with two vertices P and *Q* and a straight segment *PQ*, etc. Hence the perimeter of the 2*n*-gon is obtained from the perimeter of the original *n*-gon by means of replacing segments of broken lines by straight segments joining the same pairs of points, and thus, by the triangle inequality, the perimeter of the 2*n*-gon is less than that of the n-gon, Q.E.D.

Corollary 1. *The perimeter of any regular polygon circumscribed about a circle of radius R, is less than 8R if the polygon has more than four sides and is obtained as a result of sequential doubling the number of sides of a regular triangle or a square.*
Proof. The corollary follows immediately from the latter lemma and the expressions $P_4 = 8R$; $P_6 = 4\sqrt{3}R$.

Remark 1. If the original polygon is regular and the additional points of tangency $S_1, S_2, S_3,...$ bisect the arcs $T_1T_2, T_2T_3, T_3T_4,...$ respectively, the constructed 2*n*-gon will be regular as well.

Remark 2. When proving Lemma 10.2.3, we have never taken advantage of the suggestion that the original *n*-gon is regular. Also, it is clear from the construction used in the proof that by adding even a single new vertex to a circumscribed polygon, we replace a segment of a broken line with a straight segment, and thus the perimeter of the constructed (*n* +*1*)-gon is less than that of the original *n*-gon.

Therefore it is easy to prove, and we actually did prove that a more general result holds:

Theorem 10.2.4. *The sequence of the perimeters of circumscribed convex polygons is decreasing if each subsequent polygon is obtained from the preceding one by adding new sides that are tangent to the circle.*

Although we have mentioned this result, generalizing Lemma 10.2.3, we shall need only the statement of the latter lemma for our main goal of the section

– defining the circumference. We cannot help mentioning generalizations of the obtained results, yet for the readers' convenience we shall specify which results are essential for our discussion by calling them lemmas.

Remark 3. Think of possible generalizations of Corollary 1. For example, one can use the results of Problem 10.1.3 to show that the perimeter of a regular decagon and therefore of all its descendants obtained by the doubling procedure, is less than $8R$.

It can be proved that the conclusion of the corollary is true for any regular polygon (we do not consider such a proof). For achieving our goal of defining the circumference, we shall use a less accurate estimate ($P_n < 16R$), which will be proved below for any regular convex circumscribed polygon.

Now we are going to prove that *the perimeter of a circumscribed polygon is greater than the perimeter of any convex polygon inscribed in the same circle.*

The proof is not difficult but lengthy and it will be convenient to introduce some new terminology and notations.

We say that one polygon *includes* another if the second polygon lies entirely (except, maybe, a few points common for both polygons) within the interior of the first one. We also call such polygons the *including* and the *included*, respectively. For example, a polygon inscribed in a circle is included in any polygon circumscribed about the same circle.

We shall also need in our discussion the notions of *included* and *including* broken lines. A polygon *includes* a broken line if, with the exception of a few common points, the entire broken line is situated within the interior of the polygon. For example, in Figure 10.2.4, polygon *ABC...DEF includes* the broken line *AMNF*. (The dots in *ABC...DEF* and the dotted line in the figure mean that there may be more vertices and sides between *C* and *D*; we shall use such notations in our discussion).

If two broken lines have common endpoints and the first line is included in the polygon formed by the second line and the segment joining the endpoints, the second broken line is said to *include* the first broken line. For example, *ABC...DEF* includes *AMNF* in Figure 10.2.4; points *A* and *F* are the common endpoints of the two broken lines.

We shall say that one broken line is *greater* than the other if the sum of the sides of this line is greater than that of the second line. The second broken line will be said to be *less* than the first one.

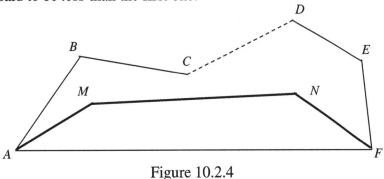

Figure 10.2.4

It is convenient to have a special notation for the sum of the sides of a broken line and for the perimeter of a polygon with given vertices. We shall denote the sum of the sides of a broken line *ABC...DEF* by S(*ABC...DEF*), i.e. S(*ABC...DEF*) = *AB* + *BC* + ... + *DE* + *EF*.

The perimeter of a polygon *ABC...DEF* will be denoted P(*ABC...DEF*), i.e. P(*ABC...DEF*) = *AB* + *BC* + ... + *DE* + *EF* + *FA*.

It follows from these definitions that
P(*ABC...DEF*) = S(*ABC...DEF*) + *FA*.

Now we are ready to prove two important lemmas (the first one is auxiliary for the second), which imply that the perimeter of a circumscribed polygon is greater than the perimeter of any convex polygon inscribed in the same circle.

Lemma 10.2.5. *A convex broken line is less than any broken line that includes it.*

Proof. We shall prove the statement using induction by the number of sides of the included broken line.

First step. Let us prove the statement for a convex broken line that consists of two sides.

Suppose a broken line *ABC...DE...F includes* some broken line *AMF* (Figure 10.2.5). We are going to prove that S(*AMF*) < S(*ABC...DE...F*).

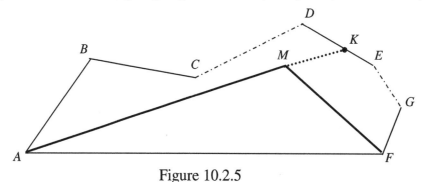

Figure 10.2.5

Let us extend *AM* beyond *M* till it intersects one of the sides of the including broken line, e.g. some side *DE* at some point *K*. As a straight segment connecting two points is less than any broken line joining the same points, we can write

$AM + MK = AK < AB + BC + ... + DK$ and $MF < MK + KE + ... + GF$. By adding these inequalities, we obtain
$AM + MK + MF < MK + AB + BC + ... + DK$. After subtracting *MK* from both sides of this inequality and taking into account that $DK + KE = DE$, it can be rewritten as $AM + MF < AB + BC + ... + DE + ... + GF$, which means S(*AMF*) < S(*ABC...DE...F*). Thus, the first step of induction has been proved.

Inductive step. Suppose the statement of the lemma holds for any included convex broken line with *n* or fewer sides. Let us show that the conclusion of the lemma will be true for an included convex line with (*n+1*) sides.

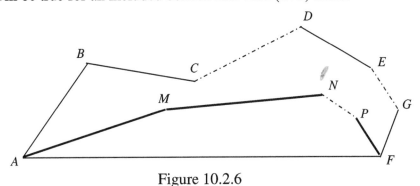

Figure 10.2.6

Suppose a convex broken line *AMN...PF*, which consists of (*n+1*) sides, is included by a broken line *ABC...DE...GF* (Figure 10.2.6).

We have to prove that $S(AMN...PF) < S(ABC...DE...GF)$.

Let us extend the first side *AM* of the included broken line beyond *M* till it intersects some side *DE* of the including polygon at some point *K* (Figure 10.2.7). The obtained convex broken line *MNP...F* consists of *n* sides, and therefore, by the inductive suggestion, $S(MNP...F) < S(MKE...GF)$, i.e.
$MN + NP + ...PF < MK + KE + ... + GF$.

Also, straight segment $AK = AM + MK$ is less than broken line *ABC...DK* which joins the same pair of points *A* and *K*, which means $AM + MK = AK < AB + BC + ... + DK$.

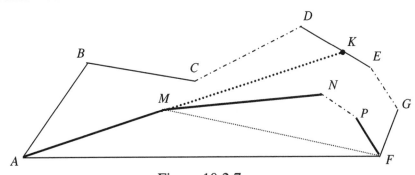

Figure 10.2.7

By adding the last two inequalities, we obtain
$AM + MK + MN + ... + PF < AB + BC + ... + DK + MK + KE + ... + GF$.

After subtracting *MK* from both sides of the inequality and taking into account that $DK + KE = DE$, the latter inequality results in
$AM + MN + ... + PF < AB + BC + ... + DE + ... + GF$, which means that.
$S(AMN...PF) < S(ABC...DE...GF)$. This completes the proof of the inductive step: if the inductive suggestion is true for *n*, it will be true for (*n+1*).

This completes the proof by induction. Thus, any included convex broken line is less than a broken line that includes it, Q.E.D.

An analogous statement is true for polygons, and its proof is based on the above lemma.

Lemma 10.2.6. *__The perimeter of any convex polygon is less than the perimeter of a polygon that includes the first polygon.__*

<u>Proof.</u> Let some polygon *A...BC...DE...F* include a convex polygon *MNP...L* (Figure 10.2.8). We have to prove that $P(MN...PL) < P(A...BC...DE...F)$.

Let us suppose that at least one of the sides of the included polygon, e.g. *MN*, lies entirely in the interior of the including polygon. Let this side be extended beyond its endpoints. These extensions will intersect two sides of the including polygon: some side *BC* at some point *Q* and some other side *DE* at some point *R*.

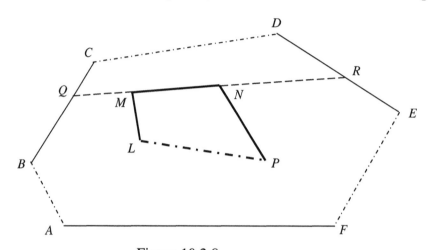

Figure 10.2.8

By applying the result of the preceding lemma to the convex broken line *NP...LM* and broken line *NRE...FA...BQM* which includes it, we can write
$NP + ... + LM < NR + RE + ... + FA + ... + BQ + QM$.

QR, which is equal to $QM + MN + NR$, is a straight segment and thus it is less than any broken line connecting *Q* and *R*; therefore
$QM + MN + NR < QC + ... + DR$. After adding the latter two inequalities, we obtain
$NP + ... + LM + QM + MN + NR < NR + RE + ... + FA + ... + BQ + QM + QC + ... + DR$
, which can be reduced to
$MN + NP + ... + LM < QC + ... + DR + RE + ... + FA + ... + BQ$. This inequality can be rewritten as $MN + NP + ... + LM < BC + ... + DE + ... + FA + ...$, which means $P(MN...PL) < P(A...BC...DE...F)$, <u>Q.E.D.</u>

In the beginning of the proof we made an assumption that at least one side of the included polygon lies entirely in the interior of the including one. Let us consider the other possibility: suppose that all vertices of the included polygon lie on the sides of the including polygon (Figure 10.2.9).

Then each side of the included polygon is the shortest distance between two neighbouring vertices. (For instance, $MN < NF + ... + EM$, etc.). By adding all the inequalities expressing this fact, we obtain the required inequality between the perimeters. This completes the proof.

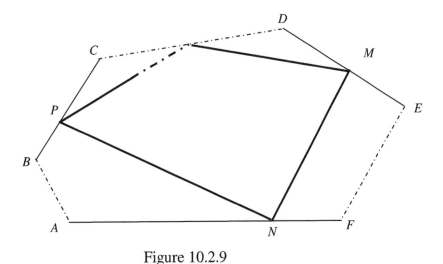

Figure 10.2.9

The following corollaries, which are crucial for defining the circumference, follow immediately from this lemma:

Corollary 1. *The perimeter of any convex polygon inscribed in a circle is less than the perimeter of a polygon circumscribed about this circle.*
Proof. A polygon inscribed in a circle is always included in any polygon circumscribed about this circle, whence the statement follows.

Corollary 2. *The perimeter of any convex polygon inscribed in a circle of radius R, is less than 8R.*
Proof. Any convex polygon inscribed in a circle is included in a square circumscribed about the circle, hence, by Th. 10.2.6, the perimeter of this polygon is less than $P_4 = 8R$, Q.E.D.

Also, we can obtain an upper estimate for the perimeter of any circumscribed regular convex polygon:
Corollary 3. *The perimeter of any convex regular polygon circumscribed about a circle of radius R, is less than 16R.*
Proof. Given a circle of radius R with its centre at some point O, draw a square with the centre of symmetry at O and a side equal $4R$ (Figure 10.2.10a). Its perimeter is $16R$. The closest to the center points of the square, such as point P in the figure, are located at the distance $2R$ from the centre; all the other points of the square will be farther away.

Let us show that all regular (convex) polygons circumscribed about the circle, are included in this square.

It is easy to see that for a circumscribed regular triangle, the distance from the centre to the vertex is $OP = OM/\cos 60^0 = 2R$. Let us prove that for any other regular n-gon it will be less.

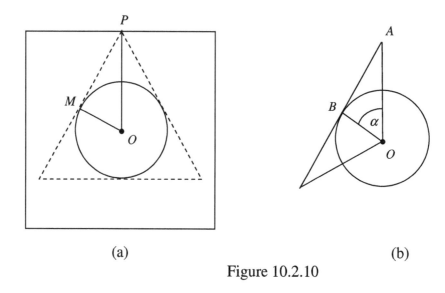

Figure 10.2.10

Let A be a vertex of a regular circumscribed polygon with n sides, and a side emanating from A is tangent to the circle at some point B (Fig.10.2.10b). Then $OB = R$, and $OA = R/\cos\alpha$. Since the number of sides n is greater than 3, angle α is less than 60^0; therefore its cosine is greater than $\frac{1}{2}$, and hence

$$OA < R/\cos 60^0 = 2R.$$

The latter inequality means that any regular convex polygon circumscribed about the given circle will be included in the square; therefore, by Lemma 10.2.6, the perimeter of any such polygon is less than the perimeter of the square, i.e. it is less then $16R$, Q.E.D.

The estimate obtained in this corollary, rough as it is, will turn out to be very useful in the proof of the following important result:

Lemma 10.2.7. *A side of a circumscribed regular polygon can be made less than any given segment by choosing a polygon with a sufficiently large number of sides.*

Proof. Let there be a sequence of regular (convex) n-gons circumscribed about a given circle and obtained as a result of doubling the number of sides of a polygon. As we have already mentioned, we shall denote the side and perimeter of a circumscribed regular polygon with n sides b_n and P_n respectively.

According to the above estimate of Corollary 3, the perimeter of any regular (convex) circumscribed polygon is less than $16R$, where R is the radius of the circle.

Then, $b_n = \frac{1}{n}P_n < \frac{1}{n}16R$, i.e. $b_n < \frac{1}{n}16R$.

Let ε be a given segment (maybe a very small one). It follows from the Archimedes Principle (I Continuity Axiom) that for any two segments ε and $16R$, there exists such a (positive) integer n that $n\varepsilon > 16R$, hence no matter how small ε is, we can find such n that $\frac{1}{n}16R < \varepsilon$.

Thus we have already established that $b_n < \frac{1}{n}16R$, and if n is sufficiently large, $\frac{1}{n}16R < \varepsilon$; therefore, $b_n < \frac{1}{n}16R < \varepsilon;$ $\Rightarrow b_n < \varepsilon$, Q.E.D.

Now we are ready to prove that, speaking in common language, the difference between the perimeters of circumscribed and inscribed regular polygons vanishes (can be made as small as desired) when the number of their sides increases.

Lemma 10.2.8. *For a given circle, the difference between the perimeters of regular polygons with the same number of sides, one of them inscribed and the other circumscribed, can be made less than any given segment by sequential doubling the number of sides of the polygons.*

Proof. Suppose a circle of some radius R with its centre at some point O is given. Let $\{P_n\}$ and $\{p_n\}$ be the sequences of the perimeters of regular circumscribed and inscribed polygons obtained by doubling the number of sides. We shall show that for any given segment ε, no matter how small it is, we can find such a natural number n that $P_n - p_n < \varepsilon$, and similar inequalities will hold for all the subsequent polygons of these sequences, i.e. $P_{n+1} - p_{n+1} < \varepsilon$, etc.

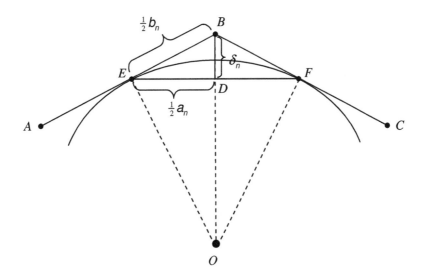

Figure 10.2.11

In Figure 10.2.11, an arc of the circle and two sides, *AB* and *BC*, of a circumscribed regular *n*-gon and one side, *EF*, of an inscribed regular *n*-gon are shown. The point of tangency bisects the side of the circumscribed regular polygon (why?), hence $EB = \frac{1}{2}b_n$. *OB* is a perpendicular bisector of *EF* (why?); therefore $ED = \frac{1}{2}a_n$. For convenience, we denote: $BD = \delta_n$.

As it follows from the triangle inequality for $\triangle BDE$, $\frac{1}{2}b_n - \frac{1}{2}a_n < \delta_n$. We shall use this estimate of the difference between the sides later on in our proof.

Now let us estimate the difference between the two perimeters, circumscribed and inscribed:

$$P_n - p_n = nb_n - na_n = n(b_n - a_n) = 2n(\tfrac{1}{2}b_n - \tfrac{1}{2}a_n) < 2n\delta_n.$$

From the similarity of triangles *BEO* and *BDE* (why are they similar?), we can write:

$$\frac{BD}{ED} = \frac{EB}{OE}, \text{ or } \frac{\delta_n}{\frac{1}{2}a_n} = \frac{\frac{1}{2}b_n}{R}; \Rightarrow \delta_n = \frac{a_n}{4R}b_n.$$

By substituting this into the expression for the difference between the perimeters, we obtain:

$$P_n - p_n < 2n\delta_n = 2n\frac{a_n}{4R}b_n = \frac{2na_n}{4R}b_n = \frac{p_n}{2R}b_n.$$

According to Corollary 2 of Lemma 10.2.6, $p_n < P_4 = 8R$; therefore

$$P_n - p_n < \frac{p_n}{2R}b_n < \frac{8R}{2R}b_n = 4b_n, \text{ i.e. } P_n - p_n < 4b_n.$$

We have proved in Lemma 10.2.7 that by choosing a sufficiently great number of sides *n*, we can make b_n less than any given segment, no matter how small this segment is. Let us choose such *n* that $b_n < \frac{1}{4}\varepsilon$.

Then $P_n - p_n < 4b_n < 4\frac{1}{4}\varepsilon = \varepsilon$, i.e. $P_n - p_n < \varepsilon$, <u>Q.E.D.</u>

Since $\{P_n\}$ is a monotonically decreasing sequence whereas $\{p_n\}$ is monotonically increasing, the difference between the corresponding terms of these sequences will decrease as *n* increases; therefore each of the subsequent differences between the perimeters will be less than ε as well.

Now we can prove the major result of this section, which is the existence of a segment that is greater than any p_n and less than any P_n, and this *separating* segment *will be defined to be the circumference*.

<u>Theorem 10.2.9.</u> *For any circle, there exists a unique segment that is greater than the perimeter of any inscribed regular polygon and less than the perimeter of any circumscribed regular polygon, where the polygons are obtained by sequential doubling their number of sides.*
This segment is called the *circumference of the circle*.

Proof. Suppose a circle of some radius R is given. For this circle, we can create the two sequences of segments, $\{p_n\}$ and $\{P_n\}$, the perimeters of inscribed and circumscribed regular convex polygons obtained by sequential doubling the number of sides.

Let MN be a ray emanating from some point M. On this ray we lay off from point M the segments congruent to the terms of the sequences $\{p_n\}$ and $\{P_n\}$, starting, e.g. with the perimeters of regular triangles, p_3 and P_3:

$MA_0 = p_3;\quad MB_0 = P_3;$

$MA_1 = p_6;\quad MB_1 = P_6;$

$MA_2 = p_{12};\quad MB_2 = P_{12};$

....

$MA_6 = p_{192};\quad MB_1 = P_{192};$

....

$MA_n = p_{3 \cdot 2^n};\quad MB_n = P_{3 \cdot 2^n};$

....

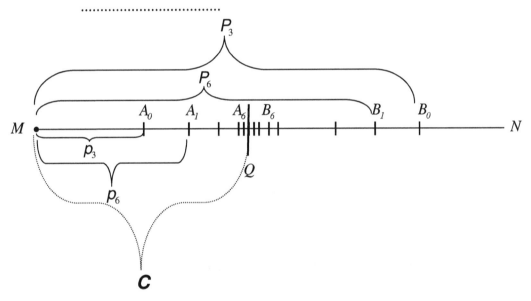

Figure 10.2.12

Thus we have created a sequence of segments A_0B_0, A_1B_1, A_2B_2, ... A_nB_n, $A_{n+1}B_{n+1}$, ... (Figure 10.2.12).

Since $p_n < P_n$ (Corollary 1 of Lemma 10.2.6), every point B_i ($i = 1,2,3,...n,...$) lies to the right from any point A_j ($j = 1,2,3,...,n,...$), i.e. every point from the A-sequence will be closer to the origin M of the ray than any point of the B-sequence.

It has been proved in Lemma 10.2.1 that $\{p_n\}$ is monotonically increasing, hence every point A_n is closer to M than its subsequent A_{n+1}, i.e. each A-point with a greater ordinal number lies to the right of the preceding one.

Similarly, since $\{P_n\}$ is monotonically decreasing (Lemma 10.2.3), each subsequent B-point lies to the left of the preceding one.

Thus, the A-points and B-points are ordered on the ray as follows: $M, A_0, A_1, A_2, ..., A_n, A_{n+1}, ..., B_{n+1}, B_n, ..., B_2, B_1, B_0$ (see Figure 10.2.12).

It means that these points form a sequence of *nested segments*: $A_0B_0 \supset A_1B_1 \supset A_2B_2 \supset ... \supset A_nB_n \supset A_{n+1}B_{n+1} \supset ...$.

According to Cantor's Principle (CONTINUITY AXIOM II), there exists a point that belongs to all these segments. Moreover, this point is unique, since, as we have proved in Lemma 10.2.8, we can choose such n that the corresponding segment $A_nB_n = P_n - p_n$ will be less than any given segment.

Therefore, there exists a *unique point* (let us name it Q) that is common to all these nested segments. This point *separates* all the A-points from the B-points. The corresponding segment $C = MQ$ *separates* the sequences $\{p_n\}$ and $\{P_n\}$ in the following sense: for every n, $p_n < C < P_n$, Q.E.D.

This *separating* segment is called the *circumference of the circle*.

<u>Remark 1.</u> It follows from Corollary 1 of Lemma 10.2.6 that the circumference of a given circle separates the perimeters of all convex (not necessarily regular) inscribed polygons from the perimeters of all circumscribed polygons.

It is not necessary to use regular polygons and the doubling the number of sides procedure in order to obtain a sequence of nested segments that contains point Q. It follows from Theorems 10.2.2 and 10.2.4 and Lemma 10.2.6 that one can construct such a sequence by means of any procedure that increases the number of vertices (and therefore the number of sides) of general (not necessarily regular) convex polygons if as a result of this procedure the sides of these polygons can be made indefinitely small (less than any segment).

Now let us estimate the circumference C of a circle of a given radius R.

It is easy to evaluate the perimeters of an inscribed regular hexagon and a circumscribed square and thus obtain the following rough estimates for the circumference:

$$6R < C < 8R.$$

More accurate estimates can be obtained, for example, by approximating the circumference by means of perimeters of inscribed regular polygons with large number of sides.

We shall use the doubling formula from Problem 10.1.2:

$$a_{2n}^2 = 2R^2 - 2R\sqrt{R^2 - \frac{a_n^2}{4}}.$$

We rewrite this expression as

$$a_{2n}^2 = 4R^2\left[\frac{1}{2}\left(1-\sqrt{1-\frac{a_n^2}{4R^2}}\right)\right]. \qquad (1)$$

Using this expression one can obtain a sequence of (lower) estimates for C:

$$C \approx p_{2n} = n \cdot a_{2n} \qquad (2)$$

where a_{2n} can be found from (1).

Let us start with an inscribed regular hexagon. The first estimate is
$$P_6 = 6 \cdot a_6 = 6R; \qquad (3)$$
the next one is

$$p_{12} = 12 \cdot a_{12} = 12 \cdot 2R\sqrt{\frac{1}{2}\left(1-\sqrt{1-\frac{1}{4}}\right)} = 2R \cdot 6\sqrt{2-\sqrt{3}} \approx 6.21R, \qquad (4)$$

followed by p_{24}, p_{48}, …. p_{192} is already a pretty accurate estimate:
$$C \approx p_{192} \approx R \cdot 6.28290494. \qquad (5)$$

By increasing n, one can estimate C within any given accuracy. One can observe from the above estimates that the approximate values of C seem to be proportional to the radius (or diameter) of the circle. Let us prove that it is so for the circumference as well. Hereafter we shall often use the word circumference instead of the length of the circumference. It is always clear from the context whether we deal with the segment or with its length.

Theorem 10.2.10. The circumference of a circle is proportional to its diameter, and the coefficient of proportionality is universal (the same) for all circles.

Proof. We shall propose two different proofs: one of them can be called "more geometrical" whereas the other one is rather "analytical".

Proof 1. We have proved in Theorem 9.6.4 that all circles are similar to each other; therefore the lengths of their corresponding parts are in proportion, whence the conclusion of the theorem follows.

Proof 2. Let us consider two circles, of radii R and r, and denote their respective circumferences C and c. We shall call these circles, for convenience, large and small, and everything that belongs to the large one we shall denote by capital letters.

Let P_n and p_n be the respective perimeters of two regular n-gons inscribed in these circles. By Th.10.1.13,

$$\frac{P_n}{R} = \frac{p_n}{r}, \quad or \quad \frac{P_n}{2R} = \frac{p_n}{2r}. \qquad (6)$$

Suppose the conclusion of the theorem is not true, i.e.
$$\frac{C}{2R} \neq \frac{c}{2r}. \qquad (7)$$

Let us suggest, for certainty, that the expression in the left-hand side of this inequality is greater than the other. It means that there exists such a positive number ε that

$$\frac{C}{2R} = \frac{c}{2r} + \varepsilon \qquad (8)$$

We can write for the circumferences:
$$C = P_n + H_n; \quad c = p_n + h_n, \qquad (9)$$
where, according to Lemma 10.2.8, each of these corrections, H_n and h_n, can be made less than any segment by choosing sufficiently large n (their respective length therefore can be made less than any positive real number).

Let us substitute (9) into (8):
$$\frac{P_n + H_n}{2R} = \frac{p_n + h_n}{2R} + \varepsilon; \quad \Rightarrow \quad \frac{P_n}{2R} + \frac{H_n}{2R} = \frac{p_n}{2R} + \frac{h_n}{2R} + \varepsilon.$$

Since $\frac{P_n}{2R} = \frac{p_n}{2r}$, the last equality can be rewritten as $\frac{H_n}{2R} = \frac{h_n}{2R} + \varepsilon$, which leads to $\frac{H_n}{2R} - \frac{h_n}{2R} = \varepsilon$.

Let us choose such n that $H_n < \varepsilon R$. Then it follows that
$$\frac{H_n}{2R} < \frac{\varepsilon R}{2R} = \frac{\varepsilon}{2}; \quad \Rightarrow \quad \frac{H_n}{2R} - \frac{h_n}{2R} < \frac{\varepsilon}{2}.$$

This contradicts to the equality $\frac{H_n}{2R} - \frac{h_n}{2R} = \varepsilon$, since number ε is positive by our assumption that $\frac{C}{2R} > \frac{c}{2r}$. Thus our assumption has been false, hence $\frac{C}{2R} = \frac{c}{2r}$, Q.E.D.

Therefore, the ratio of the circumference of a circle to the diameter of the circle is a constant, the same for all circles.

That constant is denoted π. Now we can write for the circumference of a circle of radius R:
$$C = 2R \cdot \pi, \quad \text{or} \quad C = 2\pi R.$$

The above result, one of the greatest intellectual achievements of the humankind, has been obtained by the Greek philosopher and scientist Archimedes (c.a. the III Century B.C.). He also found an estimate for π: $\pi \approx \frac{22}{7}$.

It should be noticed that π is an irrational number that cannot be expressed as a ratio but only estimated. From formula (5) we can obtain a rather accurate estimate:
$\pi \approx 3.1415$.

If an arc of a circle measures n°, the length of this arc can be expressed as
$$S = 2\pi R \cdot \frac{n}{360} = \frac{\pi R n}{180}.$$

Problem 10.3.1. *Evaluate the degree measure of an arc whose length is equal to the radius of the circle.*

Solution. Replace S by R in the above formula: $R = \dfrac{\pi R n}{180}$, whence $n = \dfrac{180}{\pi}$, i.e.

$$n^0 = \dfrac{180^0}{\pi} \approx 57°17'45''.$$

An arc whose length equals R is called a *radian*. It is convenient to measure central angles in radians: if angle θ subtends an arc of length S, its *radian measure* is

$$\theta = \dfrac{S}{R}.$$

10.2 Questions and Problems.

1. Show that, given a circle, the sequence of the perimeters of circumscribed regular polygons obtained by doubling the number of their sides is monotonically decreasing.

2. Show that, given a circle, the sequence of the perimeters of inscribed regular polygons obtained by doubling the number of their sides is monotonically increasing.

3. *Without referring to Lemma 10.2.5 or any of its implications, show that a side of a circumscribed regular polygon obtained from a regular triangle or a square by sequential doubling the number of sides can be made less than any given segment. How many times we have to apply the doubling procedure to be sure that we have obtained a polygon whose side is less then $\frac{1}{128}R$ (where R is the radius of the circle inscribed in the polygon)?

4. Prove that $3 < \pi < 4$.

5. Estimate π (find lower and upper estimates) using the perimeters of inscribed and circumscribed regular a) hexagons; b) octagons; c) decagons; d) dodecagons.

6. Before proving the fact that the circumference is proportional to the radius with the same coefficient for all circles, we had already known (Chapter 9) that all circles are similar. Then, was it necessary to prove the aforementioned statement? Does it not follow just from the similarity of all circles that $\dfrac{C}{c} = \dfrac{R}{r}; \Rightarrow \dfrac{C}{R} = \dfrac{c}{r}$?

Hereafter, in all problems implying numerical approximate answers, suggest the *approximate value* of π to be 3.14 or $\dfrac{22}{7}$ if not specified otherwise.

7. Evaluate the length of the equator, if the radius of the Earth is approximately 6400km. Evaluate the correction to the obtained estimate if π is approximated by 3.1416 (instead of 3.14).

8. An Ancient Greek scholar Eratosthenes of Alexandria, who lived in the third century B.C., evaluated the circumference of the Earth. The diagram below illustrates his method.

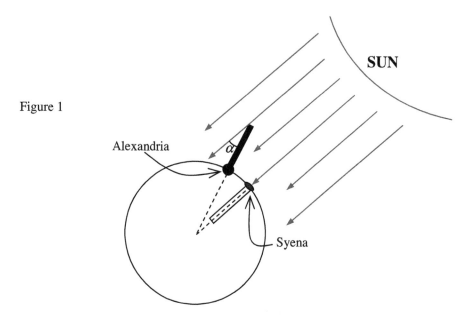

Figure 1

He knew that at noon on a day of summer solstice the Sun shone directly into a deep well in the city of Syena (modern Aswan), which was located approximately on the same meridian as Alexandria 830 km due south.

By observing the shadow cast by a high vertical obelisk located in Alexandria, Eratosthenes measured the angle α between the obelisk and the sun rays at noon of a solstice day. He found $\alpha \approx 7.5^0$.

Repeat his calculations and evaluate the Earth's circumference. Do you have to make any other approximating assumptions?

9. Find the total length of the wire spooled on a shank to make an electric reel (Figure 2), if the thickness (diameter) of the wire is 0.6mm, the diameter of the shank is 4mm, and the length of the segment of the shank occupied by the wire is 7.2mm.

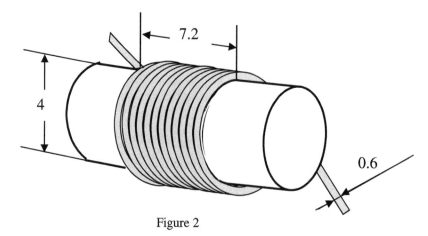

Figure 2

10. A pulley (Figure 3) is used to lift a weight at the altitude of 10m. The radius of the pulley is 70cm. How many complete revolutions will the pulley make before the goal is achieved?

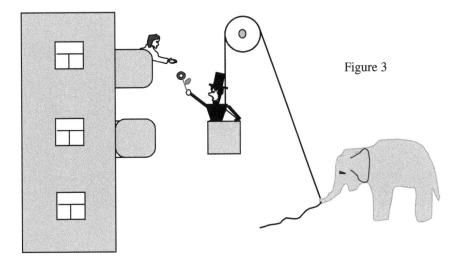

Figure 3

11. Estimate the absolute, ΔC, and relative, $\dfrac{\Delta C}{C}$, changes of the circumference if the radius, $R=20cm$, has increased by $\Delta R=2cm$. Which of the two estimated values does not depend upon the original value of R? Why?

12. A boy wishes to have the fastest possible ride on a carrousel horse. Which horse should he pick: the closest to or the farthest from the axle of the carrousel? How fast will he "race" if the horse he picks stands 5m from the axle and the carrousel makes 4 revolutions per minute (rpm)?

13. The diameter of the driving wheel (do not confuse with the steering wheel) of a locomotive is 2.1 meters, and the wheel makes 150 rpm (revolutions per minute). Evaluate the speed of the locomotive.

14. The spindle of the engine of a car drives the driving wheel by means of a driving belt, as shown in Figure 4. How many rpm (revolutions per minute) the spindle makes when the car moves at a speed of 66km/h, and the radii of the driving wheel (ρ), the used at that moment transmission sheave (R), and the spindle (r) are equal, respectively, 35cm, 9cm, and 6cm?

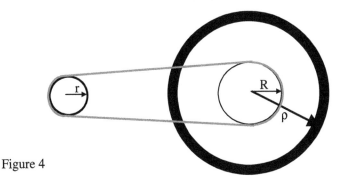

Figure 4

15. The radius of a circle equals $R=1$. Find the length of an arc that contains a) 180°; b) 90°; c) 60°; d) 45°; e) 30°; f) 1° g) 1′.

16. The radius of a circle equals 14cm. Suggest π ≈ $\frac{22}{7}$ and find the degree (also, if necessary, - minutes and seconds) measure of an arc whose length is a) 11cm;   b) 8.8cm; c) 5cm.

17. Estimate the degree measure of an arc whose length is equal to the radius.

18. Through how many degrees a wheel turns in 10 seconds if it makes 7 rpm?

19. A side of a square inscribed in a circle subtends an arc, which is 10cm long. Evaluate the radius of the circle.

20. A ship sails eastward along the equator at a speed of 15km/h. How long it will take him to change its latitude by 1°? (Suggest the radius of Earth to be approximately 6400 km)

21. Two sheaves are connected by means of a belt so that one of the sheaves can rotate another, as shown in Figure 5. Evaluate (within the accuracy of 1cm) the total length of the belt if the radii of the sheaves are *r*=2 cm and *R*=12 cm, respectively, and the distance between their axles is *L*=20 cm.

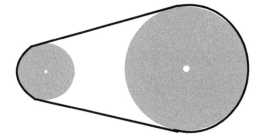

Figure 5

22. Prove that if a given angle α is subtended by an arc of length l in a circle of radius r, and an equal angle subtends an arc of length L in a circle of radius R, then $\frac{l}{r} = \frac{L}{R}$, i.e. the magnitude of the angle uniquely defines the above ratio.

Given a circle, the length of the arc subtended by a given central angle is uniquely related with the measure of the angle. That fact allows to measure angles in the lengths of corresponding arcs in relation to the corresponding radii ($\frac{L}{R}$ depends only upon the magnitude of α, as follows from the proposition of the previous problem).

That is why it is convenient to measure angle in *radians: One radian* is an angle that subtends an arc of length R in a circle of radius R.

In the *unit circle* (a circle of radius $R=1$), the radian measure of a central angle is just the length of the subtended arc (see the following problem).

23. Prove that if an angle subtends an arc of length *x* in a *unit circle* (a circle of radius *R=1*), the radian measure of that angle is *x*.

24. Find the radian measure of the following angles: a) complete revolution; b) 270°; c) 180°; d) 90°; e) 60°; f) 45°; g) 30°; h) 1°.

25. Find the degree measure of the following angles given in radians: a) ½π; b) ¼π; c) ⅔π; d) 0.1π.

A convenient characteristic of circular motion is its *angular velocity*. If a point moves along a circle, the angular velocity of the motion is the rate of change of the angle through which the radius joining the point with the centre rotates with respect to time:

$\omega = \dfrac{\Delta \theta}{\Delta t}$, where ω is the angular velocity, and $\Delta \theta$, Δt are, respectively, the increment of the angle and the period of time during which it has occurred. (Obviously, we suggest in the above definition that the motion is uniform, i.e. the angular velocity does not change during the allotted time; otherwise we would find the *average angular velocity* during that period of time).

26. Prove that for a point moving along a circle of radius R with a constant speed v, its angular velocity can be found as $\omega = \dfrac{v}{R}$.

27. Find the angular velocity of a wheel of a moving car if the radius of the wheel is 40cm, and the car moves at a speed of 100km/h. How many rpm does the wheel make?

10.3. The length of the circumference of a circle as a limit of a sequence.
(This section is optional).

One of the ways of defining the length of the circumference of a circle is based on the notion of a *limit of a sequence*.

Sequences, or *infinite sequences*, are very common in various mathematical problems. For example, one can often encounter

a sequence of natural numbers: 1, 2, 3, 4, 5, ... , n, ... , (1)
arithmetic sequence: a, $a + d$, $a + 2d$, ... , $a + (n-1)d$, ... , (2)
geometric sequence: a, ar, ar^2, ... , ar^{n-1}, ... (3)

A sequence is usually defined by a certain rule. The rule may be simple, as in the above examples, or more complicated, so that it may be uneasy or even impossible to define a sequence by a single formula expressing a term of the sequence as a function of its *index* (ordinal number): $a_n = f(n)$.

For example, when calculating consecutively lower estimates of $\sqrt{2}$ correct to within an accuracy of $\dfrac{1}{10}$, then – correct to $\dfrac{1}{100}$, $\dfrac{1}{1000}$, etc., one would create an infinite sequence:

1.4; 1.41; 1.414; 1.4142, ... (4)

Each subsequent form of that sequence improves the accuracy of the (lower) estimate of $\sqrt{2}$ tenfold, and it would be impossible to describe by a simple formula how one could obtain the next term from the previous one(s).

There is, however, a remarkable regularity in the behaviour of the terms of sequence (4): the terms approach the value of $\sqrt{2}$ *indefinitely*, i.e. they are getting closer and closer to $\sqrt{2}$, and the difference between $\sqrt{2}$ and the terms of the sequence can be done less than any given number (can be done *arbitrarily small*) by choosing the term with a sufficiently great ordinal number. One can say that sequence (4) *converges to* $\sqrt{2}$, or $\sqrt{2}$ is *the limit of sequence* (4).

Let us define the notions of *convergence* and *limit* in general.

Suppose, a sequence $\{a_1, a_2, \ldots, a_n, \ldots\}$ is given, and there exists such a number A that given any (arbitrarily small) positive number, ε, one can find such a term, a_n, of the sequence, that all terms of the sequence starting from a_n will differ from A by a magnitude less than ε (in absolute value), i.e. for any given $\varepsilon > 0$ there exists such n that

$$|a_N - A| < \varepsilon \tag{5}$$

for all $N \geq n$.

Then A is called *the limit* of the sequence $\{a_n\}$, or one can say that $\{a_n\}$ *converges to* A. It is denoted as follows: $\lim\limits_{n \to \infty} a_n = A$.

Let us consider, for example, the following sequence: 0.9; 0.99; 0.999; ..., where each subsequent term is obtained from the preceding one by adding another decimal 9 in the decimal representation of the numbers. Formally, one can write:

$$a_{n+1} = a_n + \frac{9}{10^{n+1}}. \tag{6}$$

It is easy to notice that the terms of the sequence approach *1 indefinitely*. In fact, the absolute value of the difference between a_n and *1* equals $\frac{1}{10^n}$. The latter number can be done less than any positive ε by choosing sufficiently great n.

Let us recall the definition of the length of a segment incommensurable with a unit length (it was given in Chapter 9). Now we can say that the length of such a segment is a common limit of the sequences of its upper and lower estimates with indefinitely improving accuracy. Similarly, an infinite decimal fraction, i.e. an irrational number, is a limit of a sequence of rational numbers (e.g., sequence (4)).

Theorem 10.3.1. *If a sequence has a limit, this limit is unique.*
Proof. Students can carry out the proof, having given two hints:
(1)   prove by RAA (by contradiction)
(2)   use Fig. 10.3.1.

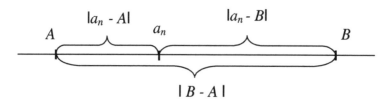

Figure 10.3.1

If in some sequence, $\{a_n\}$, $a_{n+1} \geq a_n$ (starting from some number), the sequence is called *monotonically increasing*.

If all terms of some sequence are less than a certain number, M ($a_n < M$ for n = 1, 2, 3, 4, ...), the sequence is called *bounded above*.

Theorem 10.3.2. (Weierstrass' Theorem). ***If a sequence is monotonically increasing and bounded above, it has a limit.***

The proof of the theorem is usually considered to be beyond the level of introductory courses. Yet the meaning of the theorem should be illustrated, e.g., by a picture of the terms of the sequence, $\{a_n\}$, *accumulating* near a certain point, A, on the number axis (Figure 10.3.2).

Figure 10.3.2

The notion of limit and its properties expressed in the above theorems enable us to define the length of the circumference of a circle.

Let a circle be given. We start with inscribing in the circle a regular polygon, for instance a hexagon (Fig. 10.3.3).

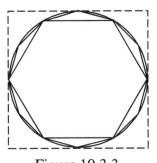

Figure 10.3.3

On some line, *MN* (Fig. 10.3.4), we lay off from some point O a segment, OP_1, equal to the perimeter of the hexagon.

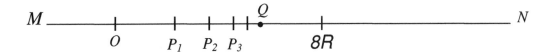

Figure 10.3.4

Then we double the number of sides of the hexagon to obtain a regular 12-gon inscribed in the circle (Fig.10.3.3). We lay off the perimeter of the 12-gon on *MN* from O to obtain another segment, OP_2. By Lemma 10.2.1, $OP_2 > OP_1$. Let us repeat the procedure indefinitely. Each time we obtain a new perimeter (at n-th step it is the perimeter of a regular $6 \cdot 2^n$ - gon), greater than the preceding one. Yet, by Corollary 1 of Lemma 10.2.6, each of these perimeters is less than the perimeter of any circumscribed polygon, for example, the perimeter of a circumscribed square. The latter equals $8R$, where R is the radius of the considered circle.

Thus, the sequence of the perimeters of regular inscribed polygons obtained by doubling the number of their sides, is monotonically increasing and bounded above; therefore, according to Weierstrass' Theorem (Th.10.3.2), the following theorem has been proved:

***Theorem 10.3.3.** Given a circle, the sequence of the lengths of the perimeters of regular inscribed polygons obtained by doubling the number of their sides converges to a certain limit.*

By definition, **this limit, C, is the length of the circumference of the circle considered.**

It is often convenient to use this definition and properties of limits to obtain various results. For example, we can propose the following succinct proof of
***Theorem 10.2.10:** The circumference of a circle is proportional to its diameter, and the coefficient of proportionality is universal (the same) for all circles.*
Proof. *Let us consider two circles, of radii* R *and* r, *and denote their respective circumferences* C *and* c. *Let* P_n *and* p_n *be the respective perimeters of two regular n-gons inscribed in these circles. By Th.10.1.13,*

$$\frac{P_n}{R} = \frac{p_n}{r}, \quad or \quad \frac{P_n}{2R} = \frac{p_n}{2r}. \qquad (9)$$

When n tends to ∞, P_n *tends to* C *and* p_n *tends to* c, *so it follows from formula (9), that expresses the equality of the two sequences, and the uniqueness of the limit of a sequence that*

$$\lim_{n \to \infty} \frac{P_n}{2R} = \lim_{n \to \infty} \frac{p_n}{2r}; \quad \Rightarrow \quad \frac{\lim_{n \to \infty} P_n}{2R} = \frac{\lim_{n \to \infty} p_n}{2r}, \text{ which means that } \frac{C}{2R} = \frac{c}{2r}. \qquad (10)$$

Therefore, the ratio of the circumference of a circle to the diameter of the circle is a constant, the same for all circles.

10.3 Questions and Problems.

1. For each of the following sequences determine, whether it converges or diverges:
    a) $1, 4, 9, 16, \ldots, n^2, \ldots$
    b) $2, -4, 8, -16, 32, \ldots$
    c) $0.3, 0.33, 0.333, 0.333\ldots$;
    d) $1, \frac{1}{2}, \frac{1}{3}, \frac{1}{4}, \ldots \frac{1}{n}, \ldots$;
    e) $\frac{2}{5}, \left(\frac{2}{5}\right)^2, \left(\frac{2}{5}\right)^3, \left(\frac{2}{5}\right)^4, \ldots \left(\frac{2}{5}\right)^n, \ldots$;
    f) $S_n = a + aq + aq^2 + aq^3 + \ldots + aq^{n-1}$;

g) $a_n = \sin\dfrac{180^0}{n}$;

2. For each of the sequences of #1 determine whether it is monotonically increasing/decreasing.

3. For each of the sequences of #1 determine whether it is bounded above/ below.

4. Prove that $\left\{\dfrac{1}{n}\right\}$ converges. Prove that it converges to 0.

5. a) Show that $\left\{\dfrac{n}{n+1}\right\}$ is a monotonically increasing sequence.

   b) Show that $\left\{\dfrac{n}{n+1}\right\}$ is a sequence bounded above.

   c) Show that the above sequence converges.

   d) Find $\lim\limits_{n\to\infty}\dfrac{n}{n+1}$.

6. $\lim\limits_{n\to\infty} a_n = L;$ find $\lim\limits_{n\to\infty} a_{n-1}$.

7. *Prove the convergence of the sequence

$\sqrt{2},\ \sqrt{2\sqrt{2}},\ \sqrt{2\sqrt{2\sqrt{2}}},\ \sqrt{2\sqrt{2\sqrt{2\sqrt{2}}}},\ \ldots$ and find the limit.

8. Prove that the sequence of the perimeters of circumscribed regular polygons obtained by means of doubling the number of the sides of the polygons is monotonically decreasing and bounded below.

9. Prove that the sequence of the perimeters of inscribed regular polygons obtained by means of doubling the number of the sides of the polygons is monotonically increasing and bounded above.

10. Show that the length of the side of an inscribed regular polygon tends to zero with indefinite increase of the number of sides, i.e. $\lim\limits_{n\to\infty} a_n = 0,$ if the radius of the circle does not change.

11. Show that the length of the apothem of an inscribed regular polygon tends to the radius of the circle when the number of sides increases indefinitely, i.e. prove that $\lim\limits_{n\to\infty} h_n = R$.

11. AREAS

> Pooh knew what he meant, but, being a Bear of Very Little Brain, couldn't think of the words.
> A.A.Milne, *The House At Pooh Corner.*

11.1 Areas of polygons

a) <u>Basic suggestions (principles)</u>.

By the *area* of a figure we mean the amount of surface contained within its bounding lines. In this chapter we consider plane figures only.

Our problem is to find numerical expressions for the areas of various figures. We require that numerical expressions of areas satisfy the following principles (axioms):

(1) *The numbers that represent areas of congruent figures are equal.*

(2) *If a figure is decomposed into a few closed figures, then the number representing the area of the whole figure is equal to the sum of the numbers representing the areas of all parts of the decomposition.*

<u>Remark 1</u>. We measure areas by positive numbers. The sum of two positive numbers is greater than any of the addends. It is necessary for the acceptance of the above principle (2) that areas of figures possess such a property. Let us explain that in more detail.

If a figure is decomposed into a few parts, these parts may be combined together in different ways so as to create different figures (see, for instance, Fig. 11.1.1).

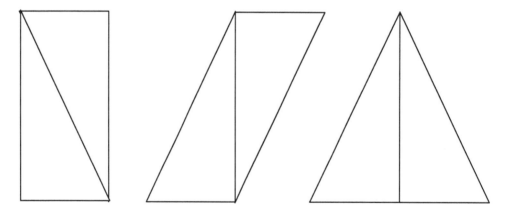

Figure 11.1.1

The question is whether one can compose the parts in such a figure that will fit entirely in the interior of the original figure.

If the answer to the above question was positive, one would obtain two figures such that one is contained within the other, and, at the same time, the numbers representing their areas, according to the principle (2), would be equal. That would mean that the number measuring the area of some part of a figure equals to the number measuring the area of the whole figure, i.e. the sum would be equal to one of the addends constituting the sum, that is impossible for positive numbers. In that case principle (2) would be unacceptable.

Yet, it has been proved by Hilbert, Schur, and others that such a hypothetical construction as described above is impossible, and, thus, principle (2) is acceptable.

Two figures of equal areas are said to be equivalent. Principle (2) is sometimes called the equivalence by finite decomposition.

It can be shown that equivalence by finite decomposition is an equivalence relation.

Congruent figures are, evidently, equivalent, whereas equivalent figures are not necessarily congruent (see, for example, the three equivalent figures in Fig. 11.1.1).

From now on we denote areas by the letter $\mathcal{A}$ and for the sake of brevity we shall use the words "the area of a figure" instead of "the number expressing the area of a figure". For example, $\mathcal{A}(\Delta ABC)$ is read "the area of triangle ABC", not "the number expressing the area of triangle ABC", even though the latter is suggested.

b) Measurement of area.

First of all we have to define the unit of the measurement. The *unit of area* is the area of a square on a side of unit length. Thus, in Fig.11.1.2 the units of a *square inch* $\equiv$ inch^2 and a *square centimeter* $\equiv$ cm^2 are shown..

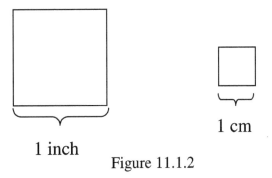

Figure 11.1.2

For the simplest figures one can determine the areas by covering each figure by as many square units as will exactly cover the whole figure. Then the area of the figure will be represented by the number of the squares covering the figure. For example, the area of the shaded figure in Fig.11.1.3 equals 34 cm^2.

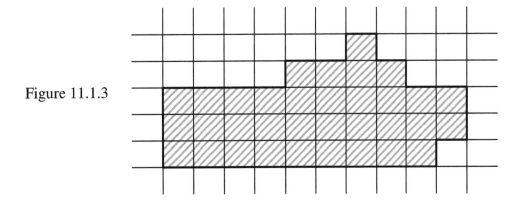

Figure 11.1.3

The aforementioned simplicity of the shaded figure depicted above (and others alike) is in its boundary: it consists of straight segments perpendicular to each other. Let us notice that such a figure may be decomposed into a finite number of rectangles. In reality figures have more complicated shapes and their areas are not found by direct superposition of units of area or their fractions onto figures.

The areas are determined *indirectly (implicitly)* by means of measuring certain segments (*linear sizes*) of figures.

When dealing with the elementary rectilinear figures, such segments are the *base* and *altitude (height)*.

The *altitude (height) of a triangle* with reference to a given side as a *base* is the perpendicular distance of the opposite vertex from the base.

The *altitude (height) of a parallelogram* with reference to a given side as a base, is the perpendicular distance between the base and the opposite side. Particularly, in a rectangle the height is the side perpendicular to a side taken as a base. The height and the base of a rectangle are called its *measurements*.

In a *trapezoid* each of the parallel sides is a base, and the *altitude (height)* is the perpendicular distance between them.

The altitude is usually denoted h, and the corresponding base (if it is not specified otherwise) is denoted b (Fig. 11.1.4).

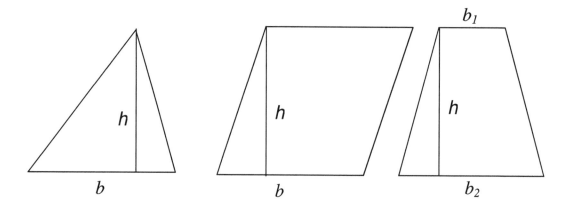

Figure 11.1.4

b) <u>Area of a rectangle.</u>

<u>Lemma 11.1.1.</u> *If a side of one square is n times (n is an integer) the side of some other square, the area of the first square is n^2 times the area of the second square.*

Proof. According to the hypothesis, n smaller squares attached sequentially by their sides one to the other form a strip which length equals to the side of the greater square (Fig. 11.1.5). The area of the strip is the sum of n areas of the smaller square. The greater square contains exactly n of these stripes, therefore its area will be equal to $n \cdot n = n^2$ areas of the smaller square, Q.E.D.

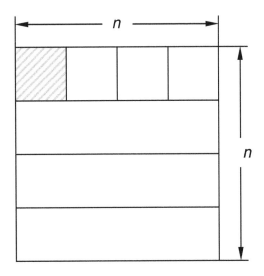

Figure 11.1.5

Theorem 11.1.2. *__The area of a rectangle equals to the product of its measurements.__*

Remark 1. Of course, we mean here: *the numerical expression of the area of a rectangle in given units of area equals to the product of the numerical expression of its sides in the corresponding linear units (units of length).*

In other words, if numbers *a* and *b* express the lengths of the height and base (width and length) of a rectangle in given linear units, the area of the rectangle is expressed in the corresponding square units as $\mathcal{A} = ab$.

Proof. There are three possibilities:
(i)   *a* and *b* are integers;
(ii)   *a* and *b* are fractions;
(iii)   *a* and *b* are irrationals.

In case (i) we just calculate the number of unit squares covering the rectangle. This number is exactly *ab*.

In case (ii) the sides of the rectangle have a common measure. Suppose, this measure to be $\frac{1}{n}$ of the unit of length, *e*. Then, each side of the rectangle contains the segment $\frac{e}{n}$ an integer number of times (α and β, respectively),

i.e.   the height $= ae = \alpha \frac{e}{n}$; the base $= be = \beta \frac{e}{n}$. That means $a = \frac{\alpha}{n}$; $b = \frac{\beta}{n}$.

Let us partition each side of the rectangle into segments of length $\frac{e}{n}$ and join the corresponding points of the partition that lye on the opposite sides (Figure 11.1.6).

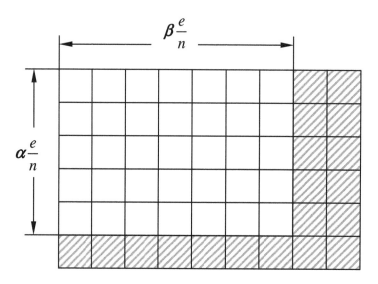

Figure 11.1.6

That will split the whole rectangle into $\alpha + \beta$ small rectangles. The area of each small rectangle (according to Lemma 11.1.1) is $\frac{e^2}{n^2}$, therefore, the area of the whole rectangle is $\mathcal{A} = \alpha\beta \frac{e^2}{n^2} = \frac{\alpha}{n} \cdot \frac{\beta}{n} \cdot e^2 = ab \cdot e^2$.

Thus, the area is expressed by the number ab, Q.E.D.

(iii) In this case the sides are incommensurable, i.e. at least one of the numbers, a, b, is irrational. Then that number can be approximated by a rational number, a_n within any given accuracy, e.g. $\frac{1}{10^n}$ if we use the decimal representations of rationals.

In this case the area can be estimated as $\mathcal{A}_n = a_n b_n$ within any given accuracy, and the precision of the estimate improves as n increases. The exact value of $\mathcal{A}$ is the limit value of $\mathcal{A}_n$ as $n \to \infty$:

$$\mathcal{A} = \lim_{n\to\infty}(a_n b_n) = \lim_{n\to\infty} a_n \cdot \lim_{n\to\infty} b_n = ab, \quad \text{Q.E.D.}$$

Remark 2. We had not proved the existence of limit of $\mathcal{A}_n$ before we used it in the last formula. It is easy to prove the existence based on the property of increasing bounded from above sequences (see section 10.3). It would be a good exercise for advanced students.

Remark 3. One can prove the validity of the formula $\mathcal{A} = ab$ by showing that the area defined by this formula satisfies principles (1) and (2), stated in the beginning of the chapter (another good exercise for students).

c)   Areas of polygons.

Theorem 11.1.3. *The area of a parallelogram equals to the product of its height by its base:* $\mathbf{A} = h \cdot b$.

Proof. In Figure 11.1.7, the base of parallelogram *ABCD* equals $AD = b$, and its height is $AE = h$. Draw $DF \perp BC$. $\triangle AEB = \triangle DFC$, hence their areas are equal, and, therefore (why?) parallelogram *ABCD* and rectangle *AEFD* are equivalent. The area of the latter is hb, whence the conclusion follows.

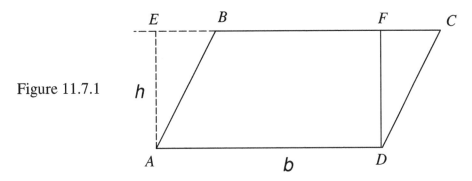

Figure 11.7.1

Theorem 11.1.4. *The area of a triangle equals half of the product of its height by the base:* $\mathcal{A} = \frac{1}{2} hb$.

Proof. $\triangle ABC$ and congruent to it $\triangle AEB$ in Fig.11.1.8 form a parallelogram (*AEBC*) with area *hb*. Then, the area of each of the triangles is ½ *hb*, Q.E.D.

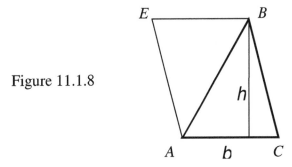

Figure 11.1.8

It is easy to prove the following corollaries (students should do that).

Corollary 1. *If a side and the corresponding altitude of one triangle are, respectively, congruent to a side and the corresponding altitude of the other, then the triangles are equivalent* (even though they are not congruent!).
Corollary 2. *The area of a right triangle is half the product its legs.*

Corollary 3. *If an angle of a triangle is congruent to an angle of some other triangle, the areas of the two triangles are in the same ratio as the products of the sides including congruent angles.*

Corollary 4. *The area of a rhombus is half the product its diagonals.*

Inasmuch as any polygon can be decomposed into a few triangles, we can determine the area of any polygon using the results of the latest theorem.

Theorem 11.1.5. *The area of a trapezoid equals the product of its height by half-sum of its bases.*

Proof. Draw a diagonal, e.g., AC (Fig. 11.1.9) to split the trapezoid into two triangles. Then the area of the trapezoid is the sum of the areas of the two triangles, that is

$$\frac{1}{2}ah + \frac{1}{2}bh = \frac{a+b}{2}h, \quad \text{Q.E.D.}$$

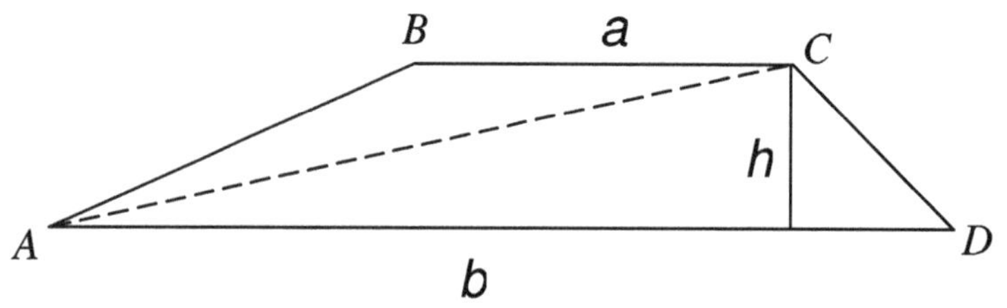

Figure 11.1.9

Corollary 1. *The area of a trapezoid equals the product of its midline by its height.*

Theorem 11.1.6. *If a polygon is circumscribed about a circle, the area of the polygon equals half the product of its perimeter and the radius of the circle:* $\mathcal{A} = \frac{1}{2}Pr$.

Figure 11.1.10

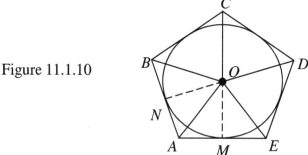

Proof. Split the polygon into a few triangles by joining the centre of the circle with the vertices (see Figure 11.1.10). In each of these triangles the height equals the inradius. Therefore, the area of the polygon is

½ $AE \cdot r$ + ½ $AB \cdot r$ + ... = ½ $(AE + AB + ...) r$ = ½ Pr, Q.E.D.

Corollary 1. *The area of a regular polygon is half the product its perimeter and the apothem.*

Proof. It follows immediately from the theorem, as one can always inscribe a circle into a regular polygon, and the apothem equals the inradius.

Problem 11.1.1. (Heron's Formula). If the sides of a triangle are *a, b, c,* its area equals

$$\mathcal{A} = \sqrt{p(p-a)(p-b)(p-c)}, \text{ where } p = \tfrac{1}{2}P = \tfrac{1}{2}(a+b+c).$$

The proof of this formula is not very easy, still it can be carried out by students, especially if they are given a few hints:
(1) start with our regular formula of the area of a triangle;
(2) express the height through a side and its projection upon the base;
(3) express the projection in terms of the sides;
(4) simplify the expression for the height using some reduction formulae (the difference of squares, the square of the sum/difference).

It follows from Heron's formula that the area of an equilateral triangle with side *a* equals $a^2 \dfrac{\sqrt{3}}{4}$.

e) <u>Pythagorean Theorem in terms of areas.</u>

As we know, for a right triangle with the legs *a, b* and hypotenuse *c,*
$$c^2 = a^2 + b^2.$$
This formula can be interpreted in terms of areas:

The area of the square constructed on the hypotenuse of a right triangle, equals the sum of the squares constructed on its legs.

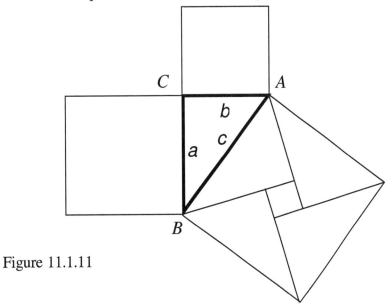

Figure 11.1.11

One can derive the conclusion of the theorem directly from the consideration of the square on the hypotenuse (Fig. 11.1.11): the area of that square equals the sum of the areas of four triangles congruent to $\triangle ABC$ plus the area of the (small) square with sides (*a-b*), i.e.
$$c^2 = 4 \cdot \tfrac{1}{2}ab + (a-b)^2 = 2ab + a^2 - 2ab + b^2 = a^2 + b^2, \underline{\text{Q.E.D.}}$$

11.1   Areas of polygons

1. Are congruent figures necessarily equivalent?

2. Adduce a few examples of figures that are equivalent but not congruent.

3. Formulate and illustrate by examples the principle of the *equivalence by finite decomposition*.

Do not use formulae but only the *principles* of the
(1) equivalence of congruent figures,
(2) equivalence by finite decomposition
to solve problems # 4 - # 9.

4. Decompose a parallelogram into two parts that constitute an equivalent rectangle.

5. *Decompose a triangle into three parts that constitute an equivalent rectangle.

6. Show that the total area occupied in the plane by two figures equals the sum of their areas reduced by the area of their intersection:
    $\mathcal{A}(F \cup G) = \mathcal{A}(F) + \mathcal{A}(G) - \mathcal{A}(F \cap G)$.

7. In Figure 1, $AB=BC$ and $MN=NP$. Show that the bold-lined rectangle is equivalent to the shaded triangle.

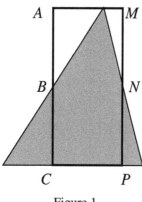
Figure 1

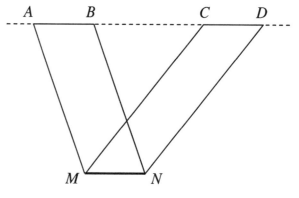
Figure 2

8. Show that parallelograms *AMNB* and *CMND* in Figure 2 are equivalent.

9. Show, how one can divide a rectangle into two equivalent figures by means of one straight cut. Are the perimeters of the obtained figures equal?

10. The area of figure M is 9dm². What is the area of M in a) square centimeters? b) square inches? c) square meters?

11. U and u are two linear units of measurement, and $U=ku$. Find the relation between the numbers representing the area of the same figure in U^2 and u^2 units of area.

12. Given a square, divide it into two equivalent parts, one of which is a square.

13. Prove that among all rectangles with given perimeter the square has the greatest area.

14. Prove that among all equivalent rectangles, the square has the least perimeter.

15. Propose geometric illustrations of the following identities (suggest that the participating numbers, *a, b, c,* and *d* are positive):
    a) $a(b+c) = ab + ac$;
    b) $a(b-c) = ab - ac$;
    c) $(a+b)(c+d) = ac + ad + bc + bd$;
    d) $(a+b)^2 = a^2 + 2ab + b^2$;
    e) $\dfrac{a}{b} = \dfrac{c}{d}$.

16. The diagonals of a convex quadrilateral are equal *D* and *d*. Find its area. Will the answer change if the quadrilateral is not convex?

17. Prove that the area of the square constructed on a diagonal of a rectangle as a side is greater or equal twice the area of the parallelogram. In which case the equality takes place?

18. Evaluate the amount of iron spent to make an iron slab in the shape of the frame shown in Figure 3, if it takes 5g of iron to produce 1sq.cm of the slab. (All sizes in the figure are in centimeters).

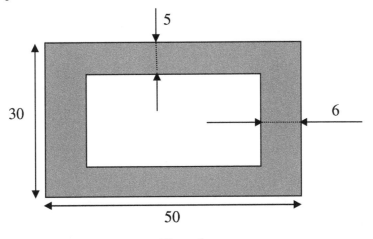

Figure 3

19. Decompose a parallelogram into a few parts and put them together to obtain a rectangle (equivalent to the given parallelogram). Use the procedure to derive the formula {height×base} for the area of a parallelogram.

20. Decompose a triangle into a few parts and put them together to obtain a rectangle (equivalent to the given triangle). Use the procedure to derive the formula {one half of height×base} for the area of a triangle.

21. Decompose a trapezoid into a few parts and put them together to obtain a rectangle (equivalent to the given trapezoid). Use the procedure to derive the formula {height×(half sum of bases)} for the area of a trapezoid.

22. May the formula for the area of a trapezoid be used for a parallelogram?

23. May a formula for the area of a trapezoid be used for a triangle?

24. The sides of a parallelogram are respectively equal to the sides of some other parallelogram. Are the two parallelograms equivalent?

25. The area of a parallelogram is half the area of a rectangle whose sides are respectively equal to the sides of the parallelogram. Determine the angles of the parallelogram.

26. Prove that the area of a parallelogram equals the product of the two adjacent sides times the sine of their included angle.

27. Show that the distances between the opposite sides of a parallelogram are inverse proportional to the length of the corresponding sides.

In the following problems, approximate the numerical estimates for areas to the nearest integer of the corresponding square units (where appropriate).

28. In a rhombus with its side $x = 10cm$, one of the angles equals 150°.
    a) Find the area of the rhombus as a function of x.
    b) Find the increment of the area of the rhombus if its side has increased by $\Delta x = 2cm$ (and the angles have not changed).
    c) The side of the rhombus grows starting with $x=10cm$ at a rate of $3cm/s$. Find the instantaneous rate of change of the area after *6 seconds.*

29. Prove that the area of a triangle equals half the product of two sides of the triangle times the sine of their included angle.

30. The sum of two sides of a triangle is 25cm, and the attitudes drawn to these sides are 6cm and 9cm long. Evaluate the area of the triangle.

31. Evaluate the area of an equilateral triangle with the side *a.*

32. Evaluate the area of a triangle whose sides equal 6cm and 10cm include an angle of 150°.

33. The projections of the legs of a right triangle upon the hypotenuse are equal 3m and 12m. Evaluate the area of the right triangle.

34. Evaluate the area of a right triangle, having given its inradius equal 3cm and one of its legs equal 8m.

35. Evaluate the area of a right triangle, having given one of its legs 15cm long and the projection of this leg upon the hypotenuse equal 9cm.

36. Evaluate the side of a rhombus whose area is 120cm², if the difference between its diagonals is 14cm.

37. Find the area of an isosceles triangle if its altitude dropped onto the base equals 30cm, and the altitude dropped onto the lateral side is 36cm long.

38. Find the base of an isosceles triangle whose area is 192cm², and the lateral side is 20cm.

39. Find the area of an isosceles right triangle with hypotenuse *c.*

40. In an isosceles trapezoid, one of its angles is 60°, the lateral side is 10cm, and the

least base is 16cm. Evaluate the area of the trapezoid.

41. Evaluate the areas of the figures shown in Figure 4 (a, b). (All sizes are given in cm). In each case use two different methods of calculation (e.g., different decompositions) to check your result.

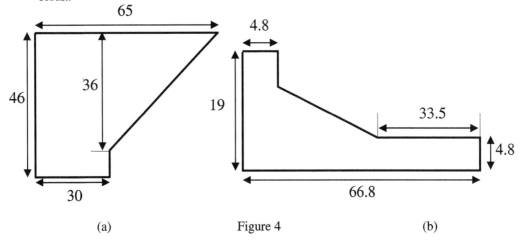

(a)　　　　　　　Figure 4　　　　　　　(b)

42. Find the area of a trapezoid whose bases are 36cm and 12 cm, respectively, and a lateral side, equal 14cm, forms a 150° angle with a base.

43. The area of a trapezoid is 54cm², and the altitude is 6cm. Find the bases if they are in a ratio of 4:5.

44. The lateral sides of a trapezoid are extended till they meet each other to form a triangle. The area of this triangle is 50 cm². Find the area of the trapezoid if its bases are in a ratio of 5:3.

45. Find the ratio of the areas of the figures into which a triangle is divided by its midline.

46. Divide a triangle into two equivalent figures by a line parallel to a side of the triangle.

47. Given a trapezoid, divide it into two equivalent parts by a line parallel to the bases.

48. A triangle is divided into two parts by a line parallel to the base and 5cm distant from the base. The areas of the parts are in a ratio of 1÷3 counting from the vertex. Find the altitude of the triangle.

49. A lateral side of a triangle is divided in a ratio of $m : n : p$ = 5:4:3 (from the base to the vertex) by two lines parallel to the base. Find the ratio of the areas of the figures into which the triangle is divided.

50. A trapezoid is divided into two equivalent parts by a line parallel to the bases. Find the segment of this line included between the lateral sides, if the bases of the trapezoid are a and b.

51. Show that a straight line drawn through the midpoints of the bases of a trapezoid, divides the trapezoid into two equivalent figures.

52. Find the locus of the vertices of triangles that are equivalent to a given triangle and have a common base with it.

53. In △ABC, three lines are drawn parallel to the sides through the centroid (the point of intersection of the medians), as shown in Figure 5. Show that the obtained three trapezoids are equivalent.

Figure 5

54. (Heron's Formula). Prove that the area of a triangle with sides *a*, *b*, and *c* equals
$$\mathcal{A} = \sqrt{p(p-a)(p-b)(p-c)},$$ where *p* is the semi-perimeter (half the perimeter) of the triangle.

55. Show that the area of a triangle with sides *a*, *b*, *c*, and inradius *r* equals
$$A = rp,$$ where *p* is its semi-perimeter ($p = \frac{a+b+c}{2}$).

56. *Show that the area of a triangle with the sides *a*, *b*, *c*, and circumradius *R* equals
$$\mathcal{A} = \frac{abc}{4R}.$$

57. *Construct a triangle equivalent to a given quadrilateral.

58. Construct a square equivalent to a given quadrilateral.

59. Propose an algorithm for the construction of a triangle equivalent to a given polygon.

60. Find the area of a regular hexagon inscribed in a circle, which is inscribed in a regular triangle with the side *a*.

61. Find the area of a regular octagon circumscribed about a circle of radius *R*.

11.2 Areas of similar figures.

Theorem 11.2.1. *The areas of two similar triangles are to each other in the same ratio as the squares of (any two) corresponding sides.*

Proof. It follows from Corollary 3 of Th.11.1.4. Suppose, △ABC ~ △A´B´C´, and

∠A = ∠A´ (Fig. 11.2.1), then the areas of these triangles are to each other as

$$\frac{AB \cdot AC}{A'B' \cdot A'C'} = \frac{AB}{A'B'} \cdot \frac{AC}{A'C'} = \left(\frac{AB}{A'B'}\right)^2 = \frac{AB^2}{A'B'^2}.$$

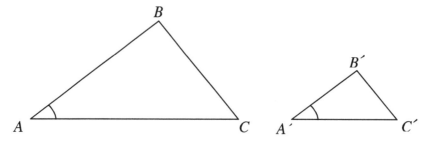

Figure 11.2.1

Theorem 11.2.2. ***The areas of two similar polygons are to each other in the same ratio as the squares of any two corresponding sides or diagonals.***

Proof. It is a trivial corollary of the preceding result and the principle of the equivalence by final decomposition. The polygons should be decomposed into triangles by, e.g., the corresponding diagonals. Then the conclusion follows. (Still, students should complete the proof in detail).

Corollary 1. ***The areas of two regular polygons with the same number of sides are in the same ratio as the squares of their sides, or their apothems, or circumradii.***

The results of this section can be briefly reformulated as follows:

The ratio of the areas of two similar figures equals k^2, where k is their similarity ratio.

Remark 1. This result is valid for any plane figures, and yet we have proved it only for rectilinear figures. The problem is in the definition of areas of arbitrary (in particular, curvilinear) figures. That definition is given in advanced calculus courses. We shall consider only the area of a circle.

11.2 Questions and Problems.

1. The sides of two squares are in a ratio of 3:4. Find the ratio of their areas.

2. The areas of two squares are in a ratio of 25:144. Find the ratio of their diagonals.

3. The areas of two rhombi whose obtuse angles are congruent are in a ratio of 36:64, and the side of the greatest rhombus is 16cm. Determine the side of the least rhombus.

4. In what ratio is the area of a triangle divided by a midline?

5. Which part of the area of a triangle makes up the area of the triangle formed by the midlines?

6. A line parallel to a side of a triangle divides it into two equivalent figures. In what ratio does it divide a lateral side?

7. Two triangles are similar, and the area of the lesser triangle is one quarter of the area of the greater. Find the distance from the centroid of the lesser triangle to the nearest vertex, if the medians of the greater triangle are, respectively, 12cm, 16cm, and 19cm.

8. A lateral side of a triangle is divided by straight lines parallel to the base in a ratio 2:3:4 (from the vertex. to the base). In what ratio the area of the triangle has been divided?

9. Given a triangle, divide it into three equivalent parts by straight lines parallel to one of its sides.

10. A trapezoid is divided into two equivalent trapezoids by a straight line. Find the segment of this line that lies within the trapezoid if the bases of the latter are *a* and *b*.

11. Construct a quadrilateral whose area is a) four times b) twice the area of a given quadrilateral.

12. Two corresponding diagonals of similar polygons are in a ratio of $3:\sqrt{5}$. What is the ratio of their areas? What is the ratio of their perimeters?

13. The area of a regular *n*-gon equals $\mathcal{A}$, its circumradius equals R, and a circle of radius h is inscribed in that polygon. Another *n*-gon is inscribed in the latter circle. Find its area and perimeter.

14. Prove that for two similar polygons, the ratio of their areas equals the square of the ratio of their *characteristic linear sizes* (see problems and comments at the end of section 9.5 - beginning of 9.6).

15. The ratio of the areas of two figures equals the ratio of the squares of their *diameters* or any other *characteristic linear sizes*. Is the above statement true for a) convex polygons? b) non-convex polygons? c) arbitrary figures?

16. On a map using a scale 1:300 000 (1cm : 3km), the area of a city is 80cm². Find the actual area of the city.

17. On a map of a city, the length of a bridge is 3cm, whereas in reality the bridge is 600m long. Find the actual area of the city if on the map it occupies 4500cm².

18. A monument to Euclid on planet X is built in a shape of a pyramid, which is 180 m high, and consists of 30 levels (including the ground level, which is assigned #1), each of the same height. What is the area of the floor on the 10th level if the area of the base of the pyramid is 9000m²? What would be the answer to the same question if the height of the pyramid were 90m?

19. Generalize the result of the previous problem. For what kind (shape) of monuments does such a problem make sense? In particular, what are the requirements if any to the shape of the lateral surface of a monument?

11.3 The area of a circle.

Lemma 11.3.1. *When the number of the sides of a regular polygon doubles indefinitely, its side tends to zero and its apothem tends to the radius.*

Remark. In this section we shall often say for brevity *the side tends to zero* instead of *the length of the side tends to zero* (like we did in the above formulation).

Proof. If n is the number of the sides of a regular polygon, and p is its perimeter, the side of the polygon equals $a_n = \frac{p}{n}$. We know that if C is the circumference of the circumscribed circle, and $n > 6$, then $6R < p < C$, therefore, $\frac{6R}{n} < \frac{p}{n} < \frac{C}{n}$, which means that $\frac{6R}{n} < a_n < \frac{C}{n}$.

Since for a given circle both R and C are constants, it follows from the latter inequality that by choosing sufficiently great n, we can make a_n less than any given segment no matter how small the given segment is.

Therefore, by choosing sufficiently high n, we can make the length of a_n less than any given positive number, i.e. as close to zero as we wish. In cases like this we say that the length of a_n tends to 0 when n tends to infinity, and we write: $a_n \to 0$ when $n \to \infty$, or (which means the same) $\lim_{n \to \infty} a_n = 0$, Q.E.D.

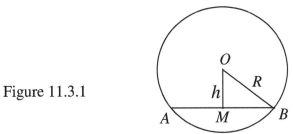

Figure 11.3.1

If in Fig.11.3.1, *AB* is a side of an inscribed regular polygon, we can apply the Pythagorean theorem to express the apothem, *h*:
$$h^2 = R^2 - (\tfrac{1}{2}a_n)^2.$$
When $n \to \infty$, $a_n \to 0$, i.e. the difference between the magnitudes of the apothem and radius can be done smaller than any given number. Therefore, one can say that when the number of sides of a regular polygon indefinitely increases (e.g., as a result of doubling), the apothem tends (indefinitely approaches) to the radius, Q.E.D.

We have shown that the area of a regular polygon equals
$$\tfrac{1}{2} ph, \qquad (1)$$
where *p* is the perimeter and *h* is the apothem.

We define *the area of a circle* as the limit of the area of an inscribed regular polygon when the number of the sides of the polygon tends to infinity.

As the number of the sides tends to infinity (as a result of sequential doubling, for instance), the perimeter of the polygon tends to the circumference (by the definition), and the apothem tends to the radius (by Lemma 11.3.1). Then, the area described by (1) tends to
$$\tfrac{1}{2} CR = \tfrac{1}{2} \cdot 2\pi R \cdot R = \pi R^2.$$
Thus, the area of a circle of radius *R* is
$$\mathcal{A} \,(\texttt{circle}) \;=\; \pi R^2.$$

Now the result of the previous section can be adopted for circles:

<u>Corollary 1.</u> ***The areas of two circles are in the same ratio as the squares of their radii (diameters).***

Proof. If *R* and *r* are the radii of two circles, the areas of these circles are to each other as
$$\frac{\pi R^2}{\pi r^2} = \frac{R^2}{r^2}, \qquad \text{Q.E.D.}$$

<u>Corollary 2.</u> ***The area of a sector is half the product of its arc and the radius.***

Proof. If the arc of a sector measures $n°$, its area is $\dfrac{n}{360}$ part of the area of the circle, i.e. it is $\dfrac{n}{360} \cdot \pi R^2 = \dfrac{\pi R n}{180} \cdot \dfrac{R}{2}$.

Let us notice that the first fraction in the latter product expresses the length of an arc that contains n degrees.

It is very convenient to express the area of a sector in terms of the *radian measure* of its central angle: if the angle measures θ radians, the length of the arc is θR, and the area of the sector is

$$\mathcal{A}(\text{sector}) = \theta R \cdot \dfrac{R}{2} = \tfrac{1}{2}\theta R^2.$$

11.3 Questions and Problems.

Note: the following problems #1 - #5 contain a procedure of defining the area of a circle.

1. Prove that the area of a regular polygon circumscribed about a circle of radius R equals half the product of this radius by the perimeter of the polygon: $S_n = \tfrac{1}{2}P_n R$.

2. Prove that the area of a regular polygon inscribed in a circle of radius R equals half the product of its perimeter by the apothem, that is $s_n = \tfrac{1}{2}p_n h_n = \tfrac{1}{2}p_n\sqrt{R^2 - \left(\dfrac{p_n}{n}\right)^2}$.

3. Prove that the sequence of areas of circumscribed regular polygons obtained as a result of consecutive doubling of the number of sides of a polygon, is monotonically decreasing and bounded below.

4. Prove that the sequence of areas of inscribed regular polygons obtained as a result of consecutive doubling of the number of sides of a polygon, is monotonically increasing and bounded above.

5. Prove that sequences S_n and s_n from the above problems #3 and #4 are *separated* by a single number which is equal to $\tfrac{1}{2}CR$, where C is the circumference. *This number, by definition, is the area of a circle of radius R.*
   This result can be reformulated in terms of limits: The sequences S_n and s_n converge to the same limit equal $\tfrac{1}{2}CR$, which is defined to be the area of the circle.

6. Prove that areas of circumscribed and inscribed in a circle of radius R regular n-gons can be find, respectively, as $S_n = nR^2 \tan\dfrac{\pi}{n}$; $s_n = \tfrac{1}{2}nR^2 \sin\dfrac{2\pi}{n}$. (Note: the angles are in radian measure, i.e. are measured in relative (to the radius) lengths of the subtended arcs).

7. Prove that for the radian measure of an acute angle θ the following inequality holds:
   $\sin\theta < \theta < \tan\theta$.

In the following problems, approximate π by 3.14 or $\frac{22}{7}$ unless specified otherwise.

8. Find the area of a circle inscribed in a square with the side equal 30cm.

9. Find the area of a circle circumscribed about a regular hexagon with the shortest diagonal equal $\sqrt{12}$.

10. Find the greatest diagonal and the area of a regular octagon inscribed in a circle whose area is 154cm².

11. Evaluate the area of the annular region enclosed between two concentric circles of radii R and r.

12. Find the radius of a circle that is located in the interior of a circle of radius R and divides it into two equivalent parts.

13. *AD*, *DC*, and *AC* are diameters of three circles that touch each other as shown in Figure 1. *BD* is a common tangent of the two interior circles. Show that the area of the shaded region equals the area of a circle with diameter *BD*.

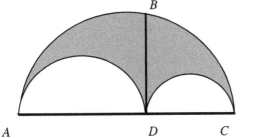

Figure 1

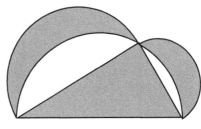

Figure 2

14. Semicircles are constructed on the legs and hypotenuse of a right triangle as on diameters, as shown in Figure 2. Prove that the sum of the areas of the shaded crescents, formed by the largest and two smaller semicircles, equals the area of the triangle.

15. *AB*, *BC*, *CD*, and *AD* are diameters of the semicircles that touch each other as shown in Figure 3. *MN* is a perpendicular bisector of diameter *AD*. The length of *MN* equals *a*. Find the area of the shaded figure.

Figure 3

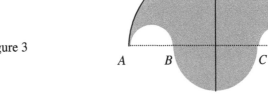

16. Find the areas of the shaded figures cut off from a square and an equilateral triangle, respectively (Figure 4).

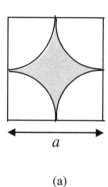
(a)

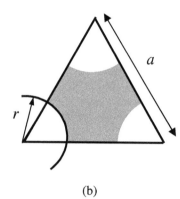
(b)

Figure 4

17. In a circle, a chord equal *a* subtends an arc twice as long as an arc subtended by a chord equal *b*. Find the area of the circle when *a*=32cm, and *b*=20cm.

18. Present the area of a circle as a function of its circumference.

19. Find the area of the base of a right circular cone with vertical axis, if the area of its horizontal cross-section located at height *h* above the base equals *A*, and the total height of the cone equals *H*.

20. What part of the area of a circle makes up a sector whose central angle contains a) 90°? b) 20°? c) 18°? d) 22°30'? e) $\frac{\pi}{3} rad$? f) $\frac{\pi}{4} rad$? g) $\frac{3}{2}\pi \, rad$? h) *1rad* ? i) *3 rad* ?

21. Find the degree measure of the central angle of a sector whose area is a) $\frac{2}{3}$; b) $\frac{1}{4}$; c) $\frac{15}{16}$ of the area of the circle.

22. Find the radian measure of the central angle of a sector whose area is a) $\frac{2}{3}$; b) $\frac{5}{4}$; c) $\frac{1}{12}$ of the area of the circle.

23. Find the area of a sector of a circle of radius *R* if the radian measure of its central angle is *θ*.

24. Find the area of a sector that subtends an arc of length *x* in a unit circle (a circle of radius equal *1*).

25. Find the area of a segment of a circle of radius *R* if the radian measure of the central angle that subtends the arc of the segment is *θ*.

26. The perimeter of a sector is 56m, and its area is 98 cm². Evaluate the radius of the circle.

27. Find the area of a segment of a circle of radius *R* if the *altitude* of the segment (the length of the lying within the segment part of the perpendicular bisector of the chord of the segment) equals *h*.

28. In a circle of radius R two chords equal a and b, respectively, are parallel to each other. Find the area enclosed between these chords.

29. A washer is a thin plate made of steel in a shape of an annular region with interior radius r=1cm and exterior radius R=3cm. The coefficient of linear heat expansion (the relative increment of length when the temperature increases by 1 Kelvin) of steel is $\alpha = 12 \cdot 10^{-6}$ K^{-1}.
a) Will the interior radius increase or decrease as a result of heat expansion?
b) Find the area of the washer (annular region) after the temperature has increased by ΔT=100 K.

11. Review

Prove propositions.

1. In a parallelogram, the distances from a point on a diagonal to the two adjacent sides are inverse proportional to the lengths of the corresponding sides.

2. The area of a trapezoid equals to the product of one of its lateral sides by the distance from the midpoint of the other lateral side to the first lateral side.

3. Two quadrilaterals are equivalent if the diagonals and an angle between them of the one are respectively equal to the diagonals and an angle between them of the other.

4. If the areas of two triangles formed by the bases of a trapezoid as bases and the segments which the diagonals cut off on each other as lateral sides are equal, respectively, p^2 and q^2, the total area of the trapezoid equals $(p+q)^2$.

5. The area of an inscribed regular hexagon is ¾ of the area of the circumscribed (about the same circle) regular hexagon.

6. In quadrilateral *ABCD*, a line drawn through the midpoint of diagonal *BD* parallel to the other diagonal, *AC*, cuts *AD* at some point *E*. Prove that *CE* bisects the area of the quadrilateral.

7. If the medians of a triangle are the sides of the other triangle, the area of the latter is ¾ of the area of the first triangle.

8. *AB* is a chord in a circle, whose centre is located at *O*. A circle is described with *OA* as a diameter. Prove that the areas of the two segments cut off by *AB* from the two circles are in a ratio of 4:1.

Computational problems.

9. Evaluate the area of a right trapezoid with one of its angles equal 60° if you know: a) two bases; b) a base and an altitude; c) a base and a non-perpendicular lateral side.

10. The bases of a trapezoid are equal a and b, and its altitude is h. Find the altitude of the triangle formed by the greater base of the trapezoid (*b*) and the lateral sides extended till their intersection.

11. A triangle of area A is given. Another triangle is formed by its midlines; the sides of the next triangle are the midlines of the second, and these process of generating new triangles continues indefinitely. Find the total area of all these triangles.

12. Given three sides of a triangle (*a, b, c*), find its inradius.

13. Given three sides of a triangle (*a, b, c*), find its circumradius.

14. Find the increment of the area of a circle of radius R generated by the increase of the radius by ΔR. Propose a simple approximate formula to calculate the increment of the area in case $\Delta R \ll R$ (which means that ΔR is considerably less than the radius).

15. Find the radius of the circle in which a segment standing on a chord equal *b* has its altitude equal *h*.

16. Evaluate the altitude and area of a segment subtended by a central angle of 60° in a circle of radius R.

17. The area of a segment on a chord *b* with altitude *h* can be estimated approximately as $\frac{2}{3}bh$. Use this formula to solve the previous problem approximately and find the relative error of the estimate.

18. Derive an approximate formula for the area of a segment whose *base* (chord) and altitude are known. Use your formula to solve problem #15 approximately and estimate the relative error.

Construction problems.

19. Through a vertex of a triangle draw two lines that divide its area in a given ratio, *m÷n÷p*.

20. Divide a triangle into two equivalent parts by a line passing through a given point on a side of the triangle.

21. Find such a point in the interior of a triangle that the straight segments joining this point with the vertices divide the area of the triangle into three equal parts.

22. Find such a point in the interior of a triangle that the straight segments joining this point with the vertices divide the area of the triangle in a given ratio, e.g. 2÷3÷4.

23. Divide a parallelogram into three equivalent parts by two rays emanating from its vertex.

24. Divide the area of a parallelogram in a given ratio (*m÷n*) by a line passing through a given point.

25. Divide a parallelogram into three equivalent parts by two lines parallel to a diagonal.

26. Divide the area of a triangle into extreme and mean ratio by a line parallel to a side of the triangle.

27. Divide a triangle into three equivalent parts by lines perpendicular to the base.

28. On a given segment as a base, construct a rectangle equivalent to a given one.

29. Construct a square whose area is ⅔ of the area of a given square.

30. Given a square, construct an equivalent rectangle with a given sum of the adjacent sides.

31. Given a square, construct an equivalent rectangle with a given difference of the adjacent sides.

32. Construct a circle equivalent to a given annular region.

33. Given two triangles, construct a triangle similar to one of them and equivalent to the other.

34. Construct an equilateral triangle equivalent to a given triangle.

35. Inscribe a rectangle of a given area (m^2) into a given circle.

36. Inscribe a rectangle of a given area (m^2) into a given triangle.

SOLUTIONS

1. ANSWERS, HINTS

> Rabbit looked at him severely.
> "I don't think you are helping," he said.
> "No," said Pooh. "I do try," he added humbly.
> A.A.Milne, *The House At Pooh Corner.*

Section 1.2.
3b. The formula is not correct; try to disprove it by a counterexample.

Section 1.3
34. A *transformation* can be explained as a passage from one object to another. In mathematics transformations are applied to figures, expressions, or sets of various nature.

In geometry we usually talk of transformations of figures. That means: as a result of a transformation, a figure is being transformed into another one or, sometimes, the same figure. In the latter case the transformation is called an *identity transformation*. Usually, when dealing with plane geometry, mathematicians prefer to talk of transformations of the plane in which the figures lie. Then, if the plane is being subjected to a transformation, the figures will change (or stay the same) as a result of this transformation.

Synonyms of the word *transformation* are *mapping*; *function*.

In general, one can view a transformation of the plane as a *correspondence* between the points of the plane to which the transformation has been applied and the points of the plane obtained as a result of the transformation. For example, when drawing a geographical map of a part of the Earth surface, we depict every point A of the surface as a point A' on the map. A' is called the *image* of A, and A is a *pre-image* of A'.

This terminology is applied for transformations in general, not only for geographical mappings. For example, in the figure below, a plane is being *transformed* into another plane in such a way that point A is *mapped* into its *image* A'.

See more about transformations in the SOLUTIONS, HINTS section.

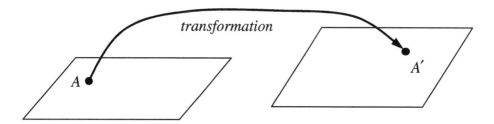

Section 4.4
5. *96°*.

Section 4.5
1. Hint: Prove by contradiction (RAA).
3. Hint: Superimpose the triangles by their congruent sides and adjacent angles; then use the property of the exterior angle to prove that the third vertices will coincide.

Sections 4.6 - 4.7
1. Hint: prove by contradiction (though it is not the only way).
2. No. Hint: prove by contradiction (RAA).
3. Hint: Auxiliary construction: extend AM beyond M by $MD=AM$. Connect D with B and C and consider obtained congruent triangles.

4. Hint: prove by contradiction.
17. Use RAA, i.e. prove by contradiction.

Section 4.9
2. The proof is similar to the proof of *SAA* condition for general triangles. Start with superimposing the triangles by their congruent legs and right angles.

Section 4.10
3. *27cm.*
4. *15cm.*

Chapter 4. REVIEW
14. Hint: consider the points symmetric to *A* in the sides of the angle.

Chapter 5. CONSTRUCTION PROBLEMS
1. Hint: Show that lengths can be transferred by a collapsing compass, or: given a point and a segment, one can use a collapsing compass and a straightedge to describe a circle of radius equal to the segment with the centre at the given point.
18. Hint: Construct the symmetric image of *M* in *AB*.

Sections 6.1-3
20. Hint: Use a) the properties of angles formed by a pair of parallels and their transversal; b) the statement of problem 18; c) the statement of problem 19;…
23. Hint: Through *B* draw a ray parallel to *AC*.
24. Hint: Through *C* draw a ray parallel to *AB*.

Section 6.5
10. 15°, 75°, 30°.
11. 99°, 15°.
13. a) 123° (or 57°); b) 135° (or 45°).
14. 104°.
15. 144°.
30. $d - 2\alpha$.
32. a) 82°; b) 105°52'.
34. Hint: use mathematical induction; suggest for the beginning that only one interior angle is reflex (greater than two right angles).

Section 6.6
2. Use central symmetry about an endpoint of the median.

Section 7.1
27. Hint: Prove by contradiction (RAA); consider the angles that constitute the obtuse angle of the parallelogram.
30. 135°; 45°.

Section 7.2
1. 25°; 65°.
5. It is convenient to use a symmetry axis.
13. Hint: Start with the construction of the two loci.
22. 30°, 60°.
25. a) 30°, 150°; b) 15°, 75°.
26. a) 60°, 120°; b) 8cm.
31. Hint: start with constructing a rhombus whose symmetry may be used to solve the problem.
32. Hint: start with constructing a rhombus whose symmetry may be used to solve the problem.

35. Two ways.
36. No.

Section 7.3
2. No.
3. Hint: Use properties of parallelograms.
4. Hint: Reduce to the preceding problem.
5. 24cm.
11. 2 cm; 14 cm; 16 cm.

Section 7.4
1. 68°; 36°.
3. a) Yes; b) No.
4. a) 130°; 60°; 120°; b) ½(*a+b*).
5. 27cm
6. *3a-b*.
10. Hint: use a *translation* (see Section 7.5).
11. $\frac{11}{34}P$.
12. $\frac{2}{7}m$.

Chapter 7.Review
37. Hint: Reduce to the construction of congruent triangles.
63. Hint: Use a translation of one of the given points along the line.

Sections 8.1-2
3. Do not forget to consider and analyze different cases.
4. Do not forget to consider and analyze different cases.
5. *2/3H*.
6. Hint: the median drawn to the hypotenuse equals half the hypotenuse (prove).
14. *1/2R; 3/2R*.

Section 8.3
2. *4R*.

Section 8.4
5. *2l; 3l*.

Section 8.5
8. *2d -2α; α; α*.

Section 8.6
26. Consider the following possible situations: (1) the sought for circle lies in the exterior of both given circles; (2) one of the given circles lies inside the sought for circle and the other lies in the exterior; (3) both given circles lie in the exterior of the sought for circle.
30. Hint: start with drawing a circle concentric with the given one and passing through the given point.

Section 8.7
5. *d/2±α*, i.e. 35° and 55°.
6. Hint: Extend *AO* into a diameter, *AF*, and connect *F* to *C*. Consider two right triangles.

Section 8.8
3. Hint: use the result of the problem #2.

4. Hint: when considering the angles of the new triangle with the vertex on a base, draw a circle with that base as a diameter and construct the symmetric in that diameter images of the other two feet of the altitudes.

Section 9.2
5. a) 9 cm; b) 1.5 cm.
6. 12 cm or 28 cm.
7. a) 13.5 cm; b) 4cm.

Sections 9.3-4
5. $3.555...cm$
6. $\dfrac{ab}{a+4b}=2.5cm;\ \dfrac{4ab}{a+4b}=10cm.$
9. 14.666…ft.
11. 150,000,000 km.
16. $\dfrac{b(2a-b)}{4a}.$
32. 4.5cm; 10cm.
35. 18cm; 12cm.
36. 22.5 cm.
38. 44mm; 77mm.
39. 112mm; 196mm.
40. 4.2 cm.
41. 5m.
42. 3m.

Section 9.5
11. 24dm; 6dm; 12dm; 18dm.
12. $k=8$.
13. 525 cm.
20. Hint: For the beginning, consider triangles and suggest that the three lines are not concurrent. Use RAA.
21. Hint: Using the similarity ratio for the two polygons, construct a polygon similar to the first one (with any point as the centre of similitude) and then prove that the constructed polygon is congruent to the second polygon.
26. Hint: Start with moving the polygons in a position where they are similarly situated.

Section 9.6
9. Hint: In the centre line find a point whose distances from the centres are in the ratio of the corresponding radii. There will be two points: one between the centres, and another – beyond the segment joining the centres, closer to the centre of the smaller circle.
12. Hint: Show that as T moves around, $P, T,$ and V are always collinear.

Section 9.8
1. 3cm; 6cm.
2. a) 15cm; b) 8cm.
4. 12cm; 3÷4.
7. One method is based on Theorem 9.8.1, and the other – on Th.9.8.2.
11. Either 10 dm or 22.5 dm.
12. 18cm.
14. 5.25cm; 6.75cm.

Section 9.9
2. 12cm.

6. (i) $c=20$, $h=9.6$, $a'=12.8$, $b'=7.2$
   (ii) $c=25$, $h=6.7$, $a'=2.0$, $b'=23.0$
   (iii) $b=12$, $h=4.6$, $a'=1.9$, $b'=11.1$
   (iv) $a=6$, $h=4.8$, $a'=3.6$, $b'=6.4$
   (v) $a=31.9$, $c=36.1$, $a'=28.1$, $b'=8$
   (vi) $a=10\sqrt{61}$, $b=12\sqrt{61}$, $h=60$, $b'=72$
   (vii) $a=\sqrt{10}$, $b=3\sqrt{10}$, $c=10$, $h=3$
   (viii) $a=4\sqrt{5}$, $b=8\sqrt{5}$, $a'=4$, $b'=16$
   (ix) $b=4$, $c=5$, $h=2.4$, $b'=3.2$
   (x) $a=9\sqrt{13}$, $b=6\sqrt{13}$, $c=39$, $a'=27$.
7. 6cm; 7.8cm; 9.4cm.
8. 13/3 m; 5/3 m.
9. 10cm; 1cm.
10. $\sqrt{61} = \sqrt{64\left(1-\dfrac{3}{64}\right)} \approx 8\left(1-\dfrac{3}{128}\right) \approx 7.8$ [cm].
11. 30cm.
12. 18m; 24m.
13. $5\sqrt{3} \approx 8.5 cm$.
14. Legs: $\sqrt{5}m \approx 2.2m$; $4\sqrt{5}m \approx 8.8m$; projections: *1m;* *4m*.
15. 44.7cm; 14.7cm.
16. $\dfrac{a\sqrt{a^2+b^2}}{a+b}$; $\dfrac{a\sqrt{a^2+b^2}}{a+b}$.
17. 2.9cm.
18. 50cm; 96cm.
19. 12.8cm; 27.2cm.
20. Triangle: $R\sqrt{3}$; $\dfrac{1}{2}R$.
21. Hexagon: R; $R\dfrac{\sqrt{3}}{2}$.
22. 9.5m, or 3.2m.
23. 25cm or 7cm.
24. $D = \dfrac{h^2 + 4a^2}{4a}$.
25. 7.8m.
26. 1cm or 7cm.
27. 5cm.
28. 7m.
29. 32.0m.
30. 5m.
31. a) obtuse; b) acute; c) right; d) obtuse; e) obtuse.
32. Hint: recall the converse to the Pythagorean theorem.
33. Hint: Use Th.9.9.5 and the Pythagorean theorem.
34. 4.8cm.
35. 20cm.
36. $50\sqrt{3} \approx 86.6 cm$.
37. 5.5cm; 3.5cm.

Section 9.10

1. $8\sqrt{15} \approx 8 \cdot 3.9 = 31.2$ [cm].
2. 8cm.

3. 30cm.
4. 1.5m.
5. $\sqrt{60} \approx 7.75$ [m].
6. $\sqrt{288} \approx 17.0$ [cm].
7. 4:3.
8. The depth is approximately 1.2+0.48=1.68 [m]. James should walk carefully.
9. 9cm.
10. 12cm.

Section 9.11
1. 5/12; 12/13; 5/12; 12/5.
2. 0.8; 0.6; 4/3; ¾.
6. The tangent is equal $2-\sqrt{3} \approx 0.27$.
9. 9cm; 12cm.
10. sinα=12/13; cosα=5/13; tanα=12/5; cotα=5/12.
17. 30°40′, 59°20′.
18. 6.2cm; 4.8cm.
19. 6.7cm.
20. 73°44′.
25. The unknown side is $\sqrt{156} \approx 12.5$ [cm]. The least angle is (approx.) 16°18′.

Section 10.1
12. $r = R(2\sqrt{3} - 3)$.
28. $a_{10} = R\dfrac{\sqrt{5}-1}{2}$. Hint: for the construction it might be more convenient to represent the side in the form: $a_{10} = \sqrt{R^2 + \left(\dfrac{R}{2}\right)^2} - \dfrac{R}{2}$.
30. Hint: $\dfrac{1}{15} = \dfrac{1}{6} - \dfrac{1}{10}$.
31. $R = \dfrac{2l}{2-\sqrt{3}}$; $h = \dfrac{\sqrt{3}\,l}{2-\sqrt{3}}$.
38. $P_n = R \cdot 2n \tan \dfrac{2d}{n}$.
39. $p_n = R \cdot 2n \sin \dfrac{2d}{n}$.

Section 10.2
3. Hint: Show that a side of each subsequent polygon is less than half of a side of the preceding one.
4. Hint: Compare the circumference with the perimeters of an inscribed regular hexagon and a circumscribed square.
9. $48\pi \approx 150.7$ [mm].
10. 2 complete revolutions.
11. 6.28cm; 10%. ΔC does not depend upon the original value of the radius, since the dependence $C(R)$ is *linear*.
12. The farthest (from the centre) horse is the fastest; 2.1m/s.
13. 990m/min ≈59.4km/h.
14. 2700 rpm.
15. f) $\pi/180$.

16. c) 24°30′; only tens of minutes should be taken into account when the value of π is approximated to within two decimals after the point.
21. Approximately 48 cm.

Section 11.1
28. 22cm².
30. 45cm².
32. 15cm².
33. 45cm².
34. 60cm².
35. 54cm².
36. 13cm.
37. 675cm².
38. 32cm, or 24cm.
39. $\frac{1}{4}c^2$.
40. 182cm².
41. 2010cm²; 503cm².
42. 168cm².
43. 8cm; 10cm.
44. 32 cm².
45. 2:3 (counting from the vertex).
46. This line divides any of the lateral sides in a ratio of $\frac{1}{\sqrt{2}-1}$. Show that and carry out the construction.
48. 10cm.
49. $m(m+2n+2p) : n(n+2p) : p^2 = 95 : 40 : 9$.
50. $\sqrt{\dfrac{a^2+b^2}{2}}$.

Section 11.3
5. Hint: On a coordinate axis, lay off from the origin the segments with lengths equal to S_n and s_n. Thus a sequence of nested segments has been constructed. Prove that this sequence contains segments that are less than any given segment. Then use Cantor's principle.
9. 12.56cm².
10. 14cm.
17. 436 cm².
25. $\frac{1}{2}R^2(\theta - \sin\theta)$. Note: this formula is also valid for segments subtended by a reflex (greater than π radians) angle, when the notion of sine is extended properly.
26. 14cm.

2. SOLUTIONS, HINTS

> "It just shows what can be done by taking a little trouble," said Eeyore. "Do you see, Pooh? Do you see, Piglet? Brains First and then Hard Work."
> A.A.Milne, *The House At Pooh Corner.*

Section 1.2.

3b.   The formula is not correct for an arbitrary quadrilateral, since a quadrilateral is not completely defined by its sides. For example, the following square and non-rectangular diamonds (rhombi) are made of the same sides. If the length of each side is a, the area of the square is a^2, whereas the area of a rhombus with the same side can be made arbitrarily small by "narrowing down" the figure, i.e. by making an angle between two sides sufficiently small, as shown in the figure below.

One can easily imagine or create such a quadrilateral by joining four straight thin planks by means of bolts and nuts (see the figure below). If the nuts are not tightened, one can change the shape of the quadrilateral from a square to a narrow diamond, making the latter enclose as little area as one pleases. A professional carpenter would instantly solve this problem *in practical terms*, since he knows that a quadrilateral with given sides *is not rigid* whereas a triangle is. Later in the course we shall see what that means in *rigorous terms*.

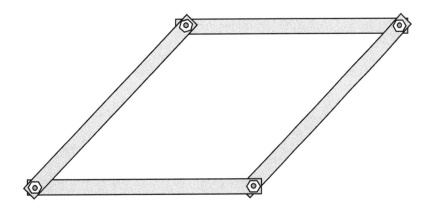

Ancient Egyptians did not notice the mistake since they were apparently dealing with approximately rectangular plots of land.
This example shows that we *should not rely on measurements when asserting general results.*

6.
  a) Only (ii) is applicable to traffic regulations.
  b) Only (iv) may be attributed to Peano axioms (see the solution to Problem 12).
  c) All four definitions would be applicable for the axiom of ancient astronomy.
     As for the Newton's first law of motion, it was not evident at all. So, only (iv) and (ii) may be applicable.

7. Syllogism is a form of reasoning in which the conclusion is derived based on two statements. The statements that form the hypothesis are called the *major premise* and the *minor premise*. For instance:

$$\left.\begin{array}{l}\text{Gentlemen do not lie}\\ \text{Robert is a gentleman}\end{array}\right\} \Rightarrow \text{Robert does not lie}.$$

In this example "Gentlemen do not lie" is the major premise, and "Robert is a gentleman" is the minor premise.

8. The square of this number is 25. $\Leftrightarrow$ The absolute value of this number is 5.

9. These are positive real numbers. Real numbers include integers, fractions (which are called *rational* numbers) and the so-called *irrational* numbers that are not integers or fractions but can be approximated by those with any prescribed accuracy. We shall consider them in detail in Chapter 9.

10. Only property (8) does not hold for magnitudes expressed in negative numbers: for instance, $-5 = (-3) + (-2)$; thus -3 is a part of -5; still $-5 < -3$. All the other properties hold according to the rules of arithmetic (properties of real numbers).

12. Leopold Kronecker (German mathematician, 1823-91) said: "God made the whole numbers, all the rest is the work of Man". To be more specific, we should say that the *set of natural numbers with the operations of addition, and multiplication* are the undefined terms of arithmetic.

The natural numbers are assumed to be an undefined notion of the arithmetic since they cannot be defined in terms of other, more primitive mathematical notions, but they can be introduced axiomatically, that is by means of axioms that describe their properties. Since we perceive the natural numbers as the ones obtained by counting, the axioms must describe (in rigorous terms) the counting procedure. For the beginning, let us summarize our experience with counting in common words.

We start counting with the number *one*, and then each natural number is followed by a number, which is by one greater then its preceding.

One is not preceded by a natural number, since it is the first counting number; otherwise every natural number is preceded by exactly one natural number.

By starting with *one* and making sequential steps from each natural number to the following one (its *successor*), we obtain the set of all natural numbers.

A rigorous description of the counting procedure has been proposed in the 19th century by a German mathematician Dedekind and an Italian mathematician Peano. They have introduced the set of all natural numbers N as an undefined notion whose properties are stated in the following five axioms (we also say: the properties are *postulated*), called the Peano Axioms.

(P1) There exists a natural number 1 (*one*).

(P2) For each natural number n there exists exactly one natural number n', called the *successor of n*.

(P3) There exists no natural number whose successor is 1.

(P4) Every natural number except 1 is a successor of exactly one number; in other words, $m' = n' \Rightarrow m = n$.

(P5) Let T be a set of natural numbers with the following properties:
(i) 1 belongs to T;
(ii) if n belongs to T, then so does n';
then T contains all the natural numbers.

The last axiom (P5) has deserved a special name "The Principle of Mathematical Induction", or "The Principle of Complete Induction". It is often used for proofs (called *proofs by induction*) of the statements that concern natural numbers. In such cases the axiom is usually used in the following formulation:

(P5$'$) *The Principle of Mathematical Induction (Complete Induction).*

Let $S(k)$ be some statement concerning the natural number k. If
(i) $S(1)$ is true, and
(ii) if S is assumed to be true for some number n, the validity of S for the successor of n follows, i.e. $S(n) \Rightarrow S(n')$,
then $S(k)$ is true for every natural number k.

It is obvious that the latter principle is just a reformulation of axiom (P5) for the case when the set T consists of all the natural numbers for which the statement S holds. It is called *complete induction*, in contrast to *incomplete induction* that draws conclusion based on the examination of a finite number of cases. We shall often use this formulation when proving *by induction* various properties of the set of natural numbers.

According to the axiom (P5′), a proof *by induction* always consists of the following two steps, which are called respectively *the first step* and *the inductive step*:
(i) *The first step*: showing that $S(1)$ holds;
(ii) *The inductive step*: showing that $S(n)$ implies $S(n')$.

The operations of *addition* and *multiplication* of natural numbers can be also introduced axiomatically, as operations with certain properties described in corresponding axioms.

The number $m+n$ (to be read "m plus n") is said to be the *sum* of the numbers m and n if
(a) $m+1 = m'$ for every natural m,
(b) $m+n' = (m+n)'$ for every natural m and every natural n,
if there exists exactly one such number for every pair m, n.
We also say that the number $m+n$ is obtained by *addition* of n to m.

The above definition is given with reservations concerning the existence and uniqueness of such a number that we called the sum of a given pair of numbers. It would not make sense to introduce a notion that cannot exist or is not *well defined* (i.e. it is not *unique*) for a given pair of numbers.

Zero and negative integers can be created by imposing certain *equivalence relations* (see the discussion following problem #51 in section 1.3) on the set of all pairs of natural numbers. Rational numbers can be introduced by imposing certain equivalence relations on the set of all possible (with some restrictions) pairs of integers.
The construction of real numbers is a sophisticated process; it will be considered in section 9.1. Those students who are interested can find a brilliant introduction to arithmetic in the book *Foundations of Analysis* by Edmund Landau (3rd edition published in 1966 by Chelsea, New York).

14. One may say (by slightly rephrasing Alfred Tarsky, an eminent logician): *a definition is a statement that tells what meaning is to be attributed to an expression (definiendum) that thus far has not occurred in a certain discipline.*

There are some restrictions on how such a statement can be constructed: the new term should be explained in terms (*definiens*) whose meaning is immediately obvious or has been explained previously. In particular, the term to be defined, or any expression previously defined with its help, cannot be used to assign meaning to the new term (*definiendum*); otherwise the definition is incorrect, it contains an error known as a VICIOUS CIRCLE IN THE DEFINITION. (Think of Euclidean "definitions" of Points, Lines, Planes)

Another restriction concerns the structure of definitions: a definition is an equivalence statement, even though usually it is not formulated explicitly as such. For example, in plane geometry (we consider only figures lying in one plane) we can give the following definition:

Two lines in a plane are *parallel* if they *do not have common points*.

Let us slightly rephrase that statement without changing the meaning:

In a plane: If *lines do not have common points,* then they are *parallel*.

If it we did not know that this is a definition, it would be justified to ask: is it necessarily true the other way around:

In a plane: if *parallel* then they *do not have common points*?

(We can imagine a plenty of examples when one statement follows from another, but the second one does not follow from the first, for example:

If I live in New York, then I live in the USA, but if I live in the USA I do not necessarily live in New York).

That is the point where the requirement of equivalence comes into play:

If a statement is a definition, we always assume that it is true in both directions:

In a plane: If *no common points,* then lines are *parallel* and

In a plane: If lines are *parallel,* then they *do not have common points*.

The second statement is not included in the original definition, but we mean it, since we always mean equivalence for definitions. Equivalence can be expressed briefly by means of the phrase "if and only if" or its abbreviation "iff" :

In a plane, lines are parallel if and only if they do not have common points,

which means: if no common points then they are parallel, but otherwise (if there are common points), they are not parallel.

To make it clear that a statement is a definition, a phrase such as "we say that" is inserted before the definition:

We say that two lines in a plane are parallel if they do not have common points.

Or: Two lines in a plane *are said to be* parallel if they do not have common points.

(We know from the phrases "*we say that*" or "*are said to*" that either of the above is a definition, and the *if* means *iff* (*if and only if*).

Compare the above statements to the following one that does sound similarly, but the phrases that would tell us that this is a definition are missing, hence it is not a definition :

If this *car* is mine, then it is white. (We understand that if a car is white it is not necessarily mine).

15.   Axioms (or postulates) describe the properties of undefined notions. We express the same by saying that we *postulate* these properties.

Section 1.3

16.   According to AXIOM 1, there is exactly one line passing through A and O. Then, all four points: $A, O, B,$ and C lie in this line, hence $A, B,$ and C are collinear.

17.   In each case, segment b seems to be greater than segment a due to an optical illusion. It is never true. The conclusion: we should not trust to what seems to be *obvious from a diagram*; if we are going to derive a result, it must be based on our axioms and results following from the axioms only.

19. There are six possible configurations of three collinear points P, Q, and M:
(1) PQM (2) MPQ (3) QMP
(4) MQP (5) QPM (6) PMQ.

Segments PQ and QP, MQ and QM, etc. are congruent, i.e. we do not make any distinction between them; hence the configurations (1) and (4), (2) and (5), (3) and (6) will lead to the same solutions, just the order in which the points follow from the left to the right will be different.

Let us consider the solutions of (a) and (b) for each of the configurations (1), (2), and (3). (Each solution is accompanied by a corresponding diagram.)

(a)

(1) $MP = MQ + QP = 16cm + 31cm = 47cm$

(2) This configuration is impossible since it is given that $MQ<PQ$.

(3) $MP = PQ - MQ = 31cm - 16cm = 15cm$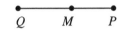

(b)

(1) The configuration is impossible since $PQ = 0.42m > PM = 3.8dm = 0.38m$.

(2) $MQ = MP + PQ = 0.38m + 0.42m = 0.8m$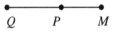

(3) $MQ = PQ - PM = 0.42m - 0.38m = 0.04m = 4cm$

21. The segment covered by a and b together will be congruent to $a + b - c$. In the first diagram below, MN is such a segment. The next diagram shows how a segment congruent to MN can be constructed: On a ray emanating from some point A, lay of $AD=a$ and then, in the same direction, $DF=b$. From F, lay off on FA a segment $FE=c$. Then, according to the definition of addition of segments, $AE = AD + DF - FE = a + b - c$.

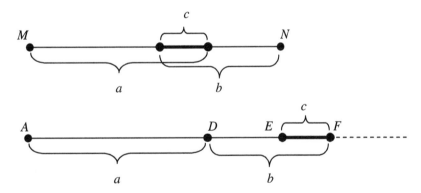

This simple result has an interpretation in *probability theory*. If the probability that some event A occurs is equal to $P(A) = a$, the probability of some event B is $P(B)=b$, and the probability that both A and B occur is $P(AB) = c$, then the probability that at least one of A or B occurs is
$P(A \text{ or } B) = P(A) + P(B) - P(AB) = a + b - c$.

23. Assuming that everyone knows the location of the Island of Twisted Luck, we can find the Island of Broken Hope, since, according to Axiom 1, there is exactly one straight line directed from the Island of Twisted Luck towards the South Pole (for small distances, such as 20 miles, we can assume the meridians to be approximately straight), and, according to Axiom 3, by laying off on that straight line a segment congruent to 20 miles, we shall find exactly one point, where the Island of Broken Hope will be situated.

If the pine marked by crossed swords exists, then, by Axiom 1, there will be exactly one ray directed from the pine to the volcano (hopefully, there is only one volcano on the island!), and by Axiom 3, we shall be able to find exactly one point on that ray by laying off on it a segment congruent to 30 steps. Of course, a step is not a very accurate measure, but the size of the chest filled with treasures is probably commensurate with the error caused by the difference in the sizes of steps of different pirates. Thus, the treasure can be found based on the information provided by the description.

24. The endline of the field (the goal line) is defined uniquely as the line passing through the bases of the goal posts (Axiom 1). From each goal post, in the direction opposite to the other post, the players should lay off a segment congruent to $(50 - 6) \div 2 = 22$ steps. This, according to Axiom 3, will determine the locations of the corners.

In practical terms the algorithm can be formulated as follows:
One player (P) takes a position that is collinear with the goal posts A and B (see the diagram below); he should stay sufficiently far (farther than 22 steps) from the nearest post.
Another player (Q) walks 22 steps from B towards P. The point C where he stops is a corner of the field.

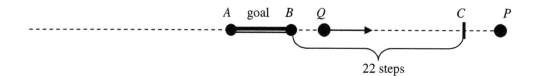

25. The idea of the solution is the following: if the segment that is to be bisected is sufficiently small, one can find the location of the midpoint approximately with a high accuracy. If, for instance, the segment is equal to 1 mm, one can mark a point just anywhere between the endpoints, and the problem will be solved within the accuracy equal to the thickness of the lead (we cannot do any better!).

If the given segment is not small, we can reduce the problem of finding its midpoint M to the problem of finding the midpoint of a smaller segment contained inside the given one.

Let AB be the given segment. From A, lay off on AB a segment AC that is less than AB. Then from B, lay off on BA a segment BD congruent to AC.

If D coincides with C, point C (or D) is the midpoint of AB. Otherwise, if C and D do not coincide, one of two cases illustrated in the diagram below will take place.

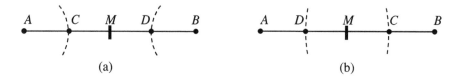

In case (a), $AC=BD$ is less than half of AB. Then, if M is the midpoint of AB, $CM = AM - AC = \frac{1}{2}AB - AC = BM - AC = BM - BD = MD$, since by subtracting equal (congruent) segments from equal, we obtain equal segments. Thus $CM = MD$, which means that M is the midpoint of CD, which is a smaller segment contained in AB.

Similarly, it can be shown that in case (b) point M is the midpoint of CD as well.

Then an analogous step can be repeated a few times until the required accuracy has been achieved.

26. $(a+b)+(a-b) = 2a$. Thus one can construct the sum of the given segments and then bisect it (use the algorithm of problem 25) to obtain a. Similarly, by constructing the difference between the given segments and bisecting it, one can find segment b: $(a+b)-(a-b) = 2b$.

27. Use a ruler to measure the thickness of the book (without the cover) and divide the result by the number of pages in the book. The thicker is the book the more accurate will be our estimate. The accuracy of a measurement is determined as a so-called relative error: the error of the measurement relative to the

size of the object to be measured. For example, the scales with the accuracy 1 kg would be suitable for measuring the weight of an elephant, but not for measuring the weight of a cat.

28.   *CD* seems to be approximately the same or even lesser than *AB*. At the same time, *CD* is much greater than *AB*. Conclusion: One can be deceived by the appearance of a diagram.

32.   Paris and Barcelona are located approximately on the same meridian (2^0 East), whereas London and Dresden are located approximately on the same parallel (51^0 North). Meridian, which is always the trace of a *great circle* (a circle cut on the surface by a plane passing through the centre of the globe) on the surface of the Earth (see diagram (a) below), is more straight ("less curved") than a parallel at such a high longitude: only the equator is the parallel described by a radius equal to the radius of Earth.

In diagrams (b) and (c) below, one can see the globe with the equator and a parallel drawn at high longitude (e.g, 51^0 N). The latter is shown in dotted line. Diagram (c) shows the view from the North Pole. The equator is described by the radius of a great circle; all the other parallels are described by smaller radii. The parallel at 51^0 North is a circle whose radius is nearly half of the radius of the equator, hence it is "more curved".

The meaning of these words is clear from the comparison of the arcs of each of the circles with the same straight segment in diagram (c): the straight segment almost coincides with the intercepted arc of the equator, and the same segment is clearly distinct from the intercepted arc of the 51st parallel.

Therefore a piece of the meridian between Paris and Barcelona is better approximated by a straight segment than a piece of the parallel between London and Dresden.

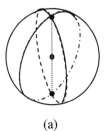
(a)

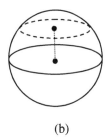
(b)

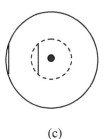
(c)

34.   A *transformation* can be explained as a passage from one object to another. In mathematics transformations are applied to figures, expressions, or sets of various nature.

Synonyms of the word *transformation* are *mapping*; *function*.

In geometry we usually talk of transformations of figures. That means: as a result of a transformation, a figure is being transformed into another one or, sometimes, the same figure. In the latter case the transformation is called an *identity transformation*. Usually, when dealing with plane geometry, mathematicians prefer to talk of transformations of the plane in which the figures lie. Then, if the plane is being subjected to a transformation, the figures will *transformed* as a result of this transformation.

A transformed figure is not necessarily different from the original one. For example, the circle in the figure below will not change (will be *mapped onto itself*) as a result of a transformation that reflects the plane in line *L* (type (iii) isometry) in the figure below.

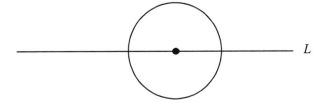

A figure that does not change under a transformation is said to be *invariant* under that transformation. A transformation that carries every point of the plane into itself is called an *identity*

transformation. One may say that an identity transformation does not change the plane. Every figure in the plane is invariant under an identity transformation.

In general, one can view a transformation of the plane as a *correspondence* between the points of the plane to which the transformation has been applied and the points of the plane obtained as a result of the transformation. For example, when drawing a geographical map of a part of the Earth surface, we depict every point A of the surface as a point A' on the map. A' is called the *image* of A, and A is a *pre-image* of A'.

This terminology is applied for transformations in general, not only for geographical mappings. For example, in the figure below, a plane is being *transformed* into a plane in such a way that point A is *mapped* into its *image* A'.

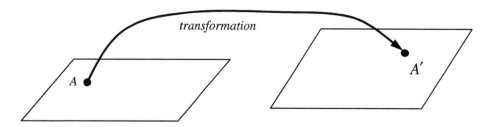

We can represent this correspondence by means of symbols and formulae. Let T denote the transformation presented in the above diagram. Then, in order to show that A is transformed into A' as a result of applying T, we can write:

$T : A \to A'$ (which is read: *T maps A into* A'), or $T(A) = A'$ (which is read: *T of A is* A').

The notation $T(A)$ means that transformation T is applied to point A.

Similar symbolic notations can be applied to figures. If, for instance, some transformation T, when being applied to some figure F, produces figure F', we shall express this fact in writing as
$$T : F \to F' \text{ or } T(F) = F'.$$

For example, in the diagram below, $T(triangle) = star$.

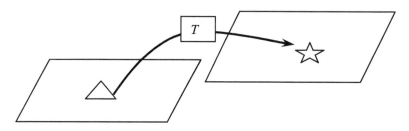

The transformed plane can be subjected to another transformation, S, which transforms the star into a new figure, as shown in the next diagram: $S(star) = face$.

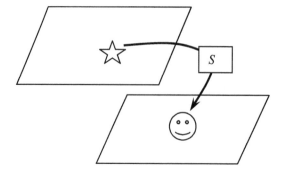

Since the triangle of the preceding diagram is transformed into the star of the latter diagram, we can also view the face of the last diagram as the image of the triangle of the preceding diagram: *T* transforms the triangle into the star, and *S* transforms the star into the face; therefore the combination "*T followed by S*" transforms the triangle into the face.

We can write this symbolically: $S(star) = S\big(T(triangle)\big) = face$.

In order to emphasize that we view "*T followed by S*" as a single transformation, we shall call it a *composition of T and S* and denote it by the symbol $S \circ T$.

Thus we can describe this transformation as

$$(S \circ T)(triangle) = S\big(T(triangle)\big) = S(star) = face, \text{ or}$$

$$S \circ T : \; triangle \to face$$

and illustrate it by the following diagram:

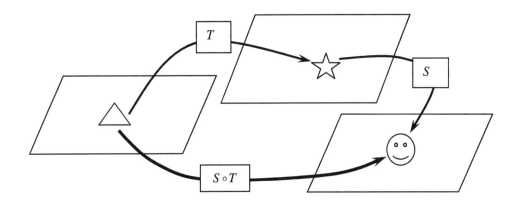

35. A few examples of possible solutions follow.

(a1)   A contraction.   (a2)   An expansion in the horizontal direction.

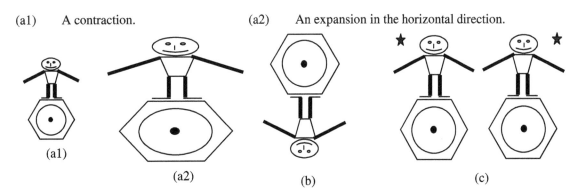

(b)   A rotation (type ii isometry) that transforms the man into his antipode, leaves the circle and the hexagon unchanged due to their symmetry.

(c)   A reflection (type iii isometry) in the vertical line passing through the centre of the circle, transforms the whole figure into itself. The same transformation moves the star from its position above the right hand into a new position, above the left hand of the man.

38.   One can use a compass to verify that the three segments are congruent, hence each of them is an isometric image of any of the two remaining: when two segments (figures) are congruent one of them can be transformed into another by means of an isometry.

We can transform each of them into any other by means of type (i) and (ii) isometries.

A word of warning: it looks like we do not need type (ii) isometry for transforming *a* into *b*: can't we just *translate* the left endpoint of *a* into the left endpoint of *b without changing the direction of the segment*, and then a will be imposed on *b*? The answer is: *we cannot*, since the notions **"to keep the same direction"** or **"to change the direction"** are neither defined nor assumed as undefined notions.

In order to define these notions we shall need another postulate, the so-called Parallel Postulate, which will be introduced later on, in Chapter 6. Then we shall define the notion of *parallel translation*.

Type (i) isometry is not a parallel translation, - it is just a translation that carries a certain point of the plane into a certain point of the plane. Thus we can certainly move segment *a* in such a way that its left endpoint falls on the left endpoint of segment *b*, but we cannot be sure that the whole segment will keep its direction (we just do not know what that means!). Thus we must assume that it will take some position *a'* (see the diagram below), to which type (ii) isometry should be applied to transform it into *b*.

39. Both (a1) and (a2) are not isometries: they change the distances between points. (b) can be performed as type (ii) or as type (iii) as a reflection in the horizontal line through the centre of the circle. (c) is a type (iii) isometry.

40. An isometry of type (iii) that leaves the line of the equator stationary, will return the map into its regular view. We can achieve the same result by various combinations of isometries; for example we can rotate the map about some point so as to return the northern semicircles into the upper position and then reflect the map in a vertical line.

In any case we shall have to use a type (iii) isometry in order to change the *orientation of the plane*.

41. For (b) and (c) we can apply a translation (type (i)) followed by a rotation (type (ii)): the translation carries a point into a point (for instance, the tip of the bow of the caravel in the position (a) can be carried into the tip of the bow of the caravel (b) or (c)), and then by means of a rotation about the tip of the bow, the caravel can be imposed onto the one in position (b) or (c), as shown in figures (b) and (c) below, where the contours of the images of the caravel after the translations are shown in dotted lines..

In order to carry the caravel into position (d) we shall have to subject it to an isometry of type (iii), maybe in combination with some other isometries. For example, we can again translate it "tip-to-tip", rotate it about the tip so as to make the caravel forming the same angle with the horizon as the caravel in (d) and then reflect in the vertical line passing through the tips.

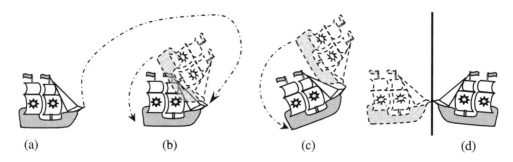

(a)   (b)   (c)   (d)

42. <u>Hint:</u> The solution is similar to the previous problem except one new feature: since the spaceship is symmetric in the axis shown below in dotted line, it is not necessary to use isometries of type (iii).

43.   a) , b) –Yes: before entering the property of the magician, the footprints had been exact copies of each other.

c) – No, it is not an isometry: the footprints have become much smaller, which means that the distances (e.g. between the heel and a toe) have not been preserved.

48.   We have to prove that any figure is congruent to itself. According to the definition, a figure is congruent to a figure if the first figure can be transformed onto the second by means of an isometry. Therefore we have to prove that a figure can be transformed onto itself by means of an isometry.

Let F be some figure in a plane, and I be a transformation that carries (maps) each point of the plane into itself. (We shall call I an *identity transformation*). Then I will transform every point of F into itself, which means that every segment will be mapped onto itself: If A and B are two points on F, then I will carry A into A and B into B, hence $I(AB) = AB$. According to Axiom 2(i), every segment is congruent to itself; therefore I maps every segment into a congruent segment, which means that I is an isometry.

Thus a figure can be mapped onto itself by an isometry. The latter means that a figure is congruent to itself, Q.E.D.

49.   Let figure F be congruent to some figure Φ. We have to show that Φ is congruent to F. In order to do this, we have to show that Φ can be mapped onto F by means of an isometry.

Since F is congruent to Φ, there must exist an isometry T that, when being applied to the plane containing F, maps F onto Φ.

Let S be a transformation that abolishes the result produced by the action of T on the plane. In other words, S is such a transformation of the plane containing Φ that carries every image $A' = T(A)$ of a point A under T into its preimage. That is $S(A') = A$.

Thus if A' and B' are two points of Φ, they will be mapped by S into their preimages A and B under T. Then $A'B'$ will be transformed by S onto the segment AB that was transformed onto $A'B'$ by T. We can write this by means of the formulae: $S(A'B') = AB$, where $A'B' = T(AB)$.

Since T is an isometry, it transforms a segment (AB) onto a congruent segment ($A'B'$); hence the last equality implies that $A'B' = AB$. Then, by Axiom 2(ii), $A'B' = AB$, and it follows from $S(A'B') = AB$ that S is an isometry.

Since F is obtained out of Φ by means of an isometry, then Φ is congruent to F, Q.E.D.
REMARK. Transformation S as described in this solution is called an *inverse* of T and denoted T^{-1}.

50.   Hint: The solution is similar to the solutions of the previous two problems. In order to show the congruence one will have to use a *composition* of two isometries. The discussion and notations will be almost identical to those of problem 49.

53.   Let $S = \{a, b, c, d, e, ..., x, y, ...\}$ be the set on which an equivalence relation is defined. Here small Latin letters (e.g., $a, b, ..., x, y, ...$) denote the elements of the set: $a \in S, b \in S$, etc.

We shall also use the following notations:

If U and V are two sets, and each element of U is also an element of V, we shall say that U is a *subset* of V and denote this by writing $U \subseteq V$ (U may also coincide with V).

$a \sim b$ will stand for "a is equivalent to b".

If a is not equivalent to b, we shall write $a \not\sim b$.

The set of all elements equivalent to some element a will be denoted $[a]$. Such a set, which is, of course, a subset of S, will be called the *equivalence class of a*.

We are going to prove that the whole set S is the union of one or more equivalence classes, and if two of these classes have a common element, then they completely coincide; in other words, two distinct classes cannot have common elements.

First of all it should be noticed that if S is not empty, there are equivalence classes in S: if a is an element of S, then by reflexivity $a \sim a$, hence $[a]$ contains at least one element, namely a.

Now let us show that if two equivalence classes have a common element, they completely coincide.

Suppose, there is such an element x that $x \in [a]$ and $x \in [b]$. That means: $x \sim a$ and $x \sim b$. Then, by symmetry, $a \sim x$, and since $x \sim b$, we can conclude that, by transitivity, $a \sim b$.

Now we shall use the transitivity property to show that any element of $[a]$ belongs to $[b]$ and vice versa. Really, if $d \in [a]$, $\Rightarrow$ $d \sim a$. Also, as we have shown, $a \sim b$; therefore, by transitivity, $d \sim b$. Hence, $d \in [b]$, which means : each element of $[a]$ is also an element of $[b]$, or $[a] \subseteq [b]$.

Similarly, using symmetry and transitivity, one can show that $[b] \subseteq [a]$. Thus, each element of $[a]$ is also an element of $[b]$ and vice versa, and therefore each of the two classes consists of the same elements: $[a] = [b]$.

We have proved that if two classes have a common element, then they coincide. In other words, two distinct classes have no common elements (they do not intersect).

55. <u>Examples from "exact sciences"</u> :

<u>I EXAMPLE</u>. Let us say that two real numbers are in relation R if their absolute values are equal: we say that xRy if $|x| = |y|$. It is easy to show that the three properties hold:

(1) For every number x, $|x| = |x|$;

(2) If $|x| = |y|$, then $|y| = |x|$;

(3) If $|x| = |y|$ and $|y| = |z|$, then $|x| = |z|$.

Thus the *equality of absolute values* is an equivalence relation on the set of real numbers. It splits the whole set of real numbers into families (equivalence classes): the equivalence class of 0 consists only of one element, which is 0 itself: $[0]=\{0\}$; all the other classes (there are infinitely many of them) consist of pairs: a number and its opposite. For example, $[5]=\{5, -5\}$; $[0.96]=\{0.96, -0.96\}$; etc.

<u>II EXAMPLE</u>. The relation "being concentric" is an equivalence relation on the set of all circles (it is easy to check that the three properties hold).

<u>An example from "political sciences"</u>:
"Being a friend" may be viewed as an equivalence relation (with certain doubts):
(1) Everyone is a friend of himself (herself);
(2) If I am your friend, then you are my friend;
(3) A friend of my friend is my friend.

56. <u>An important example from "exact sciences"</u>:

A relation of *partial order* (we shall apply it to the set of real numbers, however it can be used for more general kinds of objects):

We say that $x \geq y$ *if x is greater than y or if they are equal.*

This is not an equivalence relation since the second property (Symmetry) does not hold: if x is greater than or equal to y, then y is <u>not</u> greater than x: y is less than x or equal x. (One can check that the properties of *reflexivity* and *transitivity* hold for the relation of partial order. In general, the relation of partial order is defined as a relation that possesses the properties of

(1) REFLEXIVITY: $x \geq x$,

(2) ANTISYMMETRY: If $x \geq y$ and $y \geq x$, then $x = y$,

(3) TRANSITIVITY: If $x \geq y$ and $y \geq z$, then $x \geq z$.

<u>Another "scientific" example:</u>
Approximate equality within a given accuracy is not an equivalence relation.
For example, let us say that two segments are in relation AE (Approximately Equal) if the difference between them does not exceed 1 cm. Prove that this relation (AE) is not an equivalence relation on the set of all segments. (Hint: which property will be violated? – Propose an example showing a violation of that property)

<u>An example from "political sciences":</u>
Being an enemy is not an equivalence relation: an enemy of my enemy is not necessarily my enemy (it may be quite opposite), hence the transitivity does not hold.

64. Let two circles have equal radii r. Let us apply an isometry (type (i), translation along the line joining their centers that will move the centre of the second circle into the point O_1 where the first centre is located (we can do this, according to part (i) of the Axiom of Isometries).

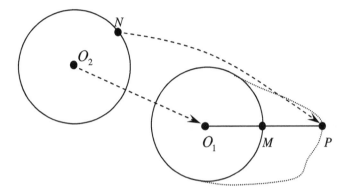

If every point of the imposed circle (or, to be exact, the isometric image of the point) falls onto the first circle, the circles are congruent by definition, and the proof is complete.

Let us suggest that some point N of the second circle is transformed by this isometry into point P, which does not lie on the first circle. Then let us draw a straight line through O_1 and P (we can do this by Axiom 1). Since O_1 is located in the interior of the circle, this line will intersect the circle at some point M.

Since O_1P is obtained by means of an isometry from O_2N, which is a radius of the second circle, it must be equal to the radius of this circle: $O_2N = r$. Also $O_1M = r$ as a radius of the first circle. Then the segments O_1M and O_1P are two congruent segments laid off on the same ray (emanating from O_1 and passing through P) from the same point O_1, and therefore their endpoints M and P must coincide, according to Axiom 3.

This contradicts to our suggestion that P lies beyond the first circle. Hence this suggestion has proven false, and each point of the second circle will lie on the first circle as a result of the isometry that makes their centers coincide. The circles coincide as a result of an isometry, therefore they are congruent, Q.E.D.

65. This statement is proved as Theorem 2.2.2 in Chapter 2 of the textbook.

Section 2.1.

2.   a)   Elephants are funny.
        b)   Elephants are mammals.
        c)   Elephants are birds.

4.
a) The inference is false: some birds cannot fly, e.g. ostrich and penguins. The conclusion is based on the observation of many birds but not all of them.

h) The inference is false: modern ships are often made of metals much heavier than water. In this case the reasoning does not include a sufficient analysis of possible situations in which an object can float (for example, an object may be hollow).

i) The statement is not true even approximately. Obviously it has been derived based on some experiments in which the resistance of air played an important role, and the falling bodies differed not only in weight but in their shapes as well. When analyzing the situation, Aristotle must have not taken the resistance forces into account.

j) The conclusion is true with a high degree of accuracy: the acceleration of free falling bodies is a constant (it is true if the height from which the objects fall is negligible comparing to the size of Earth, and therefore the weight of an object may be deemed constant). Galileo observed similar bodies (heavy balls) of various sizes, and the contribution of the resistance forces to their acceleration was reduced to a minimum.

l) This law (Newton's first law of motion) is one of the axioms of Newtonian mechanics. It postulates the existence of the so-called *inertial frames of reference*, which, by the way, can exist only in ideal world. In real world an object is always subjected to some forces, and the inference is based rather on mental experiments and contemplation than on tests and measurements involving real physical objects.

6.
- a) One line in a plane divides the plane into two parts, so the statement is true for $n = 1$. Suppose n lines divide their plane into $2n$ parts. By drawing one more line, concurrent with these n lines, we shall divide two vertical angles into 4 angles, and thus 2 more angles will be added. Now the plane will be divided into $2n + 2 = 2(n + 1)$ parts. Hence, the hypothesis holds for $n + 1$ lines, and, by induction, it has been proved for an arbitrary number of lines.

- b) One line in a plane divides the plane into two parts. $2 \leq 2^1$, hence the hypothesis holds for $n = 1$. Suppose it is true for some number n. Let us show that it will be true for $n + 1$ as well.

  According to AXIOM 1, a line may intersect another line in at most 1 point. Therefore, if a plane is divided into a few regions by n lines, a new line drawn in the plane may have with these lines at most n points of intersection, and thus will create at most $2n$ new regions (2 with each line).

  Thus the total number of the regions created by $(n + 1)$ lines will not exceed $2^n + 2n$.

  It can easily be proved by mathematical induction that $2n \leq 2^n$. As a result of this inequality, the number of the regions for n + 1 lines will not exceed $2^n + 2n \leq 2^n + 2^n = 2 \cdot 2^n = 2^{n+1}$, whence the hypothesis follows for $n + 1$, and thus for any number of lines, Q.E.D.

- c) For $n = 1$, $\frac{1 \cdot (1+1)}{2} = 1$. Therefore the statement is true for $n = 1$.

  Suppose, the hypothesis is true for n: $1 + 2 + ... + n = \frac{n(n+1)}{2}$.

  Then for $n + 1$:
  $$1 + 2 + ... + n + (n+1) = \frac{n(n+1)}{2} + (n+1) =$$
  $$= (n+1)\left(\frac{n}{2} + 1\right) = \frac{(n+1)(n+2)}{2} = \frac{(n+1)((n+1)+1)}{2},$$

  which means the validity of the hypothesis for $n + 1$.
  Hence, the formula is true for any natural number, Q.E.D.

- d) The proof is completely analogous to the previous one.

9. *Inductive reasoning*: the conclusion is based on observations.
   Deductive reasoning: the conclusion is derived from some other statements that are proved (propositions, theorems) or believed to be true (axioms, postulates).
   a) Inductive, since it is based on an observation. By the way, the conclusion is false.

b) Deductive (based on AXIOM 3 and the definition of the sum of two segments)
c) Deductive. The conclusion follows from the Newton's first law of motion, which is an axiom in classical dynamics.
d) Deductive. It is based on an axiom that may be false itself, but the choice of axioms, especially in social studies, is a matter of belief.
e) Most likely, – inductive: I have not seen them, therefore I think they do not exist. Yet, their existence is not forbidden by the principles of modern science.
f) Deductive. The conclusion follows from AXIOM 3.
g) Deductive. It is a theorem in Euclidean geometry.
h) Inductive. The conclusion is based on an observation, and it is false.
i) Deductive. The statement follows from the Newton's laws of motion.

Sections 2.2-3

1.
a) If <u>the light is red</u>, then *you must stop*.
b) If <u>an object is made of wood</u>, then *it will float*.
c) If <u>an isometry is applied to a figure</u>, then *the figure will be transformed into a congruent one*.
d) If <u>the radii of two circles are congruent</u>, then *the circles are congruent*.
e) If <u>a statement is based on empiric induction</u>, then *it is not necessarily true*.
f) If <u>an axiom of a set follows from some other axioms of the set</u>, then *the independence property is violated*.
g) If <u>a city is located in Hungary</u>, then *it is a European city*.
h) If <u>a person is decent</u>, then *this person will not cheat*.
i) If <u>two lines do not coincide</u>, then *they have at most one common point*.

7.
a) (i) If some creature is a human, then it will have one nose.
   (ii) The converse: If a creature has one nose, then it is a human.
   (iii) The converse is false: for instance, a monkey has one nose and it is not a human.
   (iv)
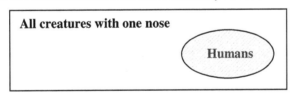

b) (i) If two circles are congruent, then their radii are equal.
   (ii) The converse: If two circles have equal radii, then they are congruent.
   (iii) The converse is true; it can be proved by means of an isometry that imposes the centre of a circle onto the centre of a circle of the same radius (before performing such a motion, it should be shown, e.g. by contradiction, that each circle has exactly one centre).
   (iv)
   > **THIS RECTANGLE REPRESENTS:**
   > The set of all circles that have the same radius as a given circle
   > ALSO:
   > The set of all circles congruent to the given circle

c) (i) If an even number is divisible by 3, then it is divisible by 6.
   (ii) The converse: If a number is divisible by 6, then it is even and divisible by 3.
   (iii) The converse is true: If a number is divisible by 6, it must contain both 2 and 3 as its factors.
   (iv) The regions denoting the set of even divisible by 3 numbers and the set of divisible by 6 numbers will completely coincide.

d) (i) If a moving object is changing the direction of motion, then this object is subjected to a force.
(ii) The converse: If a moving object is subjected to a force, it will change the direction of motion.
(iii) The converse is false: If a force is applied to an object, its velocity will change, however it may undergo a change in magnitude only, not necessarily in its direction. An example: an object decelerating under the friction force when no other forces are applied.
(iv)

```
┌─────────────────────────────────────────────┐
│ All objects to which some force is applied  │
│       ╭─────────────────────────╮           │
│       │ All objects that change │           │
│       │   the direction of motion│          │
│       ╰─────────────────────────╯           │
└─────────────────────────────────────────────┘
```

10.
(i) (a) is not necessary for (b).
(a) is not sufficient for (b). (It will be sufficient if we consider only k > 2).
(a) and (b) are not equivalent.

(ii) (a) is necessary upon (b), or $b \Rightarrow a$.
(a) is not sufficient for (b): $a \not\Rightarrow b$.
(a) and (b) are not equivalent: $a \not\Leftrightarrow b$.

(iii) (a) is not necessary upon (b): $b \not\Rightarrow a$. Let us consider the following example. Let O be a point in the plane, and T be a transformation that makes the distance between each point of the plane and O twice as big without moving O. (Such a transformation is called a *dilatation with coefficient* 2). Then let us apply to this plane another transformation that will also keep the position of the point O unchanged, but halve the distance between O and any other point of the plane (a *dilatation with coefficient* $\frac{1}{2}$). As a result of a *composition* of these two transformations (one of them followed by the other), all the distances in the plane will stay unchanged. So the composition of these transformations is an isometry, although each of them is not.
(a) is sufficient for (b): $a \Rightarrow b$. Apparently, $a \not\Leftrightarrow b$.

(iv) (a) is necessary upon (b): $b \Rightarrow a$.
(a) is not sufficient for (b): $a \not\Rightarrow b$. Apparently, they are not equivalent.

(v) (a) is necessary and sufficient for (b), i.e. they are equivalent.
Really: if (a) is true, then the number n is a product of some odd number and 5: $n = (2k+1) \cdot 5 = 10k + 5$, so the last digit is 5. Then (a) implies (b), or (a) is sufficient for (b). If, on the other hand, (b) is true, $n = 10k + 5 = 5 \cdot (2k+1)$, i.e. n contains 5 and some other odd factor(s). thus, it is odd and (a) is true. The latter means: (b) implies (a), or (a) is necessary upon (b). Therefore, (a) and (b) follow from one another, and so they are equivalent.

Sections 4.1-2

2. According to he general definition, *a figure is convex if for any two points in its interior, the segment joining these points lies entirely in the interior.*

Let us suppose that *A* and *B* are such points in the interior of a polygon bounded by a convex broken line, that some of the points of *AB* are not located in the interior of the polygon. Then *AB* should intersect the boundary of the polygon twice. Since two straight lines (or segments) can intersect in at most one point (Axiom 1), *AB* cannot intersect the same side twice; hence it will intersect two different sides of the polygon. Let us label these sides *m* and *n*.

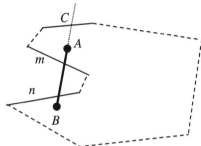

Since point *A* is in the interior of the polygon, the extension of *BA* beyond *A* in the direction opposite to *B* will intersect the boundary of the polygon. This intersection of the extension of *BA* cannot be an intersection with *m* or *n* (Axiom 1); hence the point of intersection *C* will belong to some other side of the polygon. Since *A* lies on the other side from *m* than *n*, the point of intersection *C* will also lie on the other side from *m* than side *n*. The latter means that the boundary of the polygon is not a convex broken line, which contradicts to the hypothesis.

Therefore the suggestion that *AB* has points lying in the exterior of the polygon is false. That means that every segment whose endpoints are within the polygon lies entirely within it, i.e. it is a convex figure.

6. Since a vertex can be joined by a diagonal with every other vertex except the two neighbouring ones, there will be $n-2$ triangles.

Proof by induction.

<u>First step.</u> In a convex quadrilateral, only one diagonal can be drawn, and the polygon is split into 2 triangles. $2 = 4-2$, which comletes the first step.

<u>Induction step.</u> Suppose, for every convex *n*-gon there are $n-2$ triangles. Let us prove that the formula works for $(n+1)$ –gon. In an $(n+1)$ –gon, draw the diagonal joining a vertex with one of the nearest vertices following a neighbouring vertex.

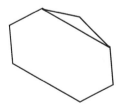

This will split the polygon into a triangle and an *n*-gon; so the total number of the triangles will be $(n-2)+1=(n+1)-2$, which confirms the inductive suggestion and thus completes the proof.

8. (The solution is based solely on the symmetry and cannot be substantiated by a rigorous argument at this point of the course. It will be substantiated in Chapter 10).

Draw a circle, and from the centre draw five rays that form $\dfrac{360^0}{5}=72^0$ angles with each other.

Join consecutively the points of intersection of the rays with the circle; this way you will obtain a convex regular pentagon, shown in the left diagram below.

By joining consecutively each point of intersection with the next to the nearest one (by drawing all the diagonals of the convex regular pentagon), we can obtain a non-convex regular polygon, which is usually called a pentagram.

Similarly, one can construct a regular decagon. Also, five vertices of a regular decagon can be obtained by bisecting the arcs subtended by the sides of a convex regular pentagon.

Section 4.3

6. Construct point M' symmetric to M in AB (this can be done by drawing through M a line perpendicular to AB and laying off on this line from the point P, at which it intersects AB, a segment PM' congruent to MP and lying on the other than MP side of AB). Draw a straight line through M' and N; it will intersect with AB at some point C.
Since PM' is symmetric to PM, $\angle ACM = \angle ACM'$, and $\angle ACN = \angle ACM'$ as vertical angles; hence $\angle ACM = \angle ACM'$, and C is the required point.

7. Let O be the centre of the circle and M be the point of intersection of AB and CD, as shown in the diagram. For convenience we denote the radius of the circle by the symbol r (for instance, $AB = 2r$, since AB is a diameter).

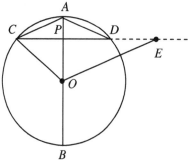

Let us perform a type (iii) isometry: a reflection in AB. $\angle CPA = \angle DPA$ since each of these angles is right; therefore PC will fall upon the line through P and D as a result of the isometry.

Suppose point C does not fall onto D, but it will be mapped into some other point E. Thus OE is obtained from OC by means of an isometry (it is an isometric image of OC); hence $OE = OC$. If point E is located in the exterior of the circle, then $OE > r$. The latter is impossible since $OC = r$, and therefore OE must be congruent to r as well. Thus we have arrived to a contradiction, which shows that C cannot fall as a result of the isometry on the extension of PD beyond D.

Similarly, we can show that the C cannot be transformed by means of the reflection in AB into a point lying in the interior of PD. Therefore, it fall exactly onto D, i.e. point D is the reflection image of C.

If C falls on D, then AC falls onto AD (according to Axiom 1, two lines through the same pair of points must coincide), and hence $AC = AD$, Q.E.D.
Remark. For brevity, we often use expressions like "C falls on E" instead of saying "The isometric image of C falls on E" or "C is mapped onto E".

8. Hint. Use type (iii) isometry: the reflection of the plane in AB. Since the angles are congruent and they have a common side AB, the second sides will fall onto one another as a result of the isometry. Then

prove by contradiction that *C* will be mapped into *D*. The solution is almost identical to the solution of problem 7 of this section.

Section 4.4

19. $\Delta DAB = \Delta CBA$ by *SAS* condition: *AB* is common, $\angle DAB = \angle CBA$, and $\Delta DBA = \Delta CAB$ by the hypothesis. Therefore, the diagonals are congruent as corresponding sides in congruent triangles.

26. $\Delta OAQ = \Delta OBQ$ by *SSS*: *OQ* is a common side, *OA=OB*, and *QA=QB* as radii of the same circles. Therefore, if ΔOAQ is reflected in *OQ*, it will fall onto ΔOBQ, which means that *B* is the reflection image of *A*, and hence they lie on a perpendicular to *OQ*. Thus, $AB \perp OQ$, Q.E.D.

Section 4.5

2. Draw a line through *M* and *B* (we can do this according to Axiom 1).

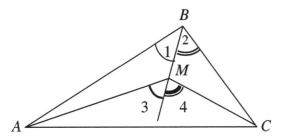

$\angle 3 > \angle 1$ since $\angle 3$ is exterior for ΔABM. Similarly, $\angle 4 > \angle 2$. Therefore, $\angle ABM = \angle 3 + \angle 4 > \angle 1 + \angle 2 = \angle ABC$.

4. Let us prove the conclusion by contradiction (RAA).
    Let *ABC* be an obtuse triangle, and $\angle A$ be obtuse. Suppose point *H*, the foot of the altitude from *B*, falls onto *AC*, in the interior of this segment. Then in ΔABH, one angle is obtuse ($\angle A$), and one angle is right ($\angle AHB$). This contradicts to a corollary from the exterior angle theorem: at least two angles in a triangle must be acute. Thus our suggestion leads to a contradiction and therefore it cannot be true: the altitude from *B* cannot fall between *A* and *C*, Q.E.D.

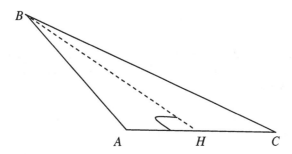

Sections 4.6 – 4.7

1. Let us join *M* with *A* and *C* by means of straight segments and extend *AM* until it intersects *BC* at some point *D* (see the figure). Since both considered perimeters include *AC*, it is enough to show that *AB+BC>AM+MC*.
    According to the *triangle inequality*:

In △ABD, AD < AB + BD, so AM + MD < AB + BD. (1)
In △MDC, MC < MD + DC   (2)
By adding the inequalities (1) and (2) we obtain:
AM + MD + MC < AB + BD + MD + DC,
whence AM + MC < AB + BC, Q.E.D.

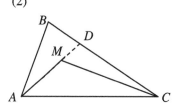

2. Let us prove by contradiction (by RAA) that it is impossible to cut a scalene (non-isosceles) triangle into two congruent ones.

Let △ABC be scalene, i.e. AB ≠ BC, BC ≠ AC, and AC ≠ AB. A line that may cut it into two triangles must pass through one of its vertices, otherwise it will split the triangle into a triangle and a quadrilateral.

Let us suggest that some line *l* passing through the vertex B splits △ABC into two congruent triangles, △ABD and △CBD, where D is the point at which *l* intersects AC.

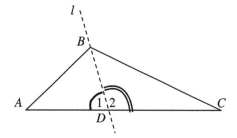

l and therefore BD may be perpendicular or not perpendicular to AC. We shall consider both cases.

If BD is not perpendicular to AC, then ∠1 and ∠2 are not congruent. Since they are supplements of each other, one of them is greater than the other. Let us suggest, for certainty, that ∠2 > ∠1. Since ∠2 is exterior for △ABD, it is greater than any of the non-adjacent interior angles, namely ∠A and ∠ABD. Then ∠2 is greater than any angle in △ABD, and therefore triangles ABD and CBD cannot be congruent: there is no angle in △ABD that is congruent to ∠2.

If BD is perpendicular to AC, line *l* splits △ABC into two right triangles whose hypotenuses (AB and BC) are not congruent (it is given). If two right triangles are congruent, their hypotenuses must be congruent (why?). Hence the obtained right triangles (ABD and CBD) are not congruent.

Therefore in any case line *l* does not divide the given triangle into two congruent triangles, Q.E.D.

3. In triangles AMB and AMC, AM is a common side, and MB=MC since AM is a median by the hypothesis. Then, according to Theorem 4.7.3(ii), ∠AMB is greater than ∠AMC as the angle opposed to the greater of the two unequal sides.

4. The idea of the solution is the following: we shall construct a triangle in which both angles, ∠BAM and ∠CAM, will be interior angles. Then we shall know that the angle opposed to a greater side is the greater of the two.

Let us extend the median beyond M and lay off on this extension MD=AM (see the figure below). Then △ABM = △DCM by SAS condition: MD=AM by construction, CM=MB by the hypothesis since AM is a median, and ∠AMB = ∠DMC as vertical. Then ∠1 = ∠3 as corresponding angles in congruent triangles, and DC=AB as corresponding sides.

In △ADC, DC = AB > AC, hence ∠2 > ∠3 as ∠2 is opposed to the greater of the two sides. Therefore, ∠2 > ∠3 = ∠1, and ∠CAM > ∠BAM (the angle between a median and the lesser side is greater than the angle between the median and the greater side).

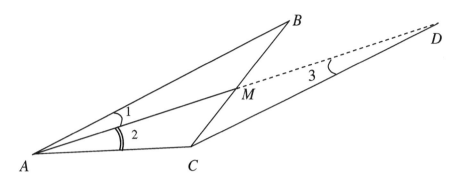

5. Suggest the opposite: in $\triangle ABC$ (see the figure) BC is the greatest side, AH is the altitude drawn to that side, and H falls on the extension of BC.

Then in $\triangle AHC$ AC is the greatest side as opposed to the right angle. Then
$AC>HC>BC$, and the latter contradicts the hypothesis. Thus, we have some to a contradiction, that proves the falsity of our suggestion.
Then, H falls between C and B.

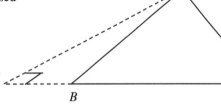

6.

> ...you must *not* forget the suspenders, Best Beloved...
> R. Kipling, *Just so stories*

How to start this problem? Usually, it is helpful to make a sketch (though we must *not forget* that a sketch itself is not a proof, Best Beloved). In the following figure, ABC is a triangle with $BC > AB > AC$. AM and AH are respectively its median and altitude drawn from the vertex A.

We have drawn the figure as accurately as we could, putting M in the middle of BC and making $\angle 3$ as right as possible. Everything looks just fine as it should be
(but do not forget that this is *not a proof*, Best Beloved!).

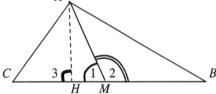

If the above figure, with H between M and C, looks natural, then some other picture, showing H located somewhere else, must look really twisted. Let us try to make such sketches (see the figures below).

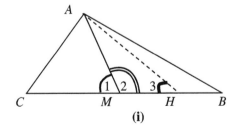
(i)

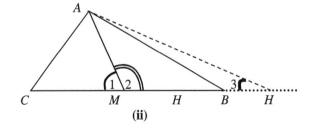
(ii)

Just look at them! Angle 3 does not look right at all! Maybe we will be able to show (we mean: to prove) that it cannot be right if H does not lie between M and C?

In cases like this, when the negation of the conclusion looks apparently wrong (though no matter how ugly it looks, it is *not a proof*, Best Beloved!), it may be helpful to assume the negation and arrive to an absurdity. In other words, the RAA method (reduction to absurdity, or a proof by contradiction) may work very well.

Let us prove the conclusion by contradiction (RAA method). Suppose *H* does not lie between *M* and *C*. Then one of the following must take place: (i) *H* lies between *M* and *B*; (ii) *H* lies on the extension of *BC* beyond *B*; (iii) *H* lies on the extension of *BC* beyond *C*.

First of all, it should be noticed that, according to Theorem 4.6.1, an angle opposed to a side that is not the greatest in a triangle is not the greatest in a triangle. Therefore such angle is necessarily acute: only one angle in a triangle can be obtuse or right (it is easy to show that existence of two obtuse or right angles would contradict the exterior angle theorem).

(i) If this is the case (*H* lies between *M* and *B*), then ∠1 is exterior for △*MBH*, and therefore ∠1 is greater than ∠3. At the same time, ∠1 is acute since it is opposite to *AC*, which is the least side in △*ABC*. ∠3 is right by the hypothesis: *AH* is an altitude. Hence, ∠1 cannot be greater than ∠3, and we have arrived to a contradiction. Thus, *H* cannot fall between *M* and *C*.

(ii) Let us consider △*AMH* in the corresponding figure. If suggestion (ii) is true, then ∠1 is exterior for △*AMH*, and therefore it is greater than ∠3. However, ∠1 is acute (see the proof above) whereas ∠3 is right (*AH* is an altitude). Thus, assumption (ii) leads to a contradiction, and we conclude that it is not true.

(iii) This situation is analogous to the previous one. Students should draw a sketch and complete the proof.

It has been shown that the assumption of *H* not lying between *M* and *C* leads to contradictions, therefore *H* is located between M and *C*, Q.E.D.

Section 4.8

1. In △ *ABC* (see the figure) *BC*>*AC*>*AB*, and *M* is a point on *BC*. Drop a perpendicular *AH* onto *BC*. If *M* is located between
H and *C*, *AM*<*AC* as an oblique with the lesser projection on *BC*.
Similarly, if *M* lies between *H* and *B* (*AM* is shown as a dotted line), then *AM*<*AB*.

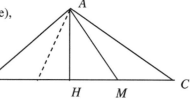

3. In △ *ABC* ∠*C*=*d*, and ∠*CAD*=∠*DAB*, i.e. *AD* is the bisector of ∠ *BAC*. On *AB* cut off a segment, *AE*=*AC*. *E* is located between *A* and *C* since *AB* is the greatest side (the hypotenuse). Join *E* with *D* by a straight segment.
Triangles *ACD* and *AED* are congruent, so ∠*AED* is right, and *CD*=*DE*. *DB* is an oblique, and *DE* is a perpendicular to *AB*; hence, *DE*<*DB*, therefore, *CD*<*DB*, Q.E.D.

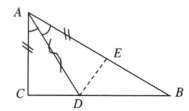

Chapter 4. REVIEW

7. For convenience, we shall use the following brief notations. In $\triangle ABC$, the sides opposed to the angles A, B, and C, will be denoted by the corresponding small letters: a, b, and c. The medians to these sides will be respectively m_a, m_b, and m_c.

Let us start with proving that the sum of the medians is less than the perimeter.

In the figure below, AM is the median from the vertex A. Let us extend it beyond M (we can do this, according to Axiom 2) and on the extension lay off a segment MN = AM (this is allowed by Axiom 3). Let us join N with C by means of a straight segment (Axiom 1).

By the triangle inequality, in $\triangle ABC$, AN < AC + CN. Triangles ABM and MNC are congruent by SAS condition: AM = MN (by construction), $\angle$ AMB = $\angle$ CMN (as verticals), and BM = MC (since AM is a median). Hence, CN = AB, and by replacing CN with AB in the above inequality, we obtain: AN < AC + AB, which can be rewritten as

$2m_a < b + c$ 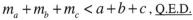. Similarly, one can obtain: $2m_b < a + c$ and $2m_c < a + b$.

By adding these inequalities and dividing the resulting one by 2, we have:

$m_a + m_b + m_c < a + b + c$, Q.E.D.

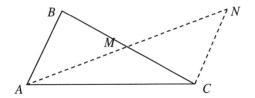

Now let us show that the sum of the medians is greater than half-perimeter.

The triangle inequality for $\triangle ABC$ can be written as AM + MB > AB, or $m_a + \frac{1}{2}a > c$. Similarly, $m_b + \frac{1}{2}b > a$, and $m_c + \frac{1}{2}c > b$. By adding the three inequalities, we shall obtain $m_a + m_b + m_c > \frac{1}{2}(a+b+c)$, Q.E.D.

10. On AB, starting from A, lay off a segment AE=AC.
Join E with D. $\triangle ADC = \triangle ADE$ by **SAS** condition.
Then $\angle ADC = \angle ADE < \angle ADB$, QED.
Also, $\angle B$ is less than $\angle ADC = \delta$ (which is exterior for $\triangle ADB$), and $\alpha > \delta$ (as an exterior angle). Thus, $\angle B < \delta < \angle DEB$. Hence, in $\triangle DEB$ DB>DE, and the latter segment equals CD.
Hence, DB>CD, Q.E.D.

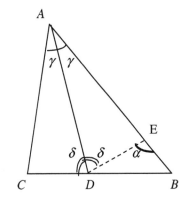

11.

> There was only one spear. I drawded
> it three times to make sure.
> R. Kipling, *Just so stories*

Let ∠A be some angle bisected by line *l*, and *P* a point not lying on the bisector.

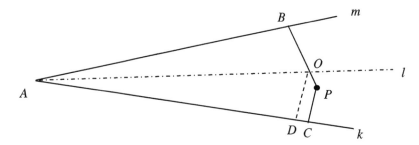

We are going to prove that the distances from *P* to the arms *m* and *k* of ∠A are not congruent. Our proof will be *direct*, i.e. it will not be a proof *by contradiction* (RAA).

Let *PB* and *PC* be the perpendiculars from P onto the arms *m* and *k*, respectively. Our task is to show that *PB* > *PC*.

PB intersects *l* at some point *O*. Let *OD* be a perpendicular dropped form *O* onto *k*. At first sight, it seems tempting to prove that *OD* (which is congruent to *OB*) is greater than *PC*. If we could manage to prove this, then the required result *PB* > *PC* would immediately follow. In the above picture, *OD* is apparently greater than *PC*.

After a few minutes of unsuccessful attempts to prove that *OD* > *PC*, we may start suspecting that this is not necessarily true. Are we deceived by our particular picture? – Yes! It happens very often when solving geometrical problems. In order to avoid this kind of mistake we should try to draw another picture, or maybe even a few more sketches that would reflect various situations.

Let us, for example, draw a picture where ∠A is obtuse. Such a situation is depicted below. In this figure, in contrast with the previous one, *PC* is apparently greater than *OD*.

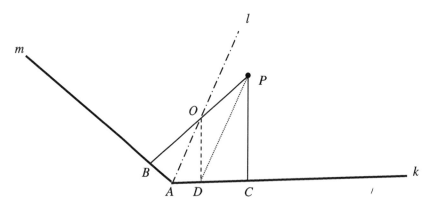

The moral of the story: pictures do not provide us with solutions, although they may give us hints. Some of these hints may be misleading; to avoid deception one should sometimes draw two or more sketches just "to make sure".

Now let us prove that *PB* > *PC*.

OB = *OD* as perpendiculars dropped from the bisector onto the arms of ∠A.
Then, *PB* = *OB* + *PO* = *OD* +*PO*, which, by the triangle inequality (for △ *OPD*), is greater than *PD*. The latter segment is an oblique drawn from *P* to *k*, and therefore it is greater than *PC*, which is a perpendicular onto *k* from *P*. Thus, we have received the following inequalities:
PB > *OD* + *PO* > *PD* > *PC*, i.e. *PB* > *PC*, Q.E.D.

Chapter 5. CONSTRUCTION PROBLEMS.

19a. <u>Analysis.</u> Let b be the base, α the given angle, and s the segment equal to the sum of the lateral sides. Suppose $\triangle ABC$ to be the required triangle.

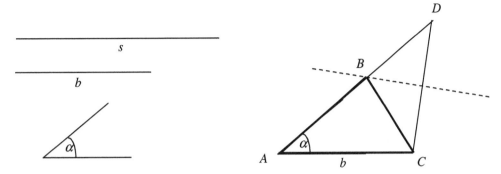

Extend AB till point D such that $AD=s$. Now $s=AB+BD=AB+BC$, whence we conclude that $BC=BD$. Then point B is located on the perpendicular bisector of CD.

<u>Construction.</u> On the sides of $\angle A = \alpha$, lay off $AC=b$ and $AD=s$. Draw a perpendicular through the midpoint of AD. It cuts AD at some point B. Connect B with C. $\triangle ABC$ is the sought for.

<u>Synthesis.</u> The proof is trivial.

<u>Research.</u> There is no solutions if $s \leq b$; moreover, if s is not substantially greater than b, the perpendicular bisector will not cut AD between A and D. Otherwise, there is a unique solution (why?).

Section 6.5
31. Extend the median beyond the point where it meets the hypotenuse by a segment congruent to the median.

Section 7.2
11. Suggest the problem to be solved; construct an auxiliary rectangle with the given segment as a side.

Section 7.3
2. Adduce counterexamples.

3. See Solution 2 in Section 7.4 of the text.

8. Let $ABCD$ be a quadrilateral and points M, N, P, and Q the midpoints of its sides. We have to prove that $MNPQ$ is a parallelogram.

Draw the diagonals AC and BD.

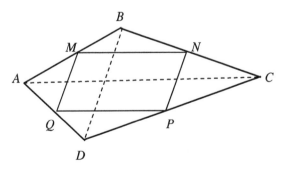

In $\triangle ABC$, MN is a midline since it joins the midpoints of two sides. Therefore, by the theorem of a midline in a triangle, $MN // AC$. Similarly, from $\triangle ADC$, we obtain that $QP//AC$. Thus, by *Corollary 2* from the Parallel Postulate, $MN//QP$.

A similar consideration of triangles ABD and CBD yields $MQ//NP$.

The opposite sides of $MNPQ$ are parallel, and then by definition it is a parallelogram, <u>Q.E.D.</u>

11. Since *MN//AC*, it follows that *KP//DL* as parts of *MN* and *AC*. Also, *K* is the midpoint of *BD*, hence *KP* is a midline in △*BDL* (*Corollary 1* from *Th.7.3.1*). Thus, $KP = \frac{1}{2}DL$. Also, *KP* = *DQ* as opposite sides of the parallelogram *KPQD* (it is given that *PQ // BD*, hence *PQ // KD*, so both pairs of the opposite sides in *KPQD* are parallel).

Therefore, $QL = DL - QD = DL - KP = DL - \frac{1}{2}DL = \frac{1}{2}DL = KP$, and thus opposite sides *QL* and *KP* in the quadrilateral *KPLQ* are congruent and parallel. Hence, *KPLQ* is a parallelogram, and its diagonals *KL* and *PQ* bisect each other. In particular, *PQ* is bisected by *KL*, Q.E.D.

The conditions *AL = LC* and *BD* ⊥ *AC* have not been used in the proof, so they are superfluous.

Section 7.4

27. Suppose the sought for trapezoid with the bases *AD=a*, *BC=b* and diagonals *AC=d*, *BD=D* to be constructed. Through *C* draw a line parallel *BD* till it cuts the extension of the other base at *E*. The sides of △*ACE* are known: *AC=d*, *CE=D*, and *AE=a+b* (why?). Then, construct △*ACE* and translate *C* parallel to *AE* by segment *b*. Thus, *B* is obtained.

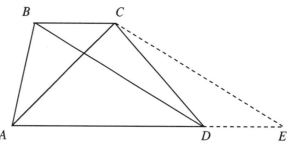

Section 7.5

3. Let the segments *a*, *b*, *c*, and *d* in the picture below be the consecutive sides of a trapezoid; *b* and *d* are the bases, and *a* and *c* are the lateral sides. Let us construct such a trapezoid.

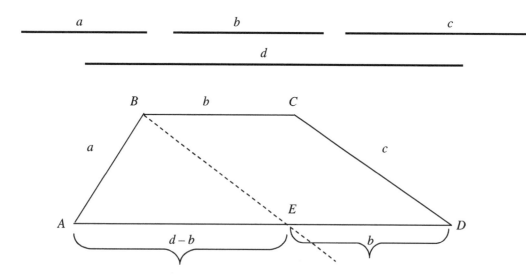

Analysis.

Suppose ABCD in the above figure is the required trapezoid. Through *B*, draw a line parallel to *CD*; it will meet *AD* at some point *E*. *ED* as part of *AD* is parallel to *BC* (*AD//BC* as the bases of a trapezoid); *BE//CD* by construction; hence *BCDE* is a parallelogram. The opposite sides of a parallelogram are congruent, therefore *ED = BC = b*, and *BE = CD = c*.

As a result of the latter, $AE = AD - ED = d - b$. Thus, we know all the sides of $\triangle ABE$: $AB = a$, $BE = c$, and $AE = d - b$.

Therefore, we can construct $\triangle ABE$ by **SSS** and then by means of a parallel translation applied to BE we shall construct the trapezoid.

Construction.

Construct $\triangle ABE$ with $AB = a$, $BE = c$, and $AE = d - b$ (**SSS** condition).

Extend AE beyond E (Axiom 2), and on the extension from point E lay off $ED = b$.

Now we shall *translate BE* along AD: Through B, draw a line parallel to AD and from B lay off a segment $BC = b$ on this line. Join C with D by a straight segment to complete the construction.

Synthesis.

By construction, BC//AD, hence $ABCD$ is a trapezoid. $AB = a$, $BC = b$, $CD = c$, and $DA = d$, so this trapezoid satisfies the given conditions, Q.E.D.

Investigation.

How many different trapezoids that satisfy the given conditions can be constructed?

$\triangle ABE$ is defined uniquely by its three sides (**SSS** condition of congruence).

There is only one straight line that passes through A and E (Axiom 1) and only one segment congruent to b can be laid off on this line (on this extension) from E (Axiom 4).

One (Th.6.1.1) and only one (Axiom 5) line parallel to AD can be drawn through B, and only one segment congruent to b can be laid off on this line from B to the right from AB (Axiom 4).

One and only one line joins C and D (Axiom 1).

Therefore, each step of the construction is defined uniquely, and so is the sought for trapezoid.

Still, could we have constructed a different trapezoid had we used a different method?

Chapter 7.Review

15. Let l be a line and P its exterior (not lying in the line) point. Connect P with a point A lying in the line. The midpoint M of AP is equidistant from A and P. Through M draw a line k parallel to l. Let us prove that k is the sought for locus.

(i) Let us prove that k possesses the required properties.

If B is any point lying in l, and N is the point where k cuts PB, then by Theorem 7.3.1, $PN = NB$, so N is equidistant from P and l, Q.E.D.

(ii) Let us prove that any point that does not lie in k does not possess the required property, i.e. is not equidistant from P and l.

Let C be some point not lying in k. Draw a straight line through P and C (Axiom 1). Suppose it meets k at E and l at D. Then, as we have proved, $PE = ED = \frac{1}{2}PD$

Since PC is part of PE, and ED is part of CD, it follows that $PC < PE = \frac{1}{2}PD = ED < CD$. Therefore, if C is not in k, $PC \neq CD$, Q.E.D.

Now we have proved that the points of k and only them are equidistant from the given point and the given line, thus k is the sought for locus.

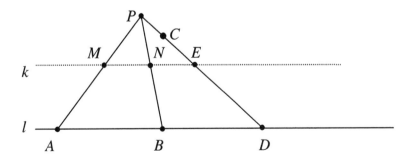

Section 8.3

3. According to Th.8.3.2, the radii of the smaller circle drawn to the points of tangency with the chords of the greater circle, are perpendicular to these chords. Thus, these chords are equidistant from the centre of the circles. Hence, by Th.8.2.2, they are congruent, Q.E.D.

6. Let a circle (its centre and radius), a line l, and an angle α be given.

Analysis. Suppose line k in the figure below is a sought for tangent, i.e. it touches the given circle and forms the given angle α with l.

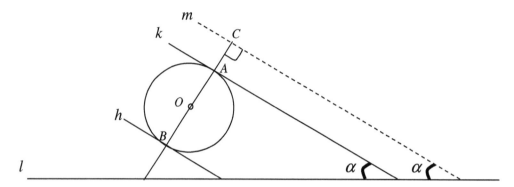

We can draw some other line that forms angle α with l, for instance, line m. Then, by the congruence of the corresponding angles, m is parallel to k.

Let OA be the radius drawn to the point A where k touches the circle. OA is perpendicular to k, hence it is perpendicular to m. Suppose the extension of the radius OA meets m at some point C. There is only one perpendicular to a given line though a given point, thus if OC is perpendicular to m and passes through O, it is the line containing the radius drawn to the point A where k touches the circle.

Construction. Draw some line m forming angle α with l. From the centre of the circle (O) drop a perpendicular upon m. This perpendicular will cut the circle at two points (A and B). Through each of these points, draw a line parallel to m. Each of these lines (h, k in the diagram) is a solution.

Synthesis. Just repeat the statements of Analysis.

Investigation. Line m can be drawn so as to cut l to the right or to the left from the centre of the circle. Therefore, there are four solutions for α acute or obtuse and two solutions if this angle is right (see the figures below).

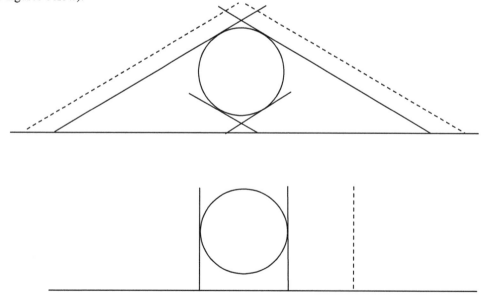

15. Let us draw the radii *OM*, *ON*, and *OP* to the three points of contact. These radii will be perpendicular to their respective tangents *AB*, *KP*, and *CD*.

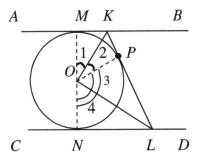

Hence, triangles *OMK* and *OPK* are right. These triangles are congruent since they legs *OM* and *OP* are congruent as radii of the same circle and the triangles have a common hypotenuse *OK*. Then, the corresponding angles $\angle 1$ and $\angle 2$ are congruent. Similarly, $\angle 3 = \angle 4$. Radii *MO* and *ON* extend each other into a diameter (why?), hence $\angle 1 + \angle 2 + \angle 3 + \angle 4 = 2d$, which means $2\angle 2 + 2\angle 3 = 2d$, wence $\angle 2 + \angle 3 = d$, Q.E.D.

Section 8.5

7. By Corollary 3 from Theorem 8.5.1, any angle inscribed in a circle and subtended by a diameter is right (figure (a) below). It is easy to show using Theorems 8.5.2 – 4 that any angle subtended by a diameter but not inscribed in the circle, is not right (figure (b) below illustrates such cases when the vertex of an angle is not located on the circle). Therefore, the circle having the hypotenuse as a diameter is the sought for locus.

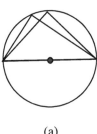 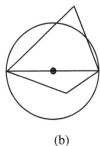

(a)   (b)

12. Since *OB*, *OC*, *PC*, and *PA* are radii, triangles *BOC* and *CPA* are isosceles with *BC* and *CA* their respective bases. $\angle OCB = \angle ACP$ as vertical, hence
$\angle O = 2d - \angle OBC - \angle OCB = 2d - 2\angle OCB = 2d - 2\angle ACP = 2d - \angle OCB - \angle CAP = \angle P$.
Since the central angles $\angle O$ and $\angle P$ are congruent, so are their subtended arcs: $\cup BmC = \cup AkC$, Q.E.D.

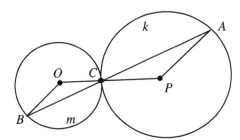

365

Section 8.6

7. <u>Analysis.</u>   Suppose the construction has been completed, and the circle with the centre at the given point O is the required one, i.e. $AB = a$, where a is the given segment (see the figure below). Then, by theorem about a diameter perpendicular to a chord, a line passing through the midpoint of the chord, will be perpendicular to the chord. Thus, if $OM \perp AB$, then M is the midpoint of AB.

<u>Construction.</u>   From the given point O lying on one side of the given angle, drop a perpendicular upon the second side of the angle. Let the foot of the perpendicular be named M. From M as a centre, describe a circle of radius equal to half of the given segment a. It will cut off on the side of the angle two segments, MA and MB, such that $AB= a$. Then, from O as a centre describe a circle of radius OA (or OB, which is the same since the perpendicular bisector of a segment is the locus of the points equidistant from the endpoints of the segment, so O is equidistant from A and B). Let us prove that this is the required circle.

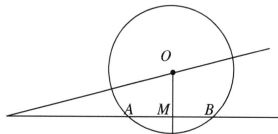

<u>Synthesis.</u>   By construction, the circle is centered at O, and $AB = AM+MB = a$.
<u>Investigation.</u>   Point M is defined uniquely since there is exactly one perpendicular onto a line from a point, and A and B are defined uniquely by AXIOM 4, thus there is exactly one solution.

11. <u>Analysis.</u>   Since the circle touches the given line , its centre must lie on the perpendicular to the line at the point of contact. Also, since it is passes through the given point of contact and another given point, the centre of the circle must be equidistant from the two points. Thus, the centre of the circle must lie on the intersection of the perpendicular to the given line at the given point with the perpendicular bisectorof the segment connecting the given point on a line with the given point.
The rest of the solution is trivial.

25. <u>Analysis.</u>   Suppose such a circle is constructed: it touches the given circle at A and the given line l.

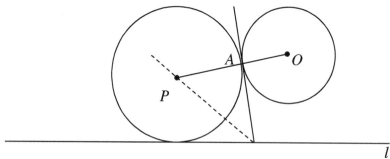

Suppose the common tangent of the circles meets l at some point C. Then, if the constructed circle touches l at B, $CB = CA$, i.e. the centre of the constructed circle lies on the bisector of $\angle ACB$. It also lies on the extension of OA (why?).

<u>Construction.</u>   Extend the segment OA joining the centre of the given circle with the point of contact into a line. Draw the tangent that touches the given circle at A. Let it meet l at some point C. Draw the bisectors of both angles formed by CA and l. Each of them will meet the extension of OA at some point. Each of these points (P or Q) will be the centre of a circle tangent to l and tangent to the given circle at A.

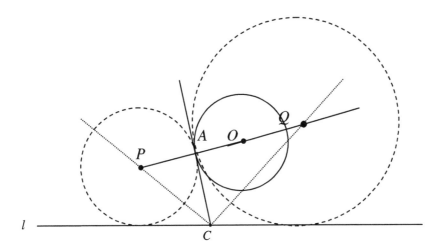

Synthesis. By Th. 8.4.3, a point of contact of two circles lies in their centre line, hence the centre of a required circle lies on the extension of *OA*. On the other hand, if two intersecting lines touch a circle, the centre of the circle must be equidistant from the lines, hence it lies on the bisector of an angle formed by the lines. *AC* forms two angles with *l*, thus the points of intersection of the bisectors of these angles with the extension of *OA* will be the centres of the required circles.

Investigation. By the preceding, the problem has two solutions: a circle touching the given one externally and internally.

32. Suppose, point *A* is the given point, and *AB* is the sought for secant. Draw the radius *BO* and extend it into a diameter *BC*. ∠*BDC* is right, and *BD=DA*. Then, △ *ABC* is isosceles, whence *AC=BC*, i.e. *AC* equals the diameter.

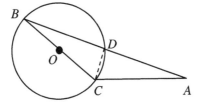

Section 8.7

4. The hypotenuse of a right triangle is twice the diameter of the circumscribed circle; it is easy to see (look at the figure) that the hypotenuse also equals
$(a-r)+(b-r)$, whence one obtain that
$a+b=2(R-r)$.
Now we can construct a triangle given a side (hypotenuse) opposite to a given (right) angle and the sum of the other two sides $2(R-r)$ (see, e.g., section 8.6).
A generalization of this problem: Construct a triangle having given one of its angles, the inradius, and the circumradius.

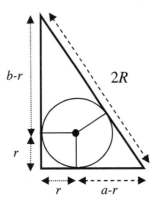

Section 8.8

12. Hint: Let AC be the given segment and O the centroid of the sought for triangle. Let M be the midpoint of AC. Draw a line through M and O, and on the extension of this line beyond O, lay off a segment $OB = 2OM$. Then ABC is the sought for triangle.

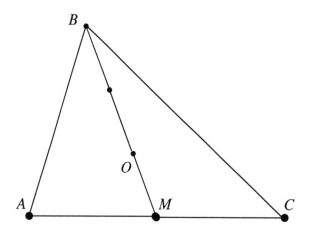

Section 9.1

4. According to the Euclidean Algorithm, GCM (A,B) = GCM (B,C). It is given that $B = 1.5\ C$, hence $B = 3\left(\frac{1}{2}C\right)$. On the other hand, $C = 2\left(\frac{1}{2}C\right)$. Thus, $\frac{1}{2}C$ is a common measure of B and C, and since the numbers 3 and 2 are relatively prime, it is their greatest common measure.

14. Since $nB=kR$, $B = k\left(\frac{1}{n}R\right)$, so $\frac{1}{n}R$ is a common measure of B and R. Then $A = mB + R = mk\left(\frac{1}{n}R\right) + R = (mk+n)\left(\frac{1}{n}R\right)$, i.e. $\frac{1}{n}R$ is a common measure of A and R. Therefore, $\frac{1}{n}R$ is a common measure of all three segments. By the way, it will be their greatest common measure if the numbers n and k are relatively prime.

15. The hypotenuse c and a leg a in an isosceles right triangle are incommensurable. Their incommensurability implies the incommensurability of any two segments that are their respective multiples, e.g. $2c$ and $5a$.

Another example: a lateral side and the base in an isosceles triangle with the angle opposed the base equal 36^0 are incommensurable (see the next problem).

16. In $\triangle ABC$, $AC=b$; $AB=BC=a$. Draw bisector AD to a lateral side. Denote $DC=r_1$. Then $a = b + r_1$. Bisect $\angle ADB = 72°$ by DF.
$\triangle ADF$ is isosceles as well as FDC; therefore $AF=DF=DC= r_1$.
Denote $FC= r_2$. Then $b= r_1 + r_2$. Etc., etc.:
$a = b + r_1$
$b = r_1 + r_2$
$r_1 = r_2 + r_3$;....
The procedure never ends, hence the common measure of the segments does not exist: when it exists it can be found in a finite number of steps.

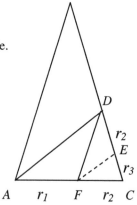

Section 9.2

8. In $\triangle ABC$, $\angle B = 120°$, and BD is the bisector.
On BA, starting from B, lay off a segment $BE=BD$.
Point E falls between A and B (why?).
Triangles AED & ABC are similar (why?).
Then, one can write a proportion and find

$$BD = \frac{ab}{a+b} = 2cm.$$

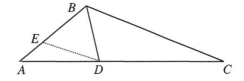

10. Extend two (lateral) sides of the given triangle and find a point whose distances from these sides (or their extensions) are equal U and 2U, where U is any unit. This can be achieved, e.g. by finding the point of intersection O of two lines, each of them // to one of the extended sides, and one is at the distance U, and another stands at the distance 2U from the line to which they are parallel. Through that point, O, draw a line perpendicular to the base of the triangle, and on that line lay off a segment equal 3U in the direction leading from the vertex of the original triangle. Through the endpoint of that segment draw a line // to the base. A new triangle is obtained.

In that triangle the distances from O to the sides are in a ratio U:2U:3U = 1:2:3. Connect O with the base vertices (the endpoints of the base) of that triangle, and in the original triangle from the base vertices draw the lines parallel to these connecting segments. They will intersect in the sought for point.

Sections 9.3-4.

12. Since $MN // AB$ as two perpendiculars to the same line, $\angle A = \angle M$ and $\angle B = \angle N$. Therefore, $\triangle ABC \sim \triangle MNC$, and $\frac{MN}{MC} = \frac{AB}{AC}$, whence $MN = \frac{AB}{AC} MC$. The segments AB, AC, and MC can be measured and thus MN can be determined.

It is not substantial for the solution that angles A and M are right, it is only important that they are congruent.

16. Let ABC be an isosceles triangle with $AB = BC = a$, and $AC = b$, and the circle inscribed in this triangle is centered at O and touches its sides at the points M, N, and P.

We have to express MN through a and b.

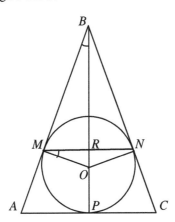

$\triangle ABC$ is isosceles, therefore it is symmetric in BP, which is the bisector of the vertex angle, and hence the median and altitude drawn to the base. Therefore, $MR = RN$, and $AP = PC = \frac{1}{2}b$. Since AP and AM are the segments of the tangents drawn to a circle from one point, $AM = AP = \frac{1}{2}b$, and thus $MB = a - \frac{1}{2}b$.

Let r denote a segment congruent to the radius of the inscribed circle; for instance, $OM = ON = r$.

∠OMR = ∠ABP as two angles with mutually perpendicular sides, hence the right triangles ABP, OBM, and OMR are similar. Hence,

$$\triangle OMR \sim \triangle OBM; \Rightarrow \frac{MR}{r} = \frac{MB}{OB}; \Rightarrow MR = r\frac{a-\frac{b}{2}}{OB};$$

$$\triangle ABP \sim \triangle OBM; \Rightarrow \frac{OB}{r} = \frac{AB}{AP}; \Rightarrow OB = r\frac{a}{2b}.$$

After substituting the latter expression for OB in the formula for MR, we find: $MR = \frac{b}{2a}\left(a-\frac{b}{2}\right)$, whence $MN = \frac{b}{a}\left(a-\frac{b}{2}\right)$.

19. Let ABC and MNP be similar triangles with ∠A = ∠M, and ∠B = ∠N. Let BD and NR be the altitudes dropped from B and N respectively.

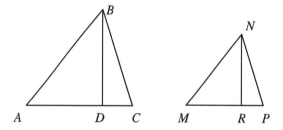

Since ∠A = ∠M, △ABD ~ △MNR as right triangles with congruent acute angles, hence $\frac{BD}{NR} = \frac{AB}{MN}$, Q.E.D.

24. Let us construct ∠A congruent to the given angle, and on the sides of the constructed angle lay off from the vertex segments AB =2a and AC =3a, where a is an arbitrary segment. Then, from A, lay off on AB (or its extension) a segment AM congruent to the given side. Through M draw a line parallel to BC; it will cut AC or its extension at some point N. △AMN is the sought for triangle (the proof is trivial).

28. Let us construct ∠A congruent to the given acute angle, and on the sides of the constructed angle lay off from the vertex segments AB =3a and AC =4a, where a is an arbitrary segment.

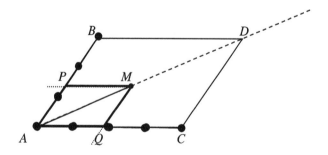

Through B and C draw two lines parallel to AC and AB respectively. They will intersect (why?) at some point D. Draw a ray that emanates from A and passes through D. On this ray, from A, lay off AM congruent to the given diagonal. Through M, draw two lines parallel to the sides of ∠A. Let the points where these lines meet the sides of ∠A, be called P and Q. APMQ is the sought for parallelogram (the proof is trivial).

Section 9.5.

9. Let *ABCD* be the given trapezoid, and let *M* be the midpoint of the greatest base (*AD*). Through *M*, draw a line parallel to *CD*; let us call this line *l*.

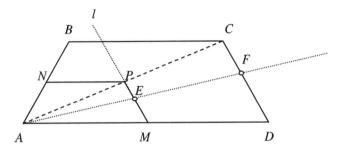

Draw *AC*. *l* will intersect *AC* at some point *P* (why are we so sure that *l* is not parallel to *AC*?). Through *P* draw a line parallel to the bases of the given trapezoid; it will meet *AB* at some point *N*.

It is easy to show that *ANPM* satisfies the requirements of the problem: since *MP//CD*, *AP*: *AC* = *AM:AD* = 1:2; also *NP//BC*, whence *AN:AB* = *AP:AC* = 1:2, thus all the sides of the two trapezoids are in the given proportion. The corresponding angles of the trapezoids are congruent since ∠*A* is common, angles *ANP* and *B* as well as *AMP* and *D* are corresponding formed by parallel lines and the angles *MPN* and *DCB* have mutually parallel sides.

(By the way, it would be even easier to show that *ANPM* is obtained by means of a homothety with the coefficient ½ applied to *ABCD*, and therefore it is similar to *ABCD* and each of its sides is half of the corresponding side of the given trapezoid. Really, any point *E* of *ANPM* is the image of some point *F* belonging to *ABCD* under a homothety with the coefficient ½ and the centre at *A*. The latter follows, for instance, from the similarity of △*AEM* and △*AFD*.)

Any other trapezoid that satisfies the given conditions will be congruent to *ANPM*; this can be proved by means of a superposition or by considering the congruent triangles constituting the trapezoids. Thus, the solution is unique.

Section 9.6

11. The circles are concentric.

Section 9.7

3. <u>Analysis.</u> Let point *P* on one side (*l*) of the given ∠*A* be equidistant from its other side (*m*) and the given point *B* in its interior. Let *C* be the foot of the perpendicular dropped from *P* upon *m*. Then, if *P* is a sought for point, *PC* = *PB*, i.e. a circle of radius *PB* with the centre at *P* on *l* is tangent to the other side (*m*) of the angle at *C*.

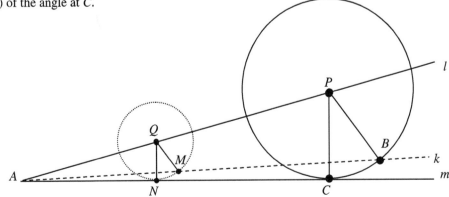

Any other circle that touches m with its centre on l will be perspectively situated with respect to this circle, with the centre of similitude at A. Let Q be the centre of such a circle, which touches m at some point N. There is a homothety that transforms the first circle into the second. As a result of such a transformation, P will be mapped into Q whereas the image of C will be N. The image of B can be found as a point of intersection of the ray k emanating from A and passing through B with the second circle. It is point M in the above picture.

The idea of construction: we shall start with an arbitrary circle like the one centered at Q at the above picture and find P as a centre of the circle obtained from the first circle as a result of a homothety with A as a centre of similitude.

<u>Construction.</u> (the beginning) Let A be the vertex and l and m the sides of the given angle. We are looking for a point lying on l and equidistant from some given point B in the interior of the angle and its other side m.

Section 9.9

3. Hint: Suggest that two given numbers with a fixed sum constitute a segment with the length equal to that sum. Consider the latter as a hypotenuse and circumscribe a circle near the triangle.

25. If the altitude dropped onto the base is h, the base is b, and the lateral side is a, then it follows from the property of a bisector that $\dfrac{h-r}{r} = \dfrac{2a}{b}$, and from the Pythagorean Theorem:

$$(h-r)^2 - r^2 = \left(a - \dfrac{b}{2}\right)^2,$$ whence one can find the following equation for a:

$$\left(1 - \dfrac{4r^2}{b^2}\right)a^2 - \dfrac{4r^2}{b}a - \left(\dfrac{b^2}{4} + r^2\right) = 0.$$

27. Hint: In the figure, the gray triangle and the shaded triangle, whose hypotenuse is a diameter, are similar (why?), whence one can obtain a proportion that relates the diameter, the diagonal, the lateral side, and the half)difference between the bases.

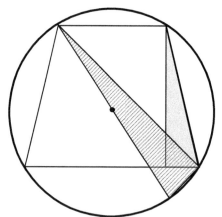

Section 9.11

6. Consider an isosceles triangle with the vertical angle of 30°=2∗15°. Draw the altitudes to the lateral sides and use the trig functions of 30° and/or the Pythagorean theorem.

14. Consider a triangle with its vertical angle α+β, where the angle is divided into α and β by the altitude drawn to the base. Let a lateral side of the triangle be equal 1 (see the figure).

$BD = BM - DM$, where
$BM = BE \cdot \cos\beta = 1 \cdot \cos\alpha \cdot \cos\beta$.
The rest is clear from the figure.

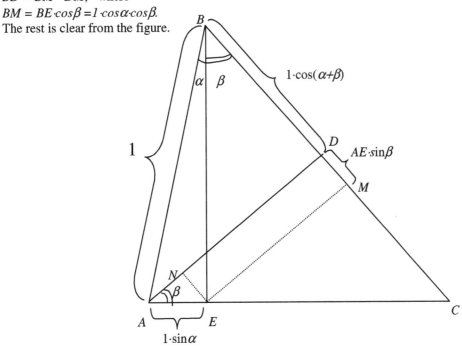

31. Hint: Use the addition formulae for trig functions from problems #13, #14 to obtain the sines of the other angles of the triangle.

Section 9.12
9. Draw a bisector of one of the base angles; it cuts the triangle into two isosceles triangles, one of which is similar to the whole triangle. From the proportion express the base in terms of the lateral side.
11. Start with the construction of the mean proportional to b and c.

9. Review
39. Hint: If ABC is a sought for triangle, and BD - the bisector, extend AB beyond B by a segment equal to BC (the other given side). Then, ...
$\dfrac{MC}{BD} = \dfrac{AM}{AB}$, whence MC can be found.

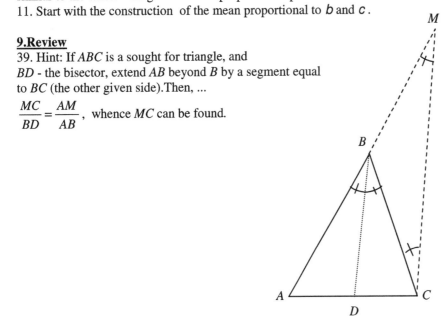

41. Hint: apply algebra.

373

Section 10.1

26. Hints: First, prove that triangles *ABC* and *AFB* are similar; then show that the marked angles are congruent, whence it follows that *FC*= to a side of the pentagon.

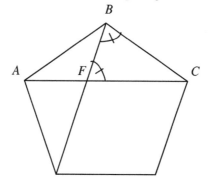

Section 10.2

3. Let us consider a sequence of regular circumscribed polygons obtained by doubling the number of sides starting with any regular polygon, not necessarily a square or a triangle.

In Figure 10.2(3), let *AB* be a side of a circumscribed regular *n*-gon; it touches the circle at *T*, which is the midpoint of *AB*; thus $AT = TB = \frac{1}{2}b_n$.

Let *Q*, *P*, ... be the points where the lines drawn from the centre of the circle to the vertices of the *n*-gon intersect the circle. Let us construct a circumscribed regular 2*n*-gon by drawing the tangents at all these points and extending them till they reach the sides of the *n*-gon. (prove that the obtained polygon is a regular 2*n*-gon or see Th. 10.1.1 for substantiation).

Thus we obtain a regular polygon *KLMN*... with 2*n* sides; $KL = LM = MN = ... = b_{2n}$. Each side of an inscribed regular polygon is bisected by the point of contact with the circle, hence $NP = PM = MT = TL = ... = \frac{1}{2}b_{2n}$.

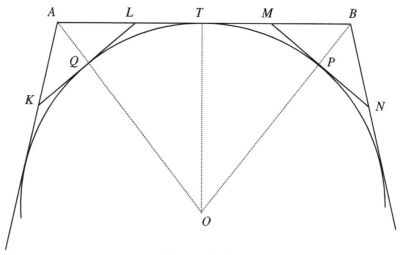

Figure 10.2(3)

Since MB is the hypotenuse in a right triangle MPB, $MB > PM = \frac{1}{2}b_{2n}$. Therefore $MB > TM$, as $TM = PM$ (each of them congruent to $\frac{1}{2}b_{2n}$). Taking into account $TM + MB = TB$, we conclude that $TM < \frac{1}{2}TB$, which means $\frac{1}{2}b_{2n} < \frac{1}{2}(\frac{1}{2}b_n)$, and therefore $b_{2n} < \frac{1}{2}b_n$.

It is easy to prove for polygons generated by doubling procedure from a triangle or a square (do it!) that for any $n > 4$, $b_n < 2R$. Thus, after applying the doubling procedure a few, e.g. $(k+1)$, times, we can obtain a regular circumscribed polygon with a side less than $\frac{1}{2}(\frac{1}{2}(\frac{1}{2}...(\frac{1}{2}2R)))= \frac{1}{2^k}R$.

Since, by choosing a sufficiently great k, 2^k can be made greater than any natural number, it follows from Archimidean Principle that $\left(\frac{1}{2^k}\right)R$ and therefore a side of a circumscribed regular polygon can be made less than any given segment, Q.E.D.

To make $b_n < \frac{1}{128}R$, start with a circumscribed square, for which $b_4 = 2R$ and apply the doubling procedure 8 or more times.

6. It does follow from the similarity, but first we have to define the length of the circumference of a circle, i.e. we have to identify the circumference with a straight segment.

21. Hint: See the diagram, where each marked angle is congruent to α (why?), and $\sin\alpha = \dfrac{R-r}{L}$.

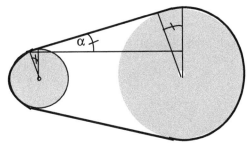

Section 11.1
5. See the figure, where ABC is the original triangle.

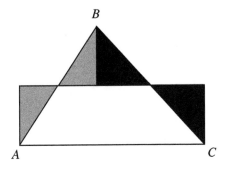

13. Suppose a rectangle and a square have the same perimeter. Let us impose the rectangle upon the square (see the figure below). Now we just have to compare the areas of the gray and white rectangles, the remainders of the areas of the square and rectangle when the common part is subtracted. The sides marked with braces are equal (why?). As for the remaining sides, the greater belongs to the gray rectangle, since the side of the square is, evidently, greater than the least and less than the greatest side of the rectangle. Then, the area of the gray rectangle is greater, and, therefore, the area of the square is greater than the area of the rectangle.

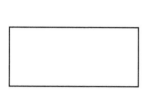

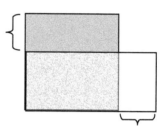

17. Hint: $a^2 + b^2 \geq 2ab$.

46. It follows from the equality of the areas that
$$hl = (H-h)(b+l),$$
where l is the base of the gray triangle, i.e. it is the segment of the sought for line that lies between the lateral sides.
From the similarity of the whole and gray triangles,
$$\frac{H}{h} = \frac{b}{l} \equiv k.$$
Then, one can obtain from the first formula: $(k-1)(k+1)=1$, whence $k = \sqrt{2}$.
Each lateral side is divided in the same ratio as the altitude (why?).

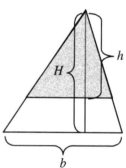

56. Hint: Let AM be a diameter of the circumcircle, and BD – the altitude dropped onto AC. Then triangles BDC and ABM are similar (by two angles), and hence
$$\frac{AB}{AM} = \frac{BD}{BC}, \text{ i.e.}$$
$$\frac{c}{2R} = \frac{h}{a}.$$
The latter equality allows to express the altitude in terms of the lateral sides and the circumradius.

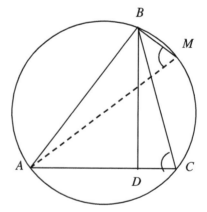

Section 11.3

7. Hint: It follows from the definition that the area of a circle is less than the area of a circumscribed polygon and greater than the area of an inscribed polygon. Therefore, the area of the sector OBC is less than the area of $\triangle OAC$ and greater than the area of $\triangle OBC$.

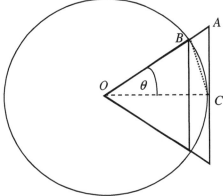

Chapter 11. Review

21. Hint: divide AC into three equal parts by points D and E. Draw a line parallel to AB through D and a line parallel to BC through E. Their point of intersection is the required point.

24. Hint: Divide a midline of the parallelogram in the ratio $m \div n$ by a point and connect this point with the given one.

26. Hint: Apply algebra.

35. Hint: Apply algebra.

36. Hint: Apply algebra.

APPENDIX:

WHY THIS TEXTBOOK IS WRITTEN THE WAY IT IS

> ...because when you are a Bear of Very Little Brain, and Think of Things, you find sometimes that a Thing which seemed very Thingish inside you is quite different when it gets out into the open and has other people looking at it.
> A.A. Milne, *The House At Pooh Corner*.

Writing a book is a personal business, even if you are retelling the stories told by many other people many times: you are retelling them in your way. So it seems natural to write an introduction that explains my motives for writing and ideas to which I have adhered in the process of doing this. And since nothing is more irritating than hearing or reading a poor interpretation of your favorite story, I cannot refrain from explaining how others misinterpret the story and why you should listen to my telling of it. As a result, this introduction consists of the following sections:

Section 1, **WHY WRITE ANOTHER TEXT?**, explains the necessity to write a decent introductory text on Euclidean Geometry and discusses, based on two popular textbooks, typical flaws that make many (if not all) modern texts unsuitable and even harmful for studying the subject.

It contains three subsections:
- **1a.** How they teach geometry in schools nowadays and a brief discussion of a typical modern textbook.
- **1b.** A critical review of another modern text, probably the most popular one.
- **1c.** Good old (really old!) textbooks.

Section 2, **HOW IS THIS TEXTBOOK WRITTEN?**, contains a detailed discussion of the principles to which I adhered when writing the text, and brief summaries of each chapter.

Section 3, **LITERATURE**, lists the books and papers mentioned in this review. These resources are worth reading, except, of course, for the two modern textbooks criticized in section 1.

It has been impossible to avoid some mathematical details in the sections 1 and 2. One can skip them at their pleasure and read only the understandable parts of the essay.

1. WHY WRITE ANOTHER TEXT?

> …But the more Tigger put his nose into this and his paw into that, the more things he found which Tiggers didn't like.
>
> A.A. Milne, *The House At Pooh Corner.*

This section contains some discourse concerning the current situation of geometry instruction in the USA and Canada and the suitability, or rather the unsuitability of modern textbooks. Two very popular textbooks will be discussed. There is also a very brief discussion of "good old" textbooks, which in general are reasonably appropriate resources for teaching the subject.

The section turned out to be fairly long, therefore the reader has an option of just reading the epigraphs (they will give you the gist of the contents) and moving to section 2, which discusses the current textbook.

1a. How they teach geometry in schools nowadays and a brief discussion of a typical modern textbook.

> On Tuesday, when it hails and snows,
> The feeling on me grows and grows
> That hardly anybody knows
> If those are these and these are those.
>
> A.A. Milne, *Winnie-the-Pooh.*

Some basic results of geometry are mentioned in the majority of modern North American high school textbooks. These results are usually presented as "facts of life", without proofs or without clear presentations of geometrical and logical underlying principles. It is not only that such texts do not improve students' understanding of mathematics and their ability to think and discourse on various subjects, which have always been important "by-products" of studying geometry. They also do not promote the knowledge of actual material: the easily acquired (without derivation or substantiation) "facts of life" are mostly forgotten immediately after or even before the diploma exams.

Our typical junior undergraduates would only vaguely remember something about the proportionality in similar triangles and the formula for the circumference, which they usually confuse with the one for the area of a circle. Moreover, the majority believes that π is a rational number equal to $22/7$, not to mention that they do not even understand why they should care whether it is rational or "…what did you call it?". It has become a common situation that in a class of 70 undergraduate students hardly anyone knows what a median of a triangle is: "– A median? – Isn't it something from statistics?"

The only way to remedy this situation is to teach Euclidean geometry as a separate subject in secondary schools as it had been done until recently for almost 2000 years with comparatively short pauses due to outbreaks of plague, obscurantism, educational reforms, and other natural or not so natural disasters. In the recent times, the periods of obscurantism have been so closely related to educational reforms that it would be hard to tell which of them have followed the others. Both would usually run under slogans that deify the use of technology and the immediate practical usefulness of any knowledge. Of course, classical subjects, including Euclidean geometry, were the first to fall victim to these fads.

As a result, it is extremely hard, if not impossible, to find a decent textbook on the subject issued in English in the last approximately 70 years. There are a few good old texts issued almost a century ago, and those will be discussed later on. Those few modern textbooks that claim to be rigorous and axiom-based are usually written in an unsystematic and even illogical manner.

The main common feature of all such books is the misinterpretation of the axiomatic method, the very method that students are supposed to master by reading these books. The principal idea of this method seems simple and has not changed much since the time of Aristotle. Any theoretical science is based on a few notions that cannot be defined in terms of more elementary notions; these are called the *undefined notions* or *primitive terms*. Their properties are described by *axioms* or *postulates,* statements that cannot be proved. All the other results of this science are derived from axioms. These results are usually called *proved statements*, *propositions,* or *theorems.*

Again, the idea seems to be simple: axioms are assumed, and they cannot be proved, whereas theorems should be proved. But it is not the case with many modern textbooks. Nothing pleases their authors more than assuming some theorems are axioms.

Let's consider for example the textbook *Essentials of Geometry for College Students* [2i], published by Pearson Education. They assume that all three tests for the congruence of triangles (angle-side-angle, side-angle-side, and side-side-side) are postulates (Postulates 2.1-2.3), whereas one of these would be enough. Moreover, even one of them would be more than enough, since earlier the authors postulate the existence of coordinates on a line and tacitly assume the existence of angle measures, and hence they could prove any statements of plane geometry by means of coordinates. Since the coordinates are postulated as are the properties of real numbers (Postulates 1.6-1.12), there is no need to introduce the segment addition postulate (Postulate 1.13) and the midpoint postulate (Postulate 1.15), – it is enough to know how to add and divide real numbers. The constructions, which are a very important component of a geometry course, are neither postulated nor proved, which should leave students perplexed. There are many other superfluous postulates and other ridiculous "marvels" in this book. The list would be too long to mention in its entirety, but I cannot help mentioning the number one prize winner of unnecessary postulates, Postulate 3.2: *A polygon has the same number of angles as sides*. It looks like the authors postulate that we use natural numbers the same way for counting angles as we do for sides. Recalling my grade one experiences, I just wonder, do we do the same things for pencils, for desks, and other countable objects?

It's interesting that this prize-winning postulate is also a feature of another textbook, *Elementary Geometry* by Gustafson and Frisk [2ii]. It goes without saying that it has about the same set of superfluous postulates as the aforementioned book. But it also has a bonus feature: superfluous undefined terms (see pg. 5 of *Elementary Geometry*).

Another typical example is the textbook *"Geometry: tools for a changing world"* by Laurie E. Bass, Basia Rinesmith Hall, Art Johnson, and Dorothy F. Wood, with contributing author Simone W. Bess, published by Prentice-Hall, 1998 [2iii]. A detailed review of this text written by Professor D. Joyce (Clarke University) can be found on line at http://aleph0.clarku.edu/~djoyce/java/elements/geotfacw.html.

The textbooks discussed above all share similar features, and there are many clones of them, which leaves us wondering why they are even written.

1b. A critical review of another modern text, probably the most popular one.

> …then this isn't an Expo – whatever it is – at all, it's simply a Confused Noise. That's what *I* say."
> A.A. Milne, *Winnie-the-Pooh*.

Another very popular text, *"Geometry" by Harold Jacobs*, published by W.H. Freeman and Company, 1987 (second edition) appears to be the lesser of two evils when compared to any of the above textbooks.

The discussion claims to be axiom-based; at least the author states which of the assertions are postulates and which of them he suggests to be theorems. However the book possesses a really amusing feature: *some postulates are derived from the others!* Anyone who has ever heard of the axiomatic approach would argue that a ***derived postulate*** is an oxymoron: it is common knowledge that postulates (axioms) are statements assumed to be true. – Not in this text.

In the lesson "The Volume of a Prism", Cavalieri's Principle is introduced as a postulate (**Postulate 14**). After a short discourse based on this postulate, the author concludes (quotation follows): "Hence, the volume of every prism can be found by the same formula. We will now state this conclusion as a general postulate. **Postulate 15**: The volume of any prism is Bh, in which B is the area of one of its bases and h is the length of its altitude." The author truly believes and teaches his students that postulates follow from one another!

The aforementioned perversion of the axiomatic method, ridiculous as it sounds, is explicitly proclaimed in Chapter 15; yet a reader can understand that something is wrong in this respect starting with the very first chapters.

It is not clear why some of the assumed axioms are not being used to prove theorems. The so-called *ruler postulate*, which states the existence of lengths and of the coordinates of the points in a line, is assumed to be one of the axioms. The introduction of this postulate in a textbook on Euclidean geometry is already a blunder, and we shall discuss it later. So far one cannot help asking: why has this axiom not been used for proving, for example, the existence of the midpoint of a segment? Such a proof would take no more than one line: the point whose coordinate is the arithmetic mean of the coordinates of the endpoints of the segment is the midpoint: just perform the subtraction and you will see. Instead, the existence and uniqueness of a midpoint of a segment is postulated. Then, why would we not postulate all the results? Why should we prove anything at all?!

Even though the ruler postulate is assumed as an axiom, it is not used in all of the proofs in which it could be useful; also, it is never used in constructions. –Why? Do we not introduce axioms in order to obtain results? Or is there a special selection rule discriminating some axioms in favour of others? – Then this rule itself must be postulated as well!

Even more mysteries are generated by the introduction of another axiom that has never been used in traditional textbooks on Euclidean geometry: the *protractor postulate* that states the existence and properties of the angle measure. The postulate is analogous to the ruler postulate for the coordinates on a line. Then, since the existence of the midpoint of a segment has been postulated it seems fair to postulate the existence of a bisector of an angle. Or maybe to state it as a theorem, which would be a good idea since it can be proved?! – Not a word about the existence of bisectors, although their existence is assumed somehow. Instead, the definition of a bisector: "*A ray bisects an angle iff it is between the sides of the angle and divides it into two equal angles*" is followed by a so-called Theorem 4: "*A ray that bisects an angle divides it into angles half as large as the angle*". It looks like the author is telling us that if we divide by two we are getting halves, not one thirds or anything else.

Now, a few words about the *ruler* and *protractor postulates*, which nowadays are present in almost every geometry text. They were originally introduced by G.D. Birkhoff in his paper "*A Set of Postulates for Plane Geometry (Based on Scale and Protractor)*", Annals of Mathematics, 33, 1932, as part of the set of axioms for plane Euclidean geometry. The set includes two more axioms: the *incidence postulate* (the existence of exactly one line through a pair of points) and a *similarity postulate* (side-angle-side criterion for similarity of triangles). The four postulates describe the undefined notions of a *point, line, distance between two points*, and *angle formed by three ordered points*. Birkhoff's system of postulates does describe the geometry of the Euclidean plane and it can be used for solving geometric problems. Yet such a choice is not suitable for a textbook on Euclidean geometry, especially for one that introduces readers to the subject. An introductory text ought to have some Euclidean spirit, which is implemented in the so-called *synthetic approach* (geometry without numbers). The synthetic approach allows the reader to view the subject with maximum generality, without dependence on a number system, and emphasizes the geometric contents of the discipline, in contrast with an approach based on arithmetic or calculus. Another important feature of a synthetic approach is the opportunity or rather the necessity of introducing the notion of *continuity*, one of the most important notions in mathematics. It was the idea of continuity that made it possible to construct real numbers and to define the notion of the length of a segment. If the existence of lengths is being postulated, the students are missing the whole story including the notion of a real number. Unfortunately, very few resources pay attention to this important issue. For interested readers I would recommend the paper "*Teaching*

Geometry According to Euclid" by Robin Hartshorne in the Notices of the AMS (American Mathematical Society), Volume 47, # 4, April 2000.

Let us come back to the H. Jacob's book. It does contain the ruler and protractor postulates, but the undefined notions and other postulates are neither the same nor equivalent to the ones used by Birkhoff. This is not even the sloppy set of SMSG (School Mathematics Study Group) postulates, fairly criticized in Morris Kline's bestseller "Why Johnny Cannot Add: The Failure of the New Math"(*Vintage Books*, New York, 1974).

Rather, it is an "Irish soup" of various approaches (Birkhoff, Euclid, Hilbert, - all in one pile, – the more the merrier) complemented with a few superfluous postulates proposed apparently by the author himself in spite of already having too many of them. As in the books [2] discussed above, both SAS (side-angle-side) and ASA (angle-side-angle) tests for the congruence of triangles are assumed to be postulates. In the Euclidean approach both tests are theorems; in the Hilbertian – one of them (SAS) is a postulate and the other is proved; in Birkhoff's system, assuming the ruler and protractor postulates, none of the tests is required as a postulate.

In rigorous theories the set of axioms is required to be *independent*: none of the axioms can follow from others. Then we do know what the underlying statements of the theory are. If the set is overfilled, as in the text discussed above, it is impossible to understand what the theory is "made of".

One could argue that a text for secondary schools cannot be completely rigorous, and I would agree with that. Yet, a textbook should not be based on a mixture of incompatible approaches as it happens with the H. Jacobs' book. The book reminds me of a children's story ("*Knights*" by V. Dragounsky). In this story, a boy, being in need of money in order to buy a present for Mother's Day, emptied two bottles of his father's drinks, a sherry and a beer, into a jar, arguing: "Wine and wine mixed together will still be wine."

The only possible reason for making such a cocktail of axioms and redundant propositions is the intention to create the appearance of high-level rigor. This pseudo-rigor does not help anything, - it only generates incomprehensible definitions such as, for instance, the definition of a *ray* (half-line) by means of *betweenness*, which does not contain explicitly the notion of the *vertex* or *origin* (later called the *endpoint*) of the ray. In the definition and the subsequent discussion of the notion of an *angle* not a word is said about its interior, not to mention the fact that angles are introduced after polygons although the latter are actually made of angles and owe them their name (in Greek: poly = many; gonio = angle) . Some axioms of betweenness are introduced (this could be easily avoided in an introductory course, - they have never been substantially used anyway) whereas the others (e.g., Pasch's axiom) are ignored. The same can be said about some incidence axioms that were implicitly meant by Euclid in his *Elements* and thus could be omitted without trouble (see the aforementioned Kline's book about the "new math" and level of rigor). Postulate 2 tells us: *If there is a line, there are at least two points on the line*. The first part of the phrase instigates the reader to ask: What if there is no line(s)? This formulation is a travesty of a postulate from the system proposed by D. Hilbert: *Every line contains at least two points*. The next postulate of incidence states that *there exist three points not lying in one line*. Why has this one been ignored by the author?

Moreover, the next axiom, called Postulate 3: *If there are three noncollinear points, then there is exactly one plane that contains them*, does seem to require the assurance that three noncollinear points do exist! So, why is the axiom of their existence not included?

The other numerous flaws of the book are minor comparing to the aforementioned ones. Let us list just a few of them.

The notion of the *locus*, which is crucial for Euclidean geometry, especially for constructions, is not present in the book.

Even though the book features a chapter on Non-Euclidean geometries, the author makes no effort to specify which results would still be true if the parallel postulate had not been assumed. Lesson 7 in the Chapter "Parallel Lines" is entitled "Two More Ways to Prove Triangles Congruent" and presents the angle-angle-side (AAS) test for general triangles and the hypotenuse-leg (HL) test for right ones. Thus readers may erroneously think (actually they are encouraged to think so) that these tests follow from the parallel postulate, and this false idea is supported by the proof given in the book. As a matter of fact, these tests are true in a geometry that assumes all Euclidean postulates except the parallel postulate (so-called neutral geometry), thus they would be true, for instance, in hyperbolic geometry.

Limits were introduced seemingly to explain the notion of circumference. "Because," as the author explains, "a mathematically precise definition of the word *limit*, would be quite difficult to understand, an informal explanation is given instead". Unfortunately, the given "explanation" is also difficult if not impossible to understand and contains a mistake for good measure (it is the *absolute value of the difference*, not the *difference*, between the terms of the sequence and the limit that "can be made as small as we wish"). Even the corrected "explanation" would not help much since the illustrating examples, consisting mostly of another set of "funny" pictures, do not possess the necessary content. The problems of the section also do no explain the mathematical contents of limits but only teach students to write the symbol replacing the words *limit when n tends to infinity*. An excerpt follows.

"Write each of the following statements in symbols.

Example: The limit of the sequence whose nth term is $\frac{1}{n}$ is 0.

A*nswer* : $\lim_{n \to \infty} \frac{1}{n} = 0$.

13. The limit of the sequence whose nth term is $\frac{n+5}{2n}$ is $\frac{1}{2}$.

14. The limit of the sequence whose nth term is $\frac{1-n}{n+4}$ is -1." (end of quotation)

The learning of a notion is replaced with a lesson on how to use the notation.

The uniqueness property and the Weierstrass' theorem, which is really crucial for defining the circumference, are not even mentioned. As a result, the circumference is defined as a limit whose existence is vaguely supported by a few numerical experiments. The number π (pi) is not defined explicitly as the universal ratio of the circumference to the radius.

Limits are never used anywhere else although they could help, for example, to provide beautiful proofs for the formulae of the volumes of pyramids and other solids.

The problems are trivial and often irrelevant, especially those that were supposed to show a "real-life connection". The latter is also apparently provided through an enormous number of pictures, cartoons, and photographs, many of them ugly and repulsive (such as a photo of a telephone booth packed with bodies illustrating the section "The Volume of a Prism) with very little relevance to the topic, some of them with erroneous commentaries. Lesson 7, "Spheres", in the Chapter "Geometric Solids", starts with a picture of an old man with a huge ball of string. The commentary states: "The photograph shows him [Mr. Roberts] standing with the result of his unusual hobby, a ball of string three feet in diameter! How heavy would a ball of string this size be? To answer this, it would be helpful to know how to find the volume of a sphere."(End of quotation). – It will not be helpful: the ball is obviously porous and non-uniform, hence knowing the volume will not help even for a rough estimate of the weight! Does that mean students should not learn about spheres and their volumes?

Speaking of the pictures (each chapter starts with one, approximately half a page in size), there is one which is particularly puzzling (not to say – irritating): Lesson 5 "Straight Segments" begins with a picture of Marilyn Monroe presented as a net of straight segments. What was the point of placing this picture? Is there any learning value? It is not easy to understand from it what the late actress looked like: this net of segments may just as well be a picture of Marge Simpson, Euclid, your neighbour Willy, or anyone at all. Does the message of the picture state that everything under the sun can be made of straight segments, and that is why they should be studied? Or is it intended to assert the triumph of science and technology by disfiguring beauty?

O'Henry wrote ("*Squaring the Circle*", The Complete Works of O'Henry, vol.II): "Beauty is Nature in perfection; circularity is its chief attribute. Behold the full moon, the enchanting gold ball, the domes of splendid temples, the huckleberry pie,…On the other hand, straight lines show that Nature has been deflected. Imagine Venus's girdle transformed into a 'straight front!'" So, maybe it would be better not to straighten out naturally occurring curves? There are already too many things to be straightened out in this book: the difference between the axioms and theorems, for instance! Maybe it would be better to leave the pictures beautiful, the axioms believable and independent, the theorems proved, the definitions defining, and the textbooks teaching something useful, – e.g. geometry?

1c. Good old (really old!) textbooks.

> "Rabbit," said Pooh to himself. "I like talking to Rabbit. He talks about sensible things. He doesn't use long, difficult words, like Owl".
>
> A.A. Milne, *The House at Pooh Corner*.

The category of "good old textbooks" includes a few geometric texts written approximately a century ago. Usually based on close-to-Euclidean sets of common notions and postulates and written in somewhat simplified language (compared to the English translations of *The Elements*), they are reasonable textbooks from which students could learn the actual geometric material as well as the methods of proofs and the ways of logic.

One of the best of such books is "*A School Geometry, Parts I-VI*" by H.S. Hall and F.H. Stevens (Macmillan and Co., Limited, London, 1908). The text includes a discussion of both the plane and solid geometries at the secondary school level.

There are a few flaws, which are, let us emphasize this again, very minor comparing to the ones in the aforementioned modern texts.

The set of postulates contains superfluous statements, such as, for example, the equality (congruence) of right angles, the existence of the bisectors of angles and of the midpoints of segments. The notion of equality (congruence) is not defined and not described as an undefined term (in "*The Elements*", it was mentioned among the common notions), which may leave students puzzled about the validity of the proofs of fairly obvious statements whereas less obvious ones are tacitly assumed.

Axioms of continuity have not been introduced and thus the notion of the length of a segment has not been discussed even though all kinds of measures, including irrational ones, are used throughout the book. It is a great loss for a school text since the students are losing the only opportunity to be introduced to the theory of real numbers. (The theory of real numbers is never discussed in high school Algebra and Analysis courses, nor in undergraduate Calculus; only a few higher level undergraduate courses include a discussion of Dedekind's cuts. Thus, students typically spend quite a few years considering *functions of real variable* without having an idea of what is a real number; it is not surprising then that they encounter great difficulties when dealing with limits and continuity).

The results of neutral geometry are not singled out, thus students cannot appreciate the role of the parallel postulate and the idea of generating different geometries by changing some postulates. It is however an important story, both historically and methodologically.

The notion of circumference does not receive any consideration. It is only mentioned that the ratio of the circumference to the radius is the same (approximately 3 and one seventh) for all circles.

Most of the exercises are technical and trivial, and thus would not excite interest in the subject, especially among strong students.

The discussion is very dry, sometimes lacking explanations, such as, for instance, why the constructions should be performed with a straightedge and a compass alone. The book loses the sense of creating and uncovering things that is inherent in "The Elements".

With all these disadvantages, the book by H.S. Hall and H.S. Stevens, in contrast with the aforementioned modern texts, is a decent book from which students can learn the basics of Euclidean geometry.

2. HOW IS THIS TEXTBOOK WRITTEN?

> *"Getting Tigger down,"* said Eeyore, *"and Not hurting anybody.* Keep those ideas in your head, Piglet, and you'll be all right."
> A.A. Milne, *The House At Pooh Corner.*

When writing the text, I attempted my best to make it satisfy the following conditions:
  a) The text discusses in consistent and sequential manner the basic principles and results of Euclidean plane geometry; the approach is synthetic, in Euclidean spirit.
  b) The discussion is rigorous but not overly formal, so that the first eight chapters could be understood by secondary school students.
  c) A curious young reader will appreciate the beauty of the subject and thus will enjoy working on the course.
  d) As a result of using the textbook, a student will acquire valuable intellectual habits that cannot be obtained otherwise than by studying Euclidean geometry.

The choice of the set of axioms constitutes the most important and difficult part of writing a geometry text. One cannot base an introductory secondary school level discussion of the subject on a set that is rigorous from a modern point of view, such as, for instance, the Hilbert's set of axioms, which is deemed as the briefest one for a rigorous treatment of the subject: this set is too abstract, and the principle of *"Not hurting anybody"* would be violated. Some rigor should probably be sacrificed in favour of clarity.

There is an excellent discussion of the problem of excessive rigor in introductory textbooks in the chapter "Rigor" of Morris Kline's *"Why Johnny can't add"*. The author's opinion has been supported by the ones of great mathematicians such as Blaise Pascal and Henri Poincare. The latter wrote: "…if the demonstration rests on premises which do not appear to him [student] more evident than the conclusion, what would this unfortunate student think? He will think that the science of mathematics is only arbitrary accumulation of useless subtleties; either he will be disgusted with it or he will amuse himself with it as a game and arrive at a state of mind analogous to that of the Greek sophists."

It is hard to disagree with this opinion when talking about some axioms from Hilbert's set. I think an introductory text can safely set aside some incidence axioms, such as *"There exist at least two points in every straight line"* and *"There exist three points that do not lie in the same straight line"* (will any young student ever question this?!) and all the betweenness axioms. Really, what would any person inexperienced in abstract mathematics think about the postulate claiming that *"If A, B, and C are points of a straight line and B lies between A and C, then B lies also between C and A"*?

Morris Kline wrote about these and some other axioms often included in modern texts for beginners: "To ask students to recognize the need for these missing axioms and theorems is to ask for a critical attitude and maturity of mind that is entirely beyond young people. If the best mathematicians did not recognize the need for these axioms and theorems for two thousand years how can we expect young people to see the need for them?" This question concerns all axioms not included in the original text of Euclid's *"The Elements"*.

I do concur with this opinion of Morris Kline, whom I hold in great esteem, related to the aforementioned axioms of incidence and betweenness. Yet, based on my personal experience as a student (first and foremost!) and later as an instructor, I would argue about the inclusion of axioms of congruence and continuity.

When I studied geometry in junior high school in Russia in the mid sixties, we followed a great textbook "*Geometria*" by A.P. Kiselyov (Uchpedgiz, Moscow, 1950, the eleventh edition). The textbook, written somewhere at the end of the 19th century, used the set of the basic axioms equivalent to the original Euclidean postulates. With that, the fifth postulate (Euclidean parallel postulate) was introduced in the discussion after all the basic results following from the first four postulates had been derived. As a result of such a sequence of the presentation, even though the existence of non-Euclidean geometries has not been discussed in the book, the readers could understand that quite a few fundamental results, such as congruence of triangles, did not require the parallel postulate in order to be obtained. The Archimedean continuity axiom was added to the set right before the discussion of proportionality of segments and similarity. Based on this axiom, and the notions of lengths and real numbers were introduced almost rigorously.

Such a discussion seems to be almost perfect. Yet I do remember experiencing a sense of discontent when our teacher proved the first basic theorem: *The bisector of the vertical angle of an isosceles triangle is also a median and an altitude, and the base angles of an isosceles triangle are equal.* The symmetry of an isosceles triangle seemed to be quite obvious, and still it was being proved by means of superposition performed as a physical motion! The comparison of figures by means of superposition had never been defined or explained by means of postulates, and still we used it for a mathematical proof?! I felt it was unfair, "not playing by the rules", especially because the rules had been proclaimed just before the proof: the teacher explained that we are going to deal with ideal, i.e. non-physical, objects for which some statements (axioms) are assumed, and then she operated with these ideal objects as if they were made of cardboard! The cardboard-likeness had not been postulated!

Later on, when reading the comments to *The Elements* in the English translation by Sir Thomas Heath, I was quite happy to learn that neither Euclid nor any of his followers really trusted the *superposition principle*. This principle had been assumed as one of the axioms or common (not particularly geometrical) notions: Common Notion 4: "*The things that coincide with one another are equal to one another*", and then used only twice, in the proofs of congruence tests for triangles. It looks like Euclid shyed away from using the superposition, since he never applied it in the theorems where it could simplify the proof (for instance, in Proposition 1.5, which states the equality of the base angles of an isosceles triangle).

Keeping all this in mind, I did postulate the existence of rigid motions (isometries) that are needed for the basic congruence theorems. The motions have been defined as segment-preserving transformations. The congruence (equality) of segments has been introduced as an undefined notion described by two axioms, and the congruence of figures has been defined through superposition by means of rigid motions. This seems to be a natural and *almost* rigorous way; one can easily remake it into a rigorous one by adding the axioms of addition of segments (I use the Euclidean common notion *the sums of equals are equal* instead) and the aforementioned "obvious" incidence and betweenness axioms, which I think are not necessary in an introductory text.

The first chapter of the book introduces the undefined terms (point, line, plane, congruence of segments), first definitions and first axioms. It also includes a brief discussion of the importance of theoretical knowledge and of the subject of Euclidean geometry as a geometry in which congruence is defined through superposition by means of rigid motions. The existence of certain types of motions has been explained at the intuitive level, by means of physical motions, and then postulated.

Chapter 2 includes an elementary discussion of the basic rules of logic, specifically applied to conditional statements as well as the general structure of axiom-based fields of knowledge. Inductive and deductive reasoning are discussed. The role of axioms and derivation of theorems in axiom-based theories is being considered.

The notions of converse, inverse, and contrapositive statements are introduced and the relations between them are discussed, both theoretically and based on simple examples. Special attention has been paid to the *proof by contradiction* (Reductio ad absurdum), both the substantiation and the use of the method. Another important method of proof, *The Principle of Math Induction*, is included in the problems section accompanying the chapter.

In Chapter 3, angles are defined and their basic properties discussed. The generation of angles by the rotation of rays and the generalization of the notion of angle for the case of non-convex interiors has been considered. The existence of right angles and their congruence is proved.

Chapter 4 contains the most substantial results of the so-called neutral geometry (Euclidean geometry without the Euclidean parallel postulate): congruence of triangles, existence of the bisectors of angles and midpoints of segments, properties of isosceles triangles, inequality in triangles, loci and axial symmetry.

Chapter 5 discusses constructions. It explains why only a straightedge and a compass are being used, and all the basic classical constructions of neutral geometry are described and substantiated in detail. The general 4-step plan for solving construction problems (analysis, construction, synthesis, and investigation) is given and illustrated in the problems section.

Chapter 6 starts with the notion of parallel lines and a few more results of neutral geometry: the proof of the existence of parallel lines and the tests for parallelism based on the angles. Then the parallel postulate in Playfair's formulation is stated, and its equivalence to the 5th Euclidean postulate is shown. The congruence/supplementary of angles with parallel and perpendicular sides has been proved. Applications to constructions are considered.

Chapter 7 contains all the basic classical results of the Euclidean geometry of parallelograms and trapezoids as well as related topics, such as central symmetry and translations and their use in constructions. The discussion closely follows that of the Kiselyov's "*Geometria*".

The basic results of the Euclidean geometry of circles (some of them post-Euclidean, still treated as classical) is discussed in Chapter 8. With a few minor deviations, the text is a translation of the section concerning circles from the Kiselyov's book. The chapter ends with the discussion of the remarkable points of concurrency: incentre, circumcentre, centroid and orthocentre of a triangle.

Chapter 9 starts with a rigorous discussion of the notion of measurement of a segment. The continuity is introduced through the Archimedes axiom and Cantor's principle of nested segments. These axioms are quite natural, and it is easy to visualize them, in contrast with the Dedekind's postulate, which does not sound very intuitive and is not easy to comprehend for students without experience in abstract mathematics (see the section *Principle of Continuity* in the comments *Other Axioms Introduced After Euclid's Time* of the Heath's translation of *The Elements*). Dedekind's principle is often employed in the university level texts for introducing lengths in geometry and real numbers in analysis. One can show the equivalence of this postulate to the Archimedes and Cantor's axioms together, and therefore the latter two, which are more intuitive, are in my opinion a better choice.

Irrational numbers are introduced as the measures of segments incommensurable with the unit of length. The existence of incommensurable segments is proved purely geometrically, and another example of such a proof is proposed as a problem for students. The ratio of segments is defined as the ratio of their measures, and it is shown that it does not depend on the choice of the unit length.

The similarity of triangles and then of arbitrary polygons is defined through the proportionality of the sides and congruence of angles and the tests are proved for general triangles, then for right triangles. Then, starting with a problem of constructing a polygon similar to a given one, the notions of a similarity transformation, centre of a homothety and magnification ratio have been introduced. These are employed for giving a new definition of similarity that is applicable for all (not necessarily rectilinear) figures and includes the original definition as a particular case. The use of similarity transformations in constructions has been discussed.

Then all the classical results concerning proportional segments follow. Those include the segments cut by parallel lines on their transversals, on parallel lines by their transversals, the properties of the bisectors of interior and exterior angles in a triangle, proportional segments in circles, and all the metric relations in right triangles, including following from them Pythagorean theorem. The same section also includes the converse of this famous theorem and its generalizations for the cases of acute-angled and obtuse-angled triangles; the latter are used in the following sections for deriving the cosine theorem for solving triangles.

Trigonometric functions of acute angles are introduced and their properties discussed. Then the cosine theorem is proved, the notion of sine is generalized for obtuse angles, and the sine theorem proved.

The chapter ends with the discussion of applications of algebra for solving geometric problems, in particular for performing constructions. A few examples, including the *golden ratio*, the *mean proportional*, the *fourth proportional*, and various expressions involving roots are considered.

Chapter 9, in contrast with the preceding chapters, does require some rudimentary knowledge of algebra, including solutions of quadratic equations in radicals.

Chapter 10 starts with a discussion of regular polygons, but the main focus of the chapter is the notion of the circumference of a circle. This is an extremely interesting and exciting topic, routinely ignored both in the secondary school and the university level texts. At the same time the notion of circumference is extremely important both from the theoretical and practical points of view, for instance for such a "minor thing" as defining trigonometric functions on the unit circle.

The circumference is introduced based on the Cantor's principle (we knew we will need it again!), and thus does not require the notion of the limit of a sequence. Still, an optional approach, by means of limits is considered, with the limits and their basic properties introduced beforehand. The chapter ends with the introduction of the radian measure of angles.

Chapter 11 deals with areas. A square with the unit side is assigned the unit area; the areas of all the other figures are defined by the principles of the equivalence of congruence figures and equivalence by finite decomposition. The existence of the area of any rectangle (with rational or irrational lengths of the sides) has been proved. Various formulae for determining the areas of parallelograms, triangles, trapezoids, and regular polygons are derived.

The Pythagorean Theorem is revisited, in terms of areas; two different proofs are proposed. The theorem claiming that the ratio of areas of similar figures is the square of their similarity ratio is proved. The derivation of the formula for the area of a circle concludes the chapter.

There is no cartoons and glitzy pictures in this book: I believe that the subject itself is so beautiful that it does not require any embellishments either in direct or ironical sense of this word. Epigraphs are a different matter. It is an old and noble tradition to start a chapter of a book with a suitable epigraph, which sheds some light on the following contents and thus helps the reader. Since the textbook is written for the children who think that they are almost adults, I found it appropriate to turn their attention to the book which many of them think they have overgrown, but they have not. Thus all the epigraphs have been borrowed from "The Complete Tales of Winnie-the-Pooh", which is really a book of wisdom for all ages. Some solutions of the problems are also preceded by epigraphs that illustrate the main idea of the solution or give a helpful hint. These are excerpts from various classical books for children and youth.

Every section of each chapter is supplied with a set of problems. Also, each chapter is concluded with a large set of *Review* problems; the latter are usually the core problems of the course whereas many of the problems concluding sections are simple technical exercises. There are over 1200 problems altogether. Some of them are quite challenging; the most difficult ones are usually marked with an asterisk and supplied with hints in the first part of the *Answers, Hints, Solutions* section and with a solution (or solutions) in the second part of the section. Also, the problems that demonstrate new methods (e.g., the use of symmetry in constructions) are provided with detailed solutions.

The students who use the book should solve problems, - it is absolutely essential. A person who solves a difficult problem experiences a sense of triumph that is unlike anything else. Solving problems and describing their solutions may be difficult in the beginning, but it must become a more customary activity for those who study the methods of proofs and constructions presented in the text. Also, in the course of their studies students will learn how to substantiate their ideas and express them in writing.

Geometry is a difficult but a very rewarding subject. I cannot imagine another one that can arouse a comparable feeling of beauty and harmony. Many of the greatest minds in the history of the humankind considered studying geometry to be the commencement of their intellectual lives. Albert Einstein could never forget the tremendous impression produced on him by his first geometry textbook. Bertrand Russell, an eminent philosopher of mathematics wrote in his autobiography: "At the age of eleven, I began Euclid… This was one of the greatest events of my life, as dazzling as first love. I had not imagined there was anything so delicious in the world."

Let us hope that modern students will not be deprived from such an experience.

3. LITERATURE: BOOKS AND PAPERS MENTIONED IN THIS REVIEW.

> "Owl looked at the notice again. To one of his education the reading of it was easy."
> A.A. Milne, *The House At Pooh Corner.*

1. *The thirteen books of Euclid's Elements translated with introduction and commentary by Sir Thomas Heath*, 2nd edition, 3 volumes, republication by Dover, 1956.
2. *(i)Essentials of Geometry for College Students*, 2/E by M.L.Lial, B.A.Brown, A.R.Steffenson, and L.M. Johnson, published by Addison-Wesley, 2003.
   (ii)Elementary Geometry, 3rd edition, by R.D.Gustafson and P.D.Frisk, published by   John Wiley and Sons, Inc.
   (iii)Geometry: tools for a changing world by Laurie E. Bass, Basia Rinesmith Hall, Art Johnson, and Dorothy F. Wood, with contributing author Simone W. Bess, published by Prentice-Hall, 1998.
3. *Geometry* by Harold Jacobs, published by W.H. Freeman and Company, 1987 (second edition).

4. G.D. Birkhoff, "*A Set of Postulates for Plane Geometry (Based on Scale and Protractor)*", Annals of Mathematics, 33, 1932.
5. *The Foundations of Geometry* by David Hilbert; authorized translation by E.J. Townsend, The Open Court Publishing Company, Le Salle, Illinois, 1950.
6. *Teaching Geometry According to Euclid* by Robin Hartshorne in the Notices of the AMS (American Mathematical Society), Volume 47, # 4, April 2000.
   This paper can be found on line at http://www.ams.org/notices/200004/fea-hartshorne.pdf
7. *Why Johnny Cannot Add: The Failure of the New Math*, by Morris Kline, *Vintage Books*, New York, 1974.
8. *Twenty years under a bed and other stories about Deniska* by Victor Dragounsky (in Russian), Rosman, Moscow, 1999.
9. *Squaring the Circle*, in The Complete Works of O'Henry, vol.II, Doubleday & Company, Inc. Garden City, NY.
10. *A School Geometry, Parts I-VI* by H.S. Hall and F.H. Stevens (Macmillan and Co., Limited, London, 1908).
11. *Geometria* by A.P. Kiselyov (in Russian), Uchpedgiz, Moscow, 1950, (the eleventh edition).
12. *The Complete Tales of Winnie-the-Pooh* by A.A. Milne with decorations by Ernest H. Shepard, Dutton Children's Books, New York, 1994.

CPSIA information can be obtained
at www.ICGtesting.com
Printed in the USA
BVHW08s1724190718
522063BV00002B/7/P